Managing the Prenatal Environment to Enhance Livestock Productivity

Paul L. Greenwood · Alan W. Bell ·
Philip E. Vercoe · Gerrit J. Viljoen
Editors

Managing the Prenatal Environment to Enhance Livestock Productivity

Springer

Editors
Dr. Paul L. Greenwood
NSW Department of Primary Industries
JSF Barker Building
Trevenna Road
University of New England
Armidale NSW 2351
Australia
paul.greenwood@industry.nsw.gov.au

Professor Alan W. Bell
CSIRO Livestock Industries
Queensland Bioscience Precinct
306 Carmody Road
St Lucia QLD 4067
Australia
alan.bell@csiro.au

Dr. Philip E. Vercoe
School of Animal Biology
Faculty of Natural and Agricultural Sciences
University of Western Australia
Crawley WA 6009
Australia
pvercoe@cyellene.uwa.edu.au

Professor Gerritt J. Viljoen
International Atomic Energy Agency (IAEA)
Division of Nuclear Techniques in Food and Agriculture
Wagramer Str. 5
1400 Wien
Austria
G.J.Viljoen@iaea.org

ISBN 978-90-481-3134-1 e-ISBN 978-90-481-3135-8
DOI 10.1007/978-90-481-3135-8
Springer Dordrecht Heidelberg London New York

Library of Congress Control Number: 2009938200

Printed on acid-free paper

Springer is part of Springer Science+Business Media (www.springer.com)

Preface

Global demand for livestock products is expanding and the role of developing countries in meeting this demand is increasing. Nutrition of the reproductive female in livestock production systems in developed and developing countries is a key to satisfying this increased demand because it influences the number, size and survival of offspring, and the frequency with which they are produced. One field of research that is attracting increasing attention is in utero nutrition of the fetus because of evidence that it influences postnatal productivity and health in the long-term.

Livestock in the developing world endure unique challenges from their environments, which are generally harsher and less managed than those faced by livestock in the developed countries. A feature of livestock production systems in developing countries is the fluctuation in the amount and quality of feed resources accessible to livestock. The fetus, therefore, is exposed to various challenges that are mostly, but not exclusively, of nutritional origin and that may influence its lifetime performance. The local genotypes within developing countries are unique in that they often have evolved, been selected for, or been exposed to trans-generational environmental effects, which dictate that survival is the overriding production objective. Often, little or no knowledge exists on whether the maternal environment influences the subsequent productive performance of offspring from these genotypes. A better understanding of how fetal development can be enhanced to improve lifetime performance in local genotypes will provide more opportunities to satisfy the increasing demand for livestock products.

The concept and outline for this publication were developed at a Consultants Meeting entitled "Improvement of animal productivity in developing countries by manipulation of nutrition in utero and indentification of future areas of research in animal nutrition" organized by the Animal Production and Health Section of the Joint FAO/IAEA Division of Nuclear Techniques in Food and Agriculture. The meeting was held in Vienna in October 2005 and participants included seven experts on in utero development and nutrition and on fetal programming from agricultural research organisations and universities in France (Prof. Jean-François Hocquette), New Zealand (Prof. Peter Gluckman), Australia (Prof. Dennis Poppi and Dr. Paul Greenwood), USA (Prof. Stephen Ford) and Denmark (Dr. Mette Olaf Nielsen and Dr. Mario Acquarone), as well as IAEA staff with expertise in livestock production (Harinder Makkar, coordinating Technical Officer). The participants addressed the extent to which nutritional challenges influence fetal development and subsequent health, growth, reproductive and lactational characteristics of offspring. The main objective of the meeting was to determine the value and scope of a new Coordinated Research Project in the field of fetal programming, focussing on nutrition-gene interactions. The goal of such a project would be to enhance livestock productivity and, within its scope, propose specific areas of research of importance to developing countries.

The consultants focussed their discussions on measurement of short- and long-term effects of challenges during pregnancy in genotypes within developing countries. They also discussed the development of production systems aimed at alleviating these constraints for instance, by supplementation, and to improve the amount and quality of outputs from production systems. A framework for a proposal highlighting the research and technical training needs required for this programme to succeed within developing countries was developed. However, it was clear from the outset that a comprehensive review of the literature for a range of livestock species was needed to support the technical training and research needs of the programme.

This book should serve as a text for any researcher with an interest in fetal programming and developmental plasticity, although it will be of most use for researchers in Member States who need a basis from which they can initiate a programme of research in this field.

Armidale, NSW	Paul L. Greenwood
St Lucia, QLD	Alan W. Bell
Crawley, WA	Philip E. Vercoe
Wien, Austria	Gerrit J. Viljoen

Contents

Contributors

R. Michael Akers Department of Dairy Science, Virginia Tech, Blacksburg, VA 24061, USA, rma@vt.edu

Lloyd L. Anderson - Department of Animal Science, College of Agriculture and Life Sciences and Department of Biomedical Sciences, College of Veterinary Medicine, Iowa State University, Ames, IA 50011-3150, USA, llanders@iastate.edu

C. Simon Bawden South Australian Research and Development Institute (SARDI) Livestock and Farming Systems, Roseworthy, SA 5371, Australia, simon.bawden@sa.gov.au

Alan W. Bell CSIRO Livestock Industries, Queensland Bioscience Precinct, St Lucia, Qld 4067, Australia, alan.bell@csiro.au

John M. Brameld Division of Nutritional Sciences, School of Biosciences, University of Nottingham, Loughborough, LE12 5RD, UK, john.brameld@nottingham.ac.uk

Anthony V. Capuco Bovine Functional Genomics Laboratory, USDA-ARS, Beltsville, MD 20705, USA, tony.capuco@ars.usda.gov

Stehphanie M. Dunn South Australian Research and Development Institute (SARDI) Livestock and Farming Systems, Roseworthy, SA 5371, Australia, stephanie.dunn@sa.gov.au

Stephen P. Ford Center for the Study of Fetal Programming and Department of Animal Science, University of Wyoming, Laramie, WY 82071-3684, USA, spford@uwyo.edu

Paul L. Greenwood NSW Department of Primary Industries, Beef Industry Centre of Excellence, University of New England, Armidale, NSW 2351, Australia, paul.greenwood@industry.nsw.gov.au

David O. Kleemann South Australian Research and Development Institute (SARDI) Livestock and Farming Systems, Roseworthy, SA 5371, Australia, dave.kleemann@sa.gov.au

Marcus Mau Research Institute for the Biology of Farm Animals, 18196 Dummerstorf, Germany, mmau@itw.uni-bonn.de

Clive J. McLaughlan South Australian Research and Development Institute (SARDI) Livestock and Farming Systems, Roseworthy, SA 5371, Australia, clive.mclaughlan@sa.gov.au

Gregory S. Nattrass South Australian Research and Development Institute (SARDI) Livestock and Farming Systems, Roseworthy, SA 5371, Australia, greg.nattrass@sa.gov.au

Pia M. Nissen Danish Institute of Agricultural Sciences, Faculty of Agricultural Sciences, Aarhus University, DK-8830 Tjele, Denmark, piam.nissen@agrsci.dk

Niels Oksbjerg Danish Institute of Agricultural Sciences, Faculty of Agricultural Sciences, Aarhus University, DK-8830 Tjele, Denmark, Niels.Oksbjerg@agrsci.dk

Charlotte Rehfeldt Research Institute for the Biology of Farm Animals, 18196 Dummerstorf, Germany, rehfeldt@fbn-dummerstorf.de

Andrew N. Thompson Department of Agriculture and Food, South Perth, WA 6151, Australia, anthompson@agric.wa.gov.au

Zehava Uni The Faculty of Agricultural, Food and Environmental Quality Sciences, The Hebrew University of Jerusalem, Rehovot 76100, Israel, uni@agri.huji.ac.il

Klaus Wimmers Research Institute for the Biology of Farm Animals, 18196 Dummerstorf, Germany, wimmers@fbn-dummerstorf.de

Matthew E. Wilson Division of Animal & Veterinary Sciences, Davis College of Agriculture, Forestry, and Consumer Sciences, West Virginia University, Morgantown, WV 26506-6108, USA, matt.wilson@mail.wvu.edu

Shlomo Yahav The Volcani Center, Institute of Animal Sciences, Bet-Dagan, Israel, yahavs@agri.huji.ac.il

Introduction

Prenatal life is the period of maximal development of organ systems and tissues in animals, and it is well recognised that factors that interfere with development can have profound effects on the embryonic, fetal and postnatal animal. However, the extent to which such factors influence the maternal, embryonic and fetal environment and how this impacts on the performance of livestock in postnatal life, requires continued research to improve both our level of understanding and our capacity to enhance productivity, health and reproductive efficiency.

In recent years, there has been a rapid increase in interest in how to manage prenatal development to enhance livestock productivity and health. To a large extent, this interest has grown as a result of the *Fetal Origins Hypothesis*, which is based on studies that show that human health in later life can be influenced by events during prenatal life. More recently, the *Thrifty Phenotype Hypothesis* has also emerged from human and animal studies, and suggests altered postnatal metabolism and responses to nutrition as a result of limited nutrient supply during prenatal development. It should be recognised, however, that scientists involved in research on livestock production have reported postnatal consequences of fetal development on aspects of productivity for decades prior to the more formal establishment of the *Fetal Origins* and *Thrifty Phenotype Hypotheses*.

Improved understanding of the mechanistic basis of postnatal consequences of prenatal development requires a synthesis of methodologies applied to the study of fetal and postnatal development. This synthesis is required at the nutritional, endocrine, metabolic, cellular and molecular levels. Many studies during fetal life and on postnatal consequences of fetal development still use radioisotopic techniques, or techniques that have evolved from radioisotopic methods.

Applications of the radioisotopic and related methodologies in developmental studies are numerous and have had a profound influence on our understanding of fetal and postnatal development. For example, radiolabelled tracers have been extensively used to study whole body or tissue specific energy and protein kinetics and metabolism, prior to the advent of stable isotopic techniques. Radioisotopic labelling and detection systems have also allowed DNA-protein interactions and tissue-specific gene and protein expression to be studied, which has expanded our understanding of the regulation of development of all organ and tissue systems. Furthermore, our understanding of endocrine regulation of development and of cell cycle activity and tissue hyperplasia in fetal and postnatal animals could not have evolved to where it is today without the advent and use of the radiolabelling techniques. Hence, it has been through the use of radiolabelling technologies that we have been able to make such positive advances in this field.

This book, *Managing the Prenatal Environment to Enhance Livestock Productivity*, is intended to provide a detailed account of phenotypic consequences of prenatal development and the mechanisms that underpin these effects in ruminants, pigs and poultry. The chapters have been divided into three parts covering fetal development and its consequences for postnatal productivity, health and efficiency. Part I deals with the quantification of prenatal effects on postnatal productivity across species; Part II is dedicated to the mechanistic bases of postnatal consequences of prenatal development in meat and fibre production in ruminants as well growth, health and carcass quality in pigs; and Part III is focused on the regulators of fetal and neonatal nutrient supply, in regard to the importance of placental vascularity as well mammary gland development and milk production.

Within the chapters, readers are provided with a comprehensive review of the literature published in this area for ruminants, pigs and poultry, an insight into the gaps in our knowledge, and an outlook into future areas of research that need to be addressed. Implicit in this is the significance of radioisotopic methodologies in enhancing our understanding of prenatal and postnatal development.

Managing the Prenatal Environment to Enhance Livestock Productivity provides a reference from which future research in developing and developed countries will evolve. It is our hope that this book will play a prominent role in helping to guide the direction of future research in this field of study that is so important to enhancing productivity, health and efficiency in livestock species.

Part I
Quantifying the Magnitude of Prenatal Effects on Productivity

Chapter 1
Postnatal Consequences of the Maternal Environment and of Growth During Prenatal Life for Productivity of Ruminants

Paul L. Greenwood, Andrew N. Thompson, and Stephen P. Ford

Introduction

There has been a recent explosion in the amount of research on consequences of fetal development for postnatal health in humans, and in the use of the sheep as a model to understand postnatal consequences of altering the prenatal environment (for review see [87]). However, there is an important need for a review that quantifies consequences of prenatal nutrition for postnatal productivity in ruminants, particularly on the extent to which traits of economic importance can be influenced or programmed by variations in the maternal and fetal environment within production systems.

Influences on growth and development can be classified according to the extent, source and type of environmental variation, and potential for permanent or transient affects on productivity traits. Consequences of development during fetal life also need to be viewed in the context of the length of the productive life of an animal which, at the extremes, varies from young animals used for meat within weeks or months of birth, to breeding animals that remain in the flock or herd until they are no longer reproductively or functionally sound. Often, it is impossible to determine whether postnatal consequences of prenatal development are permanent or transient due to the need for a prolonged "recovery" period following prenatal and/or preweaning influences. Despite this, research on consequences of fetal development conducted within relatively short postnatal time-frames may be of immense practical significance for agricultural production.

In the strictest sense, prenatal or fetal programming in the context of livestock production refers to events specific to the embryo or fetus, independent of postnatal maternal or other confounding influences, that result in permanent alterations to efficiency and outputs within the animal's productive life. However, in ruminants few studies to date have removed postnatal maternal influences on offspring by

P.L. Greenwood (✉)
NSW Department of Primary Industries, Beef Industry Centre of Excellence
University of New England Armidale, NSW 2351, Australia
e-mail: paul.greenwood@industry.nsw.gov.au

P.L. Greenwood et al. (eds.), *Managing the Prenatal Environment to Enhance Livestock Productivity*, DOI 10.1007/978-90-481-3135-8_1,

rearing them in isolation from their dams under controlled conditions [49, 115, 120]. Clearly, within production systems it is impossible to uncouple influences of fetal life from those during early postnatal life when offspring remain on their dam and are subject to maternal influences that may also be altered by the environment during pregnancy. For example, the environment encountered during pregnancy may affect the subsequent capacity of the dam to synthesise colostrum and milk, to consume or mobilise nutrients in support of nutritional requirements in later pregnancy or during lactation, and may alter maternal behaviour.

Similarly, differences in postnatal consequences of altered fetal development may exist between field and pen studies because of the apparent importance of the environment into which the young ruminant is born in determining long-term consequences of fetal development. This is particularly important because of emerging evidence in other species relating to establishment of the so-called "thrifty-phenotype" following restricted nutrition and growth during gestation which is believed to allow animals nutritionally restricted in utero to adapt metabolically to a scarcer postnatal nutritional environment [60]. Hence, where appropriate, differences in outcomes between studies conducted under pen and field conditions are reviewed in a further attempt to identify the extent to which the postnatal environment has influenced outcomes of prenatal development in ruminants.

Therefore, in this Chapter we review the literature on and quantify consequences of the fetal and maternal environment for factors relating to productivity of ruminant livestock, and attempt to determine the extent to which postnatal sources of variation and experimental conditions may exacerbate or ameliorate these effects. The reader is also referred to earlier reviews of research relating to productive outcomes of fetal development in ruminant livestock [6, 10, 11, 22, 31, 46, 48].

1.1 Managing the Prenatal Environment

A Brief Overview of Sources of Variation in Fetal Growth and Development

Factors that influence fetal growth and development include nutrition, toxins and teratogens, parity, age, live weight, stress and well-being, genotype, and prolificacy of the dam, and the thermal environment during pregnancy, which may also relate to fetal growth responses to shearing during pregnancy. Genotype of the fetus also regulates fetal growth, particularly during early- to mid-gestation prior to increasing influences of the maternal genotype [33] and the external environment. It is also important to recognise that the placenta is a powerful regulator of fetal growth and development, and plays a major role in mediating maternal influences including those of litter size on fetal growth. A comprehensive review of sources of variation that contribute to fetal growth and development is beyond the scope of this Chapter, however, the reader is referred to other Chapters within this book and earlier reviews [12, 38, 43, 48, 68, 106] for further information and references on this subject in ruminant livestock. Similarly, consequences of nutrition during pregnancy and of fetal growth for survival of ruminants have been comprehensively studied and are reviewed elsewhere [1–2, 28, 80, 90, 91].

1.2 Postnatal Consequences of the Prenatal Environment

1.2.1 Postnatal Growth and Size

Growth and mature size are of importance for animals that remain within the flock as breeding stock, and also in relation to animals slaughtered at younger ages for meat given that growth trajectories influence the time to reach market weight and, potentially, the characteristics of marketable products.

Most research on consequences of prenatal development in ruminants has been undertaken in sheep, with early research (23 studies from 1932 to 1979) on relationships between birth weight and postnatal growth of sheep reviewed by Villette and Theriez [129]. These studies showed that birth weight was correlated positively with growth to weaning, but less so or not at all with growth after weaning. In all but one study, low birth weight resulted in sheep that were smaller at ages ranging from 4 months to 3 years of age, and these findings applied whether birth weight was reduced due to litter size (singletons vs. twins), as a result of nutritional treatments during late-pregnancy, or as a result of normal variation in birth weight within flocks.

Since the Australian studies in the 1950s and 1960s on long-term consequences of severely reduced prenatal nutrition [31, 115, 116], there have been on-going attempts to elucidate influences on postnatal growth that emanate during pregnancy. While these studies initially related to agricultural production, within the last decade or so many studies have used sheep as an experimental model of fetal programming, following widespread recognition that fetal development may have long-term health implications in humans [7, 8]. Increasingly, these studies have focussed on implications of maternal nutrition during early- to mid-pregnancy when organogenesis of tissues occurs, including those of commercial importance to livestock producers.

However, it has long been recognised [30, 115, 120] that it is difficult during postnatal life to isolate specific effects on the developing fetus from those due to postnatal maternal effects in offspring suckled by their dams. Hence, cumulative effects of pregnancy and lactation on lamb growth are implicated, and artificial rearing on a standardised regimen is required to isolate influences of nutrition during pregnancy per se. This is particularly important in developing an understanding specific to influences on the fetus because of the magnitude of effects of the postnatal environment, particularly the supply of nutrients, on performance and health of the neonatal lamb.

Therefore, in the next section, we have reviewed studies that involve either artificial rearing, or rearing on the dam within pastoral and intensive systems, in an attempt to establish the extent to which environmental manipulations during pregnancy impact directly on the fetus and/or as a result of indirect effects mediated via the dam or other factors during early-postnatal life. The reader is also referred to other reviews of postnatal consequences of nutrition of ruminants during pregnancy and of fetal growth-retardation [4, 10, 11, 31, 46–48].

1.2.1.1 Studies of Sheep Reared Artificially to Weaning

Postnatal Growth Is Generally Reduced by Severe, Chronic Fetal Growth-Retardation

In sheep, it was recognised in the 1960s that postnatal growth to mature size is compromised when severe, chronic nutritional restriction is imposed throughout gestation [115] or from mid- to late-gestation until term, but not as a specific consequence of early-gestational nutritional restriction [30]. When lambs varying in birth weight are removed from their dams at or near birth and reared artificially, postnatal growth is generally reduced among small compared to larger newborns [49, 103, 115, 120, 129].

Timing, Duration and Severity of the Nutritional Insult Influence Postnatal Growth

Effects of adverse ewe nutrition during early- to mid-gestation on postnatal growth can be overcome by adequate maternal nutrition during mid- to late-gestation. In the study of [120], reduced fetal growth due to severely restricted maternal nutrition from birth to 90 days of gestation (490 vs. 548 g) [29] was completely overcome (4.2 vs. 4.3 kg birth weight) by high nutrient availability to ewes from 90 days to term. Postnatal growth during artificial rearing of lambs to 20 weeks of age was unaffected, provided that the inadequate ewe nutrition earlier in gestation was followed by adequate nutrition from 90 days of gestation to term [120].

The magnitude of postnatal effects on growth are greater due to severe maternal nutritional restriction and fetal growth retardation during late- than during early-gestation, and are greater among lambs of ewes whose nutrition is compromised throughout all of gestation. Differences in live weight due to nutrition from 90 days of pregnancy to term persisted to 20 weeks of age among artificially-reared lambs, with the magnitude of this mid- to late-gestational effect greater in lambs of ewes also undernourished during the first 90 days of pregnancy (approximately a 6 kg reduction) compared to those well-nourished during the same period (approximately a 1.5 kg difference) [120].

Further to the above findings, in comparing twins and singles that differed less in birth weight (3.5 vs. 4.6 kg) than the above study, differences in live weight were no longer evident by about 40 days of age among lambs artificially reared to day 70 of postnatal life [36].

The Magnitude of Effects of Prenatal Nutrition and Growth-Restriction for Postnatal Growth Vary with Postnatal Environment

There are additive long-term effects of inadequate prenatal nutrition throughout pregnancy and during artificial rearing to weaning on long-term growth which resulted in approximately 12 kg difference in live weight at 180 weeks of age compared to lambs well-nourished during both periods [115]. Differences of approximately 6 kg in live weight were attributable to either prenatal nutrition or postnatal nutrition alone.

Effects of fetal growth retardation on live weight at birth may be magnified by the time of weaning. For example, a 1 kg difference in birth weight of lambs reared on their dams for two days then artificially-reared resulted in a 2.6 kg difference in weaning weight at 6 weeks of age, with growth rate correlated with birth weight until 10 weeks of age following artificial rearing, but not thereafter [129]. In this study, differences in postnatal growth due to birth weight were greatest (66 g/kg birth weight/day) during the first week postpartum, and were lower over the entire period to weaning (39 g/kg birth weight/day).

Despite this, when environmental conditions from birth to weaning are optimised by ensuring adequate body weight-specific intake of high quality colostrum and thermoneutrality in a controlled environment throughout neonatal life, adverse effects on neonatal lamb growth are small and were only apparent during the immediate postpartum period [49]. These artificially-reared, severely growth-retarded lambs (2.3 vs. 4.8 kg birth weight) grew more slowly (248 g/day) than their normal birth weight counterparts (353 g/day) during the first two weeks of postnatal life, despite higher relative rates of gain. Thereafter, their absolute growth rates to 20 kg live weight (approximately 350 g/day) almost exactly matched those of the normal birth weight lambs when both groups were fed unlimited amounts of a high quality milk-replacer. In the context of fetal growth, it is also important to recognise this placentally-mediated, fetal growth-retardation model [52] produced lambs that diverged in fetal weight after 85 days of gestation [49, 50, 52].

Conclusions from Artificial Rearing Studies in Sheep

Taken overall, these findings suggest that prolonged, severe, chronic fetal growth retardation is required to adversely influence postnatal growth. They indicate that adverse effects of nutrition during early- to mid-pregnancy on growth can be overcome by adequate nutrition during late-pregnancy, however, if severe, effects of nutritional restriction and fetal growth retardation during late-pregnancy may persist. Furthermore, it appears that direct prenatal effects on capacity for growth of neonates are somewhat ameliorated or exacerbated depending upon the postnatal environment into which they are born. Hence, when maternal, social and/or nutritional and environmental disadvantages [90–91] are minimised, adverse effects of prenatal growth restriction per se on postnatal whole body growth and live weight may be minimised.

1.2.1.2 Studies of Sheep Reared to Weaning on Their Dams

Growth Rate of Lambs Reared to Weaning on Their Dams, but Not Thereafter, Is Generally Correlated with Birth Weight

Lambs reared on their dams grew 46 g/day slower per kg reduction in birth weight during the first week of postnatal life [131]. From birth to weaning at 6 weeks of age, growth rate of singletons varied by 34 g/day/kg and of twins by 28 g/day/kg difference in birth weight [129]. Weaning weight at 6 weeks of age was 2.4–2.7 kg lower per kg decline in birth weight among the singleton lambs reared on their dams, but differences in live weight stabilised from weaning to 11 weeks of age due to lambs of varying birth weight growing at equivalent rates (266 g/day) [129].

Similarly, in two studies on consequences of fetal growth retardation due to placental insufficiency, lambs of low birth weight (2.8 and 2.9 kg) reared on their dams failed to catch up in body weight to their high birth weight (4.3 and 4.4 kg) counterparts by 4 weeks (8.5 vs. 11.1 kg) or 8 weeks of age (12.7 and 12.9 vs. 15.8 and 16.4 kg) [85–86]. However, the lambs of low birth weight caught up to their high birth weight counterparts post-weaning and remained at equivalent weights as the high birth weight lambs to two years of age when they weighed 59.0 and 57.6 kg, respectively [86].

Lambs of adolescent ewes overfed during pregnancy, which resulted in reduced placental development and colostrum production, had lower birth weight (3.2 vs. 5.1 kg) and growth to weaning (weaning weight 29.3 vs. 33.9 kg) [25]. The adverse effect in males of the prenatal nutritional treatment on birth weight (2.7 vs. 5.2 kg) and pre-weaning growth rate (304 vs. 376 g/day) was more pronounced than in females (3.4 vs. 5.0 kg and 311 vs. 343 g/day), and lamb birth weight was correlated with weight at weaning irrespective of nutritional treatment. Interactions between prenatal nutrition and growth with sex were also apparent for post-weaning growth. Prenatally-growth restricted male lambs remained lighter or tended to be lighter than their well-grown counterparts until conclusion of the study at 42 weeks of age, whereas the females differed in live weight at 15, 20 and 25 weeks of age, but not thereafter.

Moderate undernutrition of ewes at 0.7–0.8 of estimated requirements during the last 6 weeks of pregnancy reduced birth weights compared to offspring of ewes fed to estimated requirements [117]. In this study, differences in offspring live weights due to nutrition of the ewe during pregnancy were evident at weaning and to one year of age when ewes were fed at low levels during lactation, but not when they received higher levels of nutriment during lactation. Post-weaning growth to 45 kg live weight was reduced, however, in lambs of ewes fed a restricted level of energy during the final third of pregnancy (198 vs. 257 g/day) compared to those of control fed ewes [40], the lambs of restricted energy fed ewes also exhibiting slower growth to weaning (200 vs. 162 g/day) [39]. Despite this, effects of nutrition during late-pregnancy were not apparent beyond 12 months of age, and differences in body condition score due to nutritional level in late-pregnancy were not evident at 12, 18 or 24 months of age [117].

Severely restricted nutrition during the final six weeks of gestation tended to reduce birth weights, but did not influence live weights of lambs reared on their dams or on foster dams to 33 weeks of age, whereas birth type and rearing type significantly affected postnatal growth [93]. Similarly, severe undernutrition of ewes for 20 days, but not for 10 days, during late-pregnancy reduced birth weights of lambs, but neither treatment had significant effects on weight at weaning at 3 months of age or at 5, 12, 22 or 30 months of age compared to lambs of ewes well-nourished during the same periods [99–100]. Enhanced ewe nutrition during the final 100 days of pregnancy increased birth weights of ewe lambs by 0.5 kg compared to those of control fed ewes. These lambs tended to be heavier at weaning, however, differences in live weight and condition score due to prenatal nutrition were not evident at 18, 30 or 42 months of age [59].

Interestingly, repeated bouts of stress of ewes during the last third of gestation increased lamb birth weights and tended to increase lamb weaning weights, but did not affect lamb weight at 8 months of age [113]. This finding suggests that conditions that result in ewes mobilising stored nutrients may enhance fetal growth, provided adequate nutrition is available to the ewe.

Moderate Undernutrition During Mid- and Late-Pregnancy Has Little Long-Term Effect on Growth, Unless Pre-weaning Growth Is Also Compromised

Sub-maintenance nutrition of ewes from days 50 to 140 of gestation resulting in ewes being 12.1 kg lighter than control ewes reduced birth weight by 0.5 kg (5.0 vs. 5.5 kg) compared to control-fed ewes [74]. Lamb weights were significantly lower to weaning at 12 weeks of age (30.9 vs. 32.5 kg) but were not significantly different during growth to 6 years of age (80.5 vs. 81.4 kg). In a second experiment in which ewes were fed at sub-maintenance levels during pregnancy and lactation [74], lamb birth weights were again reduced by 0.5 kg relative to control-fed ewes (4.2 vs. 4.7 kg). In this case, weaning weights were substantially lower for lambs of ewes fed at sub-maintenance levels (19.9 vs. 29.7 kg), and live weights remained significantly lower during growth to 4 years of age (70.2 vs. 72.6 kg). At the same age within the study specific to nutrition during pregnancy, live weights did not differ between offspring of sub-maintenance and control-fed ewes (72.6 vs. 73.7 kg). As mentioned previously, post-weaning growth to 45 kg live weight was reduced in lambs of ewes fed a restricted level of energy during the final third of pregnancy compared to those of control fed ewes, and these lambs also grew more slowly to weaning [39–40].

Nutrition During Early- to Mid-Pregnancy Has Little Direct Effect on Postnatal Growth

Since the studies of Everitt [29–31], there have been numerous experiments on influences of nutrition during early- to mid-pregnancy. Effects may or may not be evident during early postnatal growth when lambs are reared to weaning on their dams, and in most instances adverse effects of restricted nutrition do not persist during later postnatal life.

Lambs of ewes fed at 70% of nutritional requirements from 30 days prior to mating until 100 days of pregnancy had higher neonatal mortality rates, averaged 0.6 kg lighter at birth and grew more slowly after weaning than those fed at 100% of estimated requirements [97–98]. These factors, particularly reduced post-weaning growth, contributed to lambs of restricted ewes taking 18 days longer to reach the target weight of 58.5 kg. Pre-conception nutrition resulting in low maternal body condition at conception (≤ 2 vs. ≥ 3) did not affect birth weight compared to control ewes (3.8 vs. 3.5 kg), whereas restricted nutrition (50 vs. 100% of estimated metabolisable energy requirements) during early- to mid-pregnancy resulted in heavier newborns than controls (4.7 vs. 3.5 kg) [45]. Postnatal growth rates and weight at 6 months of age of these lambs did not differ significantly due to ewe pre-mating condition score (29.4 kg) or early- to mid-pregnancy nutritional restriction (31.4 kg) compared to lambs of control ewes (33.5 kg).

Nutritional supplementation for the first 50 days or from days 50 to 100 of pregnancy did not affect pre- or post-weaning growth to 60 days of age and to 30 kg live weight [92, 142]. Lamb birth weights and postnatal growth rates to slaughter at 35 kg live weight were not significantly affected in a series of experiments in which ewes were severely undernourished or adequately fed from mating to day 70 of gestation [81]. Similarly, nutrition of ewes from mating to 95 days of gestation at 50% or 100% of estimated metabolisable energy requirements had no effect on birth weight (4.0 vs. 4.4 kg, respectively) or weight at 3 years of age following growth at pasture (75.6 vs. 75.0 kg) [44]. Furthermore, severe nutritional restriction at 50% of estimated energy requirements from mating to 95 days of pregnancy did not significantly affect live weights of male or female lambs at birth, 6 weeks, or 20 months of age, nor body condition score at 20 months of age within a pastoral system [105]. Nutrition of pregnant ewes at pasture to reduce ewe condition score from 3 to 2 or to maintain condition score at 3 during the first 3 months of pregnancy had no affect on birth weight and lamb live weight at 8 months of age [23].

Recent large-scale Australian experiments [121] established the effects of a continuum of ewe nutritional levels during early or late gestation on lamb birth weights [32]. At sites in Victoria and Western Australia about 750 and 400 Merino ewes, respectively were exposed to 10 nutritional treatments in each of two years and their live weight profile during different stages of pregnancy was related to the birth weight of their lambs. This work confirmed that the effects of poor nutrition up until day 90 of pregnancy could be completely overcome by improving nutrition during late pregnancy. A loss of 10 kg in ewe live weight between joining and day 90 of pregnancy reduced lamb birth weight by about 0.3 kg, whereas gaining 10 kg from day 90 to lambing increased birth weight by about 450 g. The responses were consistent across birth rank, experimental sites and years (Table 1.1). There was no evidence that birth weight was consistently related to ewe live weight change during critical 'windows' of a few weeks during pregnancy. Few studies have reported the relative effects of ewe live weight change during different periods of pregnancy on lamb birth weight. However, from data reported in 13 studies using a range of breeds and nutritional interventions it is estimated that a 10 kg change in ewe live weight during the entire pregnancy changes birth weight by about 0.5 kg.

In the experiments described by Thompson and Oldham [121], differences in progeny live weight due to nutrition during early and mid-pregnancy or late pregnancy and lactation persisted to about 3.5 years of age (Table 1.2). When evident, the effects on mature live weight of ewe nutrition to day 90 of pregnancy and between day 90 and weaning were additive, as with earlier findings for sheep grown to younger ages [120]. Effects of rearing type on live weight also disappeared at older ages.

As with more chronic nutritional restriction during early pregnancy, acute, short-term early-gestational nutritional restriction (between days 1 and 35) did not affect

Table 1.1 Regression coefficients (± s.e.) that predict lamb birth weight (kg) as affected by ewe live weight at mating, ewe live weight change during early or late pregnancy (kg), birth type (single or twin) and sex. Data represents a combined analysis for two years (2001 and 2002) at research sites in Victoria and Western Australia (C.M. Oldham and A.N. Thompson unpublished results)

	Birth weight (kg)	
Factor	Victoria site ($n = 1995$)	Western Australia site ($n = 688$)
Constant	3.70 ± 0.16^a	3.90 ± 0.24^b
Ewe live weight at mating (kg)	0.03 ± 0.003	0.03 ± 0.004
Ewe live weight change from mating to 90 days (kg)	0.03 ± 0.004	0.03 ± 0.007
Ewe live weight change from 90 days to lambing (kg)	0.05 ± 0.004	0.05 ± 0.007
Twin lamb (kg)	-1.10 ± 0.03	
Female lamb (kg)	-0.19 ± 0.031	-0.26 ± 0.050

[a]Birth weight constant is for singleton males.
[b]Birth weight constant is for males.

birth weights, post-weaning growth or weight of ewes and rams at two to three years of age [102]. However, in lambs reared as twins, severe nutritional restriction for 10-day periods during the first 30 days of pregnancy, reduced weaning weights at 90 days of age by 4.7 kg (9.4 kg per ewe) compared to lambs of ewes better nourished throughout the first 30 days of gestation [69]. This latter finding suggests that nutrition during early pregnancy may be more important for postnatal growth of offspring of ewes that have greater maternal nutrient requirements due to multiple fetuses.

Nutrition of pregnant hoggets at pasture on low, medium or high levels of availability had no affect on lamb birth weight or survival to weaning, with live weight at weaning reduced in the offspring of the ewes fed low levels of pasture by 1.6 and 1.7 kg compared to offspring of those fed medium or high levels, respectively [95]. These findings, in which high levels of feeding did not adversely affect fetal growth and lamb birth weight are in contrast to those of Wallace et al. [132–133] who found that overfed pregnant adolescent ewes had retarded placental and fetal growth. As discussed [95], the differences between the findings probably reflect differences in the age and live weight of the ewes, maternal growth rates, pasture- vs. concentrate-based nutrition, and the use of pre-pubertal ewes and embryo transfer [95, 132–133].

Postnatal Growth Responses to Nutrition During Early- to Mid-Pregnancy May Vary Depending Upon the Severity of the Nutritional Regimen, Subsequent Nutrition, and Activity

In contrast to the above studies, ewes fed at 50% of estimated nutrient requirements from days 28 to 78 of pregnancy produced wether lambs that did not differ in weight

Table 1.2 Regression coefficients that predict live weight (kg) of progeny from weaning to 63 months of age as affected by ewe live weight at mating (kg), ewe live weight change (kg) during early pregnancy, late pregnancy or lactation and rearing type. Data represents a combined analysis for two years (2001 and 2002) at a research site in Victoria (A.N. Thompson and C.M. Oldham unpublished results)

	Age (months)					
Factor	Weaning	15	27	39	51	63
Number of offspring	1,324	1,103	1,021	768	608	380
Ewe live weight at mating (kg)	0.18	0.22	0.31	0.28	0.14	0.24
Ewe live weight change from mating to day 90 (kg)	0.24	0.14	0.18	0.18	n.s.	n.s.
Ewe live weight change from day 90 to lambing (kg)	0.19	0.13	0.12	0.17	n.s.	n.s.
Ewe live weight change from lambing to weaning (kg)	0.08	n.s.	n.s.	n.s.	n.s.	n.s.
Twin reared as singleton[a]	−1.7	−0.6	−0.3	−1.2	n.s.	n.s.
Twin reared as twin[a]	−4.8	−2.1	−2.1	−1.8	−1.6	n.s.

n.s. = not significant.
[a]Comparison with singleton born reared as singleton.

at birth but which tended to be heavier at 63 days of age (20.8 vs. 17.0 kg) and were significantly heavier at weaning at 120 days of age (26.6 vs. 21.8 kg) [34]. This difference in weight was maintained to the conclusion of the study at 8 months of age when the lambs weighed 61.7 and 56.8 kg, respectively [34, 141]. These studies investigated the impacts of early gestational undernutrition and realimentation on fetal and offspring growth and development under conditions of limited exercise, as a model for the so-called "couch potato" effect that resulting in increased fatness. The restricted nutrition resulted in the ewes losing approximately 10% of their body weight from days 28 to 78 of gestation due to nutrient restriction, then compensatory weight gain when ewes were realimented onto a control diet thereafter. The fact that fetal weights on day 78 of gestation were approximately 30% less than those of fetuses of control fed ewes, while birth weights of fetuses from nutrient restricted-realimented ewes and control fed ewes were similar, indicates compensatory growth of fetuses in utero in nutrient restricted ewes following realimentation, which may have continued into postnatal life. The ewes were maintained in individual pens throughout pregnancy which limited exercise and may also have contributed to their compensatory weight gain. Further, the wether lambs were allowed to experience maximal growth rates after weaning by allowing them unlimited access to creep feed from birth to weaning while being reared in small pens, and ad libitum concentrate-based feeding from 4 to 8 months of age.

Sheep Genotype May Influence the Postnatal Growth Response to Prenatal Nutrition

Recent research has identified marked differences between sheep genotypes in the postnatal response of their offspring to maternal nutrient restriction during early- to mid-pregnancy. Whereas a sheep genotype selected within a modern intensive production environment exhibited altered prenatal and postnatal growth characteristics [15, 34], those of similar breeding and subjected to a nomadic lifestyle that survive under conditions of limited nutrition were not affected by restricted nutrition from days 28 to 78 of pregnancy with respect to fetal growth and birth weight, or growth to and weight at 12 months of age [15].

In further studies on interactions between genotype of the sire and ewe nutrition during pregnancy, ewes mated to high growth or high meat production sires and given a low level (60% below total energy intake of high nutrient level ewes) of nutrients during late-gestation (day 105 to term) produced smaller lambs than those fed high levels in two experiments (3.1 and 3.3 vs. 4.0 and 4.2 kg) [127]. However, only within one experiment were differences between sire-genotypes in birth weight evident, and then only within ewes fed low nutrient levels, the high meat-sired lambs weighing less than high growth-sired lambs (2.6 vs. 3.4 kg). Lambs of ewes fed low levels of nutrient during late-gestation grew more slowly to and weighed less to weaning (18.0 vs. 23.9 kg) at day 56 than those of well-nourished ewes in both studies. However, they grew at similar rates from weaning to conclusion of one study at day 130, resulting in a trend towards lighter lambs from the poorly-nourished ewes (43.3 vs. 47.9 kg). In the other study, they exhibited some compensation, resulting in lighter lambs at day 141 only within the high meat-sired genotype.

Shearing During Early- to Mid-Pregnancy May Increase Birth Weight and Growth to Weaning

In contrast to the generally limited effects of moderate nutritional restriction during early to mid pregnancy on birth weight, shearing during early to mid gestation enhances mobilisation of maternal body tissues [72], and has increased the birth weight of single or twin lambs by up to 17%, although results have been inconsistent [72, 75, 94, 107–108]. This has led to the proposal that the fetal growth response to shearing occurs in ewes that would otherwise give birth to a low birth weight newborn, and the ewe must have adequate maternal reserves and/or be fed adequately to support increased fetal growth [76]. More recent larger-scale studies have supported this assertion and shown that mid-pregnancy shearing of ewes carrying multiple fetuses increased birth weights by 0.13–0.44 kg, lamb survival rates to weaning by 5.5%, and weight at weaning by 1.07 kg [77]. By contrast, hoggets shorn in mid-pregnancy produced 0.32 kg heavier newborn singleton lambs than their unshorn counterparts, however, birth weight of multiple-born lambs was unaffected [78]. In this latter study, neither mid-or late-pregnancy shearing influenced live weight at or survival of lambs to weaning. To date, no published studies have assessed longer-term impacts of shearing during pregnancy on performance of offspring.

Other Nutritional and Metabolic Treatments

Increasingly, studies are being conducted that seek to limit intrauterine growth retardation by altering the uterine environment in which the embryo resides and by increasing placental blood flow and/or the amount and quality of nutrients supplied to the fetus. These studies which have been recently reviewed [139] have included: maternal supplementation with arginine and other factors that increase placental blood flow and fetal growth; supplementation with selenium due to its antioxidant properties; and administration of somatotropin and other growth promotants or repartitioning agents to the pregnant dam, with a view to enhancing embryonic, placental and fetal growth and muscle development. Supplementation with factors that increase placental blood flow have had positive effects in utero however, postnatal consequences have not been reported to date.

Treatment of ewes with somatotropin (i.m. injection of 25 mg of bovine GH) at mating increased birth weight compared to lambs of control treated ewes (5.1 vs. 4.6 kg). Growth rate of singleton but not of multiple reared lambs of somatotropin treated ewes was greater to 75 days compared to those of control ewes, suggesting some capacity alter conceptus development by treatment of ewes at mating with metabolic modifiers [21].

Conclusions from Studies Where Lambs Are Reared on Their Dam

Restricted fetal growth that results in low birth weight generally causes slower growth to weaning, but not necessarily thereafter; however if the restriction in fetal growth is chronic and severe, reduced mature size may result. Maternal nutrition prior to and during early- to mid-pregnancy has variable effects on postnatal growth. Differences between studies in the effects of maternal nutrition during early- to mid-pregnancy on postnatal growth are difficult to reconcile. However, they probably relate to factors including the magnitude, timing and duration of the nutritional restriction; the extent to which they impact on the dam and the fetus; nutrition during the recovery period and potential for compensatory weight gain by the dam and compensatory growth by the fetus; carry-over effects of the nutritional regimens during pregnancy on lactational performance of the dams; the availability of pasture or creep feed from birth to weaning; the extent of confinement under which the lambs are reared; and potential for postnatal confounding because of factors such as litter size. Interpretation in relation to consequences of fetal programming of studies in which lambs are reared to weaning by their dams is difficult because of the potential for confounding by effects such as post-treatment nutritional regimens during later pregnancy, and because of postnatal factors such as milk supply and intake.

1.2.1.3 Studies of Cattle

Overall, birth weight of calves did not significantly influence live weight gain of calves reared artificially to weaning on a high plane of nutrition [125]. However, male calves of dams fed a low plane of nutrition during the final trimester had significantly higher growth rates during artificial rearing to weaning than those born to dams on a high plane during the same period, while the opposite occurred for female calves [124–125]. It is not clear from these studies, whether calves differing in birth weight and/or prenatal nutrition differed in their growth after weaning.

Birth weight of calf is positively correlated with calf growth rate to weaning and live weight at weaning when calves are reared on their dams [14, 17, 20, 54, 63–64, 114, 137]. For example, calves that had birth weights 5.4 and 5.9 kg lower than those from cows well-nourished during late-pregnancy were 16.5 and 17.2 kg lighter at weaning [63–64]. As with the sheep studies described previously, it is not possible to fully separate out consequences of nutrition during pregnancy on the fetus when offspring remain on their dams to weaning due to carry-over effects on maternal performance (see [49] for an example of a rearing system designed to uncouple prenatal and postnatal influences). The extent to which reduced calf growth to weaning is due to reduced calf growth capacity, or to factors including maternal milk production as affected by prenatal nutrition and cow body composition and, hence nutrient reserves, and differences in sucking stimulus and milk removal between small and large newborns is unknown. However, the findings of Tudor et al. [126] with artificially-reared calves, coupled with studies in which calves are reared on their dams [17], show that the postnatal maternal contribution to pre-weaning growth of calves is substantial. Irrespective, the net effects of maternal nutrition during pregnancy on the calf remain of practical significance to livestock producers.

Our recent studies have demonstrated that fetal growth restriction resulting in 10.2 kg or 26% lower birth weight (Table 1.3) may limit the capacity of cattle to exhibit compensatory growth. Cattle significantly growth-retarded during fetal life due to factors including severely restricted maternal nutrition from day 80 or so of pregnancy to parturition remained smaller at any given postnatal age to 30 months compared to their well-grown or better nourished counterparts (Table 1.3 and [53]). Growth of very low birth weight cattle was significantly slower than that of high birth weight cattle at all stages of postnatal growth, although pre-weaning growth was influenced by maternal nutritional status during pregnancy [17]. Whether this represents a permanent stunting or simply a delay of attainment of mature size of cattle is open to conjecture.

It should also be noted that differences in weight of calves at birth following three levels of maternal nutrition during late-pregnancy disappeared by weaning when postnatal nutrition was of high quality and availability, although residual effects of the previous years nutrition did influence calf growth [65]. Similarly, effects of variable nutrition during mid and/or late pregnancy on weight at birth can be overcome by adequate nutrition postpartum, and resulted in no differences in calf body weights at 58 days of age [35].

Twin cattle are lighter at birth and grow more slowly on their dams to weaning [62]. They may grow more slowly [56], at a similar rate [112], or more rapidly [19, 62, 134] post-weaning than singletons, depending upon the rearing system and subsequent nutrition. However, twin cattle tended to consume less feed in feedlot than singletons, due primarily to their lower live weight [112].

It is concluded from the above studies in cattle, that severe, prolonged intra-uterine growth retardation may result in slower growth throughout postnatal life, resulting in smaller animals at any given postnatal age. However, provided the

Table 1.3 Consequences of growth in utero for growth and live weight characteristics of beef cattle to 30 months of age [from 48, 53]

Variable	Prenatal growth/birth weight Low ($n = 120$)	High ($n = 120$)	Significance of difference (P)
Birth weight (kg)	28.6	38.8	<0.001
Pre-weaning ADG (g)	670	759	<0.001
Weaning (7 months) weight (kg)	174	198	<0.001
Backgrounding ADG (g)	571	603	<0.001
At equivalent age (26–30 months)			
Feedlot entry (26 months) weight (kg)	481	520	<0.001
Feedlot ADG (g)	1,480	1,617	<0.001
Feedlot exit (30 months) weight (kg)	647	703	<0.001
At equivalent feedlot entry live weight (500 kg)			
Age at feedlot entry (day)[a]	797	715	<0.001
Feedlot ADG (g)	1,515	1,583	0.019
Feedlot exit weight (kg)	671	679	0.017
Age at feedlot exit (day)[b]	914	833	<0.001

Values are predicted means from REML analyses including effects of birth weight, pre-weaning nutrition, sex/year cohort, sire-genotype and their interactions. Maternal nutritional treatments commenced between days 30 and 90 of pregnancy (refer to [17, 15] for details of pasture-based nutritional treatments and selection criteria for calves used to study long-term consequences of growth early in life).
[a]Predicted from mean average daily gain (ADG) during backgrounding.
[b]Predicted from mean ADG during background and mean feedlotting period.

maternal nutritional insult during pregnancy does not limit the capacity of the dam to lactate, restricted prenatal nutrition resulting in restricted fetal growth and low birth weight may be overcome during growth to weaning by adequate nutrition during lactation.

1.2.2 Nutrient Intake, Digestibility and Efficiency of Nutrient Utilisation

The extent to which the prenatal environment and/or fetal growth retardation affects nutrient intake, digestibility and efficiency of nutrient utilisation has received relatively little attention, with most studies focussing on consequences of birth weight. Total nutrient availability during the pre-weaning and post-weaning periods are highly correlated with nutrient intake, digestibility and efficiency of nutrient utilisation, hence the potentially confounding role of variation in the supply of nutrients to lambs fed on their dams is an important consideration for studies of prenatal impacts on these parameters.

Low birth weight lambs (2.3 kg) take longer to achieve positive weight balance and a net gain in weight takes than normal birth weight lambs (4.8 kg) immediately postpartum, even when reared in an optimal environment [49]. During the early postpartum period, digestibility of feed and various feed components are positively correlated with birth weight [66], although amounts and activity of digestive enzymes were not strongly associated with birth weight in young ruminants [57].

Normal birth weight lambs consume more nutrients than those of low birth weight to any given age during the pre-weaning period, primarily due to differences in weight at birth [49, 129]. However, low birth weight lambs consume more nutrients to any given live weight because of the greater time required to reach a given weight [39–40, 49]. Low birth weight lambs ate 13 and 20% more feed than normal birth weight lambs, respectively, when reared rapidly or slowly to 20 kg live weight [49].

Low birth weight lambs consumed 45 g more feed per kg live weight per day which equated to 20% higher relative intake than normal birth weight lambs fed ad libitum during the early postpartum period, and higher relative intakes were evident until about 25 days of age [49]. Furthermore, offspring of ewes with experimental placental restriction feed more frequently than those of ewes with normal placental development [26]. However, at equivalent live weights during ad libitum or restricted feeding to 20 kg live weight, differences in relative feed intake due to birth weight were not evident [49]. In contrast, others have found that relative intake was not affected significantly by birth weight during the first five weeks of postnatal life, although birth weight and relative feed intake were correlated during week six [129]. In contrast to the above findings in artificially reared lambs [49], relative intake did not differ due to birth weight in lambs reared on their dams during the first week of postnatal life [131].

Feed to gain ratios were lower and hence efficiency of utilisation of feed was higher in small newborn lambs, overall, during milk-feeding to 20 kg live weight than in normal birth weight lambs [49]. This was due at least in part, however, to differences in efficiency during the early postpartum period when the ratio of gain to maintenance was higher in the small compared to normal size lambs. Hence, feed to gain ratios differed little from 10 to 20 kg live weight [49]. However lambs reared on ewes fed low levels of nutrients for six weeks pre-partum were more efficient (lower milk intake per g body gain) from days 14 to 21, but not days 0–14, postpartum than those of ewes well-nourished during late-pregnancy [127]. Again, these findings may relate to differing size of lambs at the same age. Retention of ingested dry matter, nitrogen, lipid, ash and energy from birth to 20 kg live weight did not differ between low and normal birth weight lambs reared artificially, although slowly reared low birth weight lambs tended to deposit more fat per unit of fat ingested than the normal lambs [49]. Feed efficiency did not differ with birth weight during the first week of life during rearing on the dam [131] or from birth to weaning during artificial rearing [129].

During the period from weaning to 35 kg live weight, birth weight was not related to feed intake and feed efficiency [129]. Furthermore, voluntary feed intake from weaning to 2 years of age was not affected by moderate maternal nutrient restriction at 0.7–0.8 of estimated energy requirements during the final 6 weeks of pregnancy that resulted in somewhat reduced birth weights (3.2 vs. 3.5 kg) compared to ewes fed to requirements [117]. In lambs of ewes fed a restricted level of energy during the final third of pregnancy, post-weaning feed intake per day was not affected compared to lambs of control fed ewes (765 vs. 757 g/day, respectively) [40]. Similarly, differences in feed intake and efficiency were not evident in lambs grown to 17 and 24 weeks of age following restricted nutrition of ewes to 50% of predicted requirements during early- to mid-pregnancy (days 30–70) [24].

Recently, feed intake and efficiency was measured in 24 adult Merino wethers from ewes that experienced extreme differences in nutrition during pregnancy and lactation [123]. The average condition score of the ‘Low’ and ‘High’ ewes was 2.7 vs. 2.6 at joining, 2.3 vs. 2.8 at day 90 of pregnancy, 2.4 vs. 3.4 at lambing and 2.1 vs. 3.1 at weaning. Single-born lambs from the ‘Low’ group were lighter at birth (4.6 vs. 5.9 kg) and at weaning (14.7 vs. 22.2 kg) than those from the ‘High’ group. Both groups grazed together after weaning and there were no significant differences in live weight at 3.5 years of age when housed indoors and fed ad libitum for 8 weeks following an adaptation period. Those from poorly fed ewes tended to grow

Table 1.4 Consequences of growth in utero for feed intake and efficiency of beef cattle during feedlotting from 26 to 30 months of age [48]

	Prenatal growth/Birth weight		
Variable	Low ($n = 77$)	High ($n = 77$)	Significance of difference (P)
Birth weight (kg)	28.1	38.4	<0.001
At equivalent age (26–30 months)			
Feedlot entry (26 month) weight (kg)	466	513	<0.001
Feedlot ADG (kg)[a]	1.279	1.396	0.004
Feed intake (kg/day)[a]	13.21	14.63	<0.001
Feed efficiency (Feed DM intake/Feedlot ADG)	10.00	10.38	0.263
Residual Feed Intake (kg)[b]	−0.005	0.003	0.986
At equivalent feedlot entry live weight (490 kg)			
Feedlot ADG (kg)[a]	1.317	1.361	0.463
Feed intake (kg/day)[a]	13.86	14.01	0.546
Feed efficiency (Feed DM intake/Feedlot ADG)	10.25	10.15	0.888

Values are predicted means from REML analyses including effects of birth weight, pre-weaning nutrition, sex/year cohort, sire-genotype and their interactions, with feedlot entry weight as a covariate (linear and, where significant, quadratic) to predict means at equivalent feedlot entry weight. Difference in significance of feedlot average daily gain (ADG) between Tables 1.3 and 1.4 is due to the number of cohorts studied (three cohorts in Table 1.3 vs. two cohorts in Table 1.4) and the duration of the measurement period (average of 117 days in Table 1.3 vs. 70 days in Table 1.4).

[a]During 70-day period in feed intake pens.

[b]At mean metabolic live weight.

more slowly (131 vs. 173 g/day), eat less (1.51 vs. 1.65 kg dry matter/day) and be less efficient at converting feed into live weight gain (11.4 vs. 10.0 kg intake/kg gain) over the 8-week period than those from well fed ewes, although these differences were not statistically significant [123]. While it was not possible from this work to separate out the effects of late gestation versus lactation on nutrient intake and utilisation, the potential for early life nutrition to influence appetite and feed conversion efficiency during adulthood requires confirmation in sheep.

Feed intake and efficiency of cattle have also been recently measured at 26–30 months of age in a large study of heifers and steers sired by diverse genotypes (Wagyu, lower birth weight and high intramuscular fat and Piedmontese, higher birth weight and high muscling). Two cohorts comprising steers and heifers from low and high nutritional pasture-based systems that were divergent in birth weight and growth rate to weaning at 7 months were studied [48]. Differences in feed intake and feed efficiency were not evident, nor were there differences in net feed intake due to differences in growth during fetal life (Table 1.4), nor were there interactions between prenatal growth and pre-weaning growth or sire-genotype. Similarly, differences in post-weaning efficiency of utilisation of feed by offspring have not been evident in studies of maternal supplementation during late pregnancy, or in twin compared to singleton offspring (for review see [48]).

1.2.3 Body and Carcass Composition

1.2.3.1 Sheep

Severe fetal growth-retardation of lambs resulting in substantially reduced birth weight results in increased fatness of lambs. During growth to 20 kg live weight, low birth weight lambs (2.3 vs. 4.8 kg) accreted fat and energy at greater rates, and ash, nitrogen and muscle at slower rates, than normal birth weight lambs, resulting in greater fatness [49, 51]. This was due primarily to greater relative feed intakes during the first 2–3 weeks of postnatal life coupled with continued accretion of fat at greater rates during subsequent postnatal growth compared to normal birth weight lambs. Consequently, low birth weight was associated with increased fatness and reduced bone mass during postnatal growth to 35 kg live weight [130] and to 2.3 years of age [86].

Restricted energy supply to ewes during the final third of pregnancy resulted in lambs with increased omental, mesenteric, perirenal and pelvic fat, but not inguinal fat at 45 kg live weight, compared to lambs of control fed ewes [41]. Furthermore, lambs of ewes restricted to 50% of requirement from day 110 of pregnancy to term had significantly more perirenal and omental fat and relative fat mass at one year of age compared to lambs of control ewes and those of ewes restricted to 50% of requirements from days 0 to 30 of pregnancy [37]. These findings are consistent with those following restriction of fetal growth due to placental insufficiency [26]. In contrast, the effect of more acute maternal nutritional restriction during late-pregnancy on fatness was not significant at 30 months of age, with female progeny of pregnant ewes undernourished for 20 days tending to weigh less and have reduced subcutaneous fat depth [99]. Furthermore, there was no effect on fatness of offspring at 30 months of age when ewes were undernourished for 10 days during late-pregnancy [99].

In many instances, severe nutritional restriction during early pregnancy does not adversely influence body composition during postnatal life. At 58 kg live weight and 29 kg carcass weight, muscle mass and myofibre characteristics were not significantly affected in offspring of ewes were undernourished from 30 days prior to conception until 100 days post-conception [97–98]. Furthermore, in three experiments in which ewes were undernourished for the first 70 days of pregnancy, effects on size and other characteristics of muscle in offspring at 35 kg live weight and 15–16 kg carcass weight were not evident [81]. In these studies, postnatal growth rate was not affected [81] or was slightly reduced [97–98] by adverse nutrition of ewes. Similarly to these studies, various measures of fatness and body composition of offspring during growth up to 3 years of age were not significantly affected by severely restricted maternal nutrition during days 0–30 [37], days 28–80 [42, 45], days 0–90 [101] and days 0–95 [44] of pregnancy. This is despite evidence indicating reduced myofibre number [30, 119] and altered secondary to primary myofibre ratios in fetuses of ewes severely undernourished early in pregnancy [104, 140], although effects on myofibre number, size and/or types were not evident at market weights [81, 98]. In this regard, growth of sheep from birth to weaning has been shown influence myofibre characteristics during the post-weaning period, although these effects did not persist [55].

However, in studies in which growth to weaning and weaning weight were substantially greater in offspring of nutrient-restricted ewes from days 28 to 78 of gestation, offspring remained heavier (61.7 vs. 56.8 kg) and had significantly greater subcutaneous fat depth and more kidney and pelvic fat at 8 months of age [34, 141]. Similarly, evidence has been provided of a tendency towards a shift in the ratio of fat to lean in lambs grown to 24 weeks of age, which was not evident at 17 weeks of age, following severe maternal restriction from days 30 to 70 of pregnancy [24]. In contrast to the results of the above studies [34, 141], in this case lambs from nutritionally-restricted ewes were lighter at 24 weeks of age [24]. An increase in indices of fatness and poorer confirmation was observed in lambs from ewes fed at 60% of predicted metabolisable energy requirements compared to those from ewes fed at 100 and 175% of requirements from days 0 to 39 of pregnancy, but not in those from ewes that were undernourished from days 40 to 90 days of pregnancy [96]. In this latter study, the lambs from all maternal nutritional treatments had similar postnatal growth and carcass weights.

Prolonged maternal underfeeding during mid- to late-pregnancy and lactation may also increase fatness of sheep in the long-term. Although little difference in eye muscle and back fat depth were evident at about 30 kg live weight [101] or at 42–45 months of age [123] due to these factors, at maturity a sub-set of these lambs from poorly fed ewes were significantly fatter than their counterparts from ewes well-nourished during the same period [123]. While body composition of adult wethers measured using dual energy x-ray absorptiometry was most closely related to their live weight, after correction for live weight the proportion of fat was greater (34.6 vs. 27.4%) and of lean was less (65.8 vs. 71.9%) for the low than the high maternal nutritional groups, while there were no significant differences in bone mineral content. In this work, more than 80% of the variance in the proportions of total fat and lean was explained by differences in live weight of progeny, ewe live weight at joining and changes in ewe live weight between joining and day 90 of pregnancy and day 90 and lambing. The model coefficients suggested that a loss of 10 kg in ewe live weight between joining and day 90 of pregnancy or between day 90 of pregnancy and lambing increased fatness and reduced percentage lean by 5–7%. As the effects on total body fat were not evident from measurements of back fat depth measured by ultrasound, the extra fat that resulted from nutritional stresses early in life may well have been located in the abdominal region, consistent with findings in younger sheep [26, 37, 41], which is significant because central obesity has been linked to increased incidence of metabolic, cardiovascular and other diseases [87].

Despite the above findings, supplementation of pregnant ewes with corn and soybean from days 1 to 50, days 51 to 100 or days 101 to term compared to ewes fed pasture only throughout pregnancy had no effect on commercial carcass characteristics, including yield and indices of conformation, of offspring reared in confinement to approximately 30 kg live weight [142] although specific measures of fatness were not reported.

It can be concluded from the above studies that prolonged, severe growth retardation resulting in low birth weight or prolonged, severe restricted maternal nutrition during late pregnancy predisposes sheep to increased fatness during postnatal life. As with effects on postnatal growth, maternal nutrition prior to and during early- to mid-pregnancy often has little or no effect on body composition at equivalent postnatal live or carcass weights. However, some studies have shown effects on adiposity during postnatal life. These conflicting results are difficult to reconcile, and research is required to determine whether they result from direct effects on the fetus during or after the treatment period that may alter myogenesis and the potential for muscle hypertrophy, or from indirect effects on adipocyte development mediated via the dam between birth and weaning. In this regard, primary myogenesis occurs at days 32–38 in the sheep and myogenesis continues until about day 110, after which, muscle hypertrophy commences [50, 88, 136]. Certainly, severe nutritional effects on the fetus during myogenesis may influence subsequent muscle development; however, very severe maternal restriction is required to affect fetal growth during this period compared to later pregnancy when fetal muscle satellite cell activity can be markedly altered by maternal nutrition [50]. It is also worthy of note that the effect on postnatal fatness and muscle mass of nutrition during the milk-feeding phase is greater than the effect of prenatal growth restriction [49, 51], and it is possible that the postnatal factors described previously may have contributed to affects on body composition. Clearly, the importance of nutritionally mediated effects on early-life programming of body composition during growth and adulthood in the context of developing practical ewe feeding systems and marketing systems requires further investigation.

1.2.3.2 Cattle

Large reductions in birth weight due to factors including severe maternal nutritional restriction did not influence indices of fatness, apart from P8 (rump) fat, in carcasses of Wagyu- or Piedmontese-sired steers and heifers at 30 months of age, beyond that normally attributable to differences in live or carcass weight (Table 1.5). Low birth weight cattle had a similar retail yield, fat trim and bone content at equivalent carcass weight, suggesting little overall difference in carcass composition than their high birth weight counterparts. However, ossification score was higher in low compared to high birth weight calves, suggesting an impact of prenatal growth on maturity [54]. Among cattle less variable in birth weight, there were no effects of birth weight on postnatal growth and carcass compositional characteristics at 30 months of age [16].

Similarly, differences were not evident in dressing percentage, weight of organs, hide, various anatomical muscles and bones, height, and in gross chemical compositional in the whole body or the carcass, of Hereford steers or heifers grown to 370–400 kg live weight following restricted or adequate nutrition of their dams from 180 days of pregnancy to parturition that resulted in a 22% or 6.8 kg difference in calf birth weight [126].

The above findings are consistent with research on twin cattle which has demonstrated that despite significantly lower birth weights and reduced pre-weaning growth, compositional differences at equivalent slaughter weights or ages are small

Table 1.5 Consequences of growth in utero for carcass characteristics of beef cattle at 30 months of age [48, 54]

Variable	Prenatal growth/Birth weight Low ($n = 120$)	High ($n = 120$)	Significance of difference (P)
At equivalent age (30 months):			
Carcass weight (kg)	364	396	<0.001
Retail yield (kg)	239	257	<0.001
At equivalent carcass weight (380 kg):			
Eye muscle area (mm^2)	90.4	88.9	0.25
P8 fat depth (mm)	21.3	19.6	0.048
Rib fat depth (mm)	11.5	11.8	0.35
Aus-Meat marble score	1.83	1.86	0.56
USDA marble score	447	444	0.98
Ossification score	206	195	0.009
Retail yield (kg)	249	247	0.20
Bone (kg)	66.9	67.6	0.10
Fat trim (kg)	54.6	56.0	0.58

Values are predicted means from REML analyses including effects of birth weight, pre-weaning nutrition, sex/year cohort, sire-genotype and their interactions, with carcass weight as a covariate (linear and, where significant, quadratic) to predict means at equivalent carcass weight. Refer to Table 1.3 for growth characteristics of the cattle.

and not significant, with twins generally having similar or leaner carcasses than singletons [19, 56, 112, 134].

1.2.4 Meat Quality

Effects of nutritional restriction during early- to mid- or late-pregnancy on eating quality of sheep meat and/or myofibre characteristics at market weights were not evident [81, 98, 128]. Similarly, there were little or no effects on postnatal myofibre characteristics observed following early- to mid-gestational undernutrition [24] or following nutritional supplementation during early, mid or late gestation [92].

Consistent with the above findings, there were no adverse effects on objective measurements of beef quality in the *longissimus* or *semitendinosus* muscles of cattle or on myofibre characteristics at 30 months of age due to restricted growth in utero from early pregnancy to parturition that resulted in low birth weights (Table 1.6) [48, 54].

Table 1.6 Consequences of growth in utero for objective measurements of *m. longissimus* and *m. semitendinosus* quality in beef cattle at 30 months of age [48, 54]

	Prenatal growth/Birth weight		
Variable	Low ($n = 120$)	High ($n = 120$)	Significance of difference (P)
Longissimus			
Peak force (Newtons)[a]	39.2	40.5	0.26
Compression (Newtons)[a]	13.9	14.4	0.19
Cooking loss (%)	21.6	21.7	0.57
Ultimate pH	5.47	5.48	0.50
Colour L (lightness)	39.5	40.0	0.21
Colour a (red/green)	26.3	26.7	0.20
Colour b (yellow/blue)	13.6	13.8	0.15
Semitendinosus			
Peak force (Newtons)[a]	46.2	46.4	0.81
Compression (Newtons)[a]	22.6	22.7	0.97
Cooking loss (%)	21.5	21.3	0.52

Values are predicted means from REML analyses including effects of birth weight, pre-weaning nutrition, sex, sire-genotype and their interactions. Refer to Tables 1.3 and 1.5 for growth and carcass characteristics of the cattle.
[a]Objective measures of meat texture.

1.2.5 Wool Production and Quality

The quantity and quality of wool is related to characteristics of follicles in the skin [67, 71]. The development of the follicle population is affected by nutrition of the ewe during pregnancy and by lamb growth during the first few weeks following birth [1, 30, 70, 115–116]. Mechanisms of fetal development relating to follicle formation and postnatal fibre production were reviewed by [9]. Formation of secondary wool follicles is initiated between days 80 and 135 of gestation. Secondary follicles mature and commence fibre production just prior to birth until four to five weeks after birth [61, 116]. Nutritional restriction during late pregnancy can restrict secondary follicle initiation, and this effect is exacerbated by poor nutrition during the early postnatal period [73–74, 115]. The general consensus is that it is more difficult to influence the formation of primary follicles. Some studies have shown that progeny born to underfed ewes grow less wool that has a greater fibre diameter than that of progeny born to ewes fed adequately during pregnancy [27, 73] and that these differences persist throughout the life of the animal [74]. This is consistent with a reduction in the number of secondary follicles that are initiated and mature to produce a wool fibre, and these effects are similar to differences between single- and twin-born lambs [84].

Most studies of the effects of nutrition on fetal growth and development in relation to wool production have focused on late pregnancy or have considered a limited number of nutritional regimens, which may not be applicable to nutrition in more practical situations. Recently, we studied the effects of a continuum of ewe nutritional levels during early- to mid-pregnancy or late pregnancy and lactation on the amount and quality of wool produced by the progeny (Table 1.7). At sites in Victoria and Western Australia, between 400 and 750 Merino ewes were exposed to 10 nutritional treatments in each of two years and their live weight profiles

Table 1.7 Regression coefficients (± s.e.) that predict progeny wool characteristics at second shearing as affected by ewe live weight, rearing type and sex of progeny at research sites in Victoria and Western Australian. Data represents a combined analysis for 2001 and 2002 (A.N. Thompson and C.M. Oldham unpublished results)

Sites	Factor	Clean fleece weight (kg)	Fibre diameter (μm)
Victoria ($n = 1{,}100$)	Constant	2.87 ± 0.382	17.3 ± 0.52
	Ewe live weight at mating (kg)	0.010 ± 0.0028	n.s.
	Ewe live weight change from mating to 90 days (kg)	0.019 ± 0.0039	−0.03 ± 0.009
	Ewe live weight change from 90 days to lambing (kg)	0.019 ± 0.0035	−0.04 ± 0.009
	Twin reared as singleton[a]	−0.143 ± 0.0396	0.13 ± 0.103
	Twin reared as twin[a]	−0.274 ± 0.0316	0.48 ± 0.081
	Female[b]		0.29 ± 0.071
Western Australia ($n = 460$)	Constant	3.77 ± 0.133	19.0 ± 0.61
	Ewe live weight change from mating to 90 days (kg)	n.s.	−0.04 ± 0.015
	Ewe live weight change from 90 days to lambing (kg)	0.021 ± 0.0051	−0.04 ± 0.015
	Female[b]	−0.153 ± 0.0474	0.50 ± 0.120

[a]Comparison with singleton born reared as singleton for clean fleece weight; Comparison with male singleton reared as singleton for fibre diameter.
[b]Comparison with male.

during various stages of pregnancy and or lactation were related to the wool production capacity of their lambs. The results showed that improving the nutrition of Merino ewes during pregnancy and lactation increases the fleece weight and reduces the fibre diameter of their progeny's wool during their lifetime, and that the effects of poor nutrition up to day 90 of pregnancy could be completely overcome by improved nutrition during late pregnancy. A loss of 10 kg in ewe live weight between joining and day 90 of pregnancy reduced fleece weight by 0.2 kg and increased fibre diameter by 0.35 microns at the hogget shearing, whereas an increase of 10 kg in ewe live weight from day 90 to lambing had the opposite effect. The responses were consistent across birth rank, sire genotype, experimental site and year (A.N. Thompson and C.M. Oldham unpublished results), and there was no evidence that progeny wool production could be significantly manipulated by managing ewe live weight during acute windows of a few weeks' duration during pregnancy.

1.2.6 Reproductive Performance

1.2.6.1 Female Reproductive Performance

Sheep

Potential mechanisms relating to reproductive performance as a result of altered fetal development have recently been studied and reviewed [83, 109, 111] in sheep. The concept that nutrition during prenatal life may influence reproductive performance of offspring was investigated by Gunn [58], who demonstrated reduced numbers of lambs born, but not number of lambs marked, because of a low compared to a high nutritional environment from 6 weeks pre-partum until 12 months of age. Ovulation rate was reduced because of the low nutritional environment during this period but only among ewes receiving a low level of nutrition during adulthood. In a similar study [82], lifetime fertility was not affected in ewes from low (high available nutrition) and high (low available nutrition) stocking rate systems from conception to 15 months of age, although those exposed to low stocking rates weaned 50% more lambs in total, but weaned less lambs per hectare grazed, than those in the high stocking rate system. However, fecundity was affected by early-life nutrition. Lambs of ewes grazed at low stocking rates from conception to weaning at 3 months of age produced 15 more lambs per 100 ewes lambing compared to those of ewes grazed at high stocking rates, but only within a low stocking rate adult environment [82], which contrasts the earlier findings [58]. Although not specific to pregnancy, these findings suggest that female reproductive performance may be altered by fetal growth and development. Despite these results, there were no significant relationships between nutrition of Merino ewes during pregnancy and lactation and reproductive performance of the daughters at 5–6 years of age within the Lifetime Wool study (A.N. Thompson and C.M. Oldham unpublished results).

The timing of maternal nutritional treatments during pregnancy appears to influence reproductive performance of female offspring. Nutrition of pregnant ewes at

0.5 and 1.5 times estimated maintenance requirements from day 1 to day 35 of pregnancy did not affect ovulation rates during their first or second breeding seasons [102]. Similarly, during the second breeding season, when ewes were treated with pregnant mare serum gonadotropin, differences in ovulation rates because of maternal nutrition during early pregnancy were not apparent [102]. However, undernutrition of ewes from mating to day 95 of pregnancy reduced ovulation rate in female offspring at 20 months of age compared to offspring of ewes that were fed adequately during the same period [105]. This regimen did not affect birth weight or live weight, condition score or reproductive hormonal profiles of the 20 month old offspring. The proportion of ewe offspring bearing multiple lambs was increased by supplementary nutrition of their dams during the final 100 days of pregnancy [59]. This finding was attributed to differences in embryonic or fetal mortality because ovulation rate and ewe live weight and condition score were unaffected by nutrition during this phase of gestation. Despite these findings, fetal growth retardation of ewe lambs as a result of overfeeding adolescent ewes throughout pregnancy had no affect on live weight at puberty and time of onset of puberty, duration of the first ovarian cycle, incidence and number of ovarian cycles, aberrant ovarian cycles and duration of the first breeding season [25].

While some studies indicate that maternal nutrition during pregnancy affects reproductive performance of offspring [59, 105], it is important to recognise that nutritional restriction during the pre-weaning period may also influence reproductive performance of offspring [5, 59, 110] and may exacerbate the consequences of prenatal nutrition on ewe reproductive performance. Again, however, significant relationships between nutrition of Merino ewes during pregnancy and lactation and reproductive performance of the daughters to 5–6 years of age were not evident within the Lifetime Wool Project (A.N. Thompson and C.M. Oldham unpublished results).

Cattle

Recently it has been shown pregnancy rate of heifers (93 vs. 80%) and percentage of heifers calving during the first 21 days of the calving season (77 vs. 49%) were higher in heifers of dams supplemented with protein during the last trimester of pregnancy [89]. However, the percentage of heifers cycling at the beginning of the breeding season, calving dates, calf birth weights and percentage of unassisted births were not affected by maternal nutritional treatment [89]. Furthermore, various aspects of reproductive performance in nulliparous heifers and primiparous cows were not affected by birth weight [118]. These included services per conception, success at first service, age at and days to first service and to conception, and conception rates.

We studied ovarian characteristics at slaughter in 30 month old heifers that had undergone slow or rapid growth in utero following divergent nutritional treatments commencing from days 30 to 90 of gestation until parturition (birth weights 27.3 vs. 35.9 kg) and/or slow or rapid growth from birth to weaning (average daily gain 501 vs. 806 g/d). Mean ovarian weight (16.6 vs.19.4 g) and mean size of large (>9 mm) follicles (14.0 vs. 16.7 mm) were lower among those cattle that had undergone restricted growth in utero compared to their well-grown counterparts, while effects on ovarian characteristics due to pre-weaning growth were non-significant and of lesser magnitude than for prenatal growth [135]. The practical significance of these findings remains to be established.

1.2.6.2 Male Reproductive Performance

There does not appear to be any published information on the effects of altered prenatal development of male ruminants on fertility rates. However, studies with sheep suggest that adverse effects on reproductive performance of rams are possible after chronic fetal growth retardation, but not as a result of maternal nutrient restriction during early- to mid-pregnancy.

Undernutrition of ewes from mating until day 95 of pregnancy resulted in rams with higher follicle-stimulating hormone concentrations and mean basal and luteinising hormone-induced testosterone concentrations that tended to be higher at 20 months of age than those of well-nourished ewes [105]. Birth weight, live weight, condition score, testes circumference, indices of semen quality and sex hormone profiles in response to gonadotropin releasing hormone or luteinising hormone at 20 months of age were unaffected.

Male lambs restricted in growth during fetal life as a result of overfeeding adolescent ewes throughout pregnancy had lower live weight, delayed age at puberty, reduced testicular volume per unit live weight at 35 weeks of age, lower peak testosterone concentrations, and a delay in the timing of the seasonal increase in testosterone concentration compared to their counterparts that were well-nourished during fetal life [25]. Furthermore, male lambs from ewes that maintained weight during late pregnancy had a significantly lower number of sertoli cells at birth than lambs from ewes that gained about 20% of their live weight during the same period [13].

1.2.7 Lactational Performance

The reader is referred to the Chapter within this book on management and environmental influences on mammary gland development and milk production [18]. There appear to be no studies on long-term effects of fetal growth and development on lactational performance of sheep. However, while it has been suggested that modulation of early mammary-duct growth during fetal development could affect milk production [79], birth weight did not influence subsequent lactational performance of heifer cattle [118].

1.3 Conclusions

Severe, chronic growth retardation of sheep in utero resulting in very low birth weight may reduce mature size and increase fatness of sheep. However, the effects of moderate maternal nutritional restriction or fetal growth retardation may not persist in the long-term. Consequences of ovine fetal growth retardation may be exacerbated or ameliorated depending on the early-postnatal environment. Maternal nutrition during early- to mid-pregnancy may or may not affect postnatal growth and composition of lambs, and it remains unclear whether effects during this period of pregnancy are caused entirely by effects on the fetus during or after the period of altered nutrition, or whether postnatal activity of offspring and indirect effects mediated via the dam during lactation also contribute. The results of a limited number of studies with sheep have not shown long-term effects of nutrition during pregnancy, or of fetal growth, on postnatal efficiency of nutrient utilisation. Similarly, long-term effects of altered maternal nutrition or fetal growth on meat quality have not been demonstrated. However, it is well established that nutrition during pregnancy, particularly during late-pregnancy, can alter wool production characteristics. Furthermore, there is some evidence that nutrition during early life may influence subsequent reproductive performance. As yet, there does not appear to be evidence to suggest that lactational performance of ruminant offspring is affected by maternal nutrition during pregnancy or by fetal growth.

Hence, the practice of feeding ewes adequately during late pregnancy and lactation to ensure that fetal and neonatal growth is not seriously compromised will help to optimise lamb performance. However, the potential of nutritional and other treatments during early pregnancy to enhance productivity has yet to be realised.

Severe, chronic growth retardation of cattle early in life is associated with reduced growth potential, resulting in smaller animals at any given age. The capacity for long-term compensatory growth appears to diminish as the age of onset of severe nutritional restriction resulting in prolonged growth retardation declines, such that more extreme intrauterine growth retardation can result in slower growth throughout postnatal life. However, within the normal limits of beef production systems restricted growth in utero does not appear to influence efficiency of nutrient utilisation later in life.

Retail yield from cattle severely restricted in growth during pregnancy is reduced compared to cattle well grown early in life, when compared at the same age later in life. However, retail yield and carcass composition of low and high birth weight calves are similar when compared at the same carcass weight. Restricted prenatal nutrition and growth have not adversely affected measures of beef quality including shear force, compression, cooking loss and colour. Similarly, bovine myofibre characteristics are little affected in the long-term by growth in utero, despite specific myofibre type-related effects at birth. Hence, economic benefits resulting from adequate maternal nutrition during pregnancy, to optimise growth of offspring to market weights are primarily due to advantages in carcass weight and retail beef yield at a given age, reduced feed costs to reach a given market weight, stocking rates and subsequent reproductive rates of breeding cows, but not due to differences in beef quality characteristics [3].

We propose that within pasture-based production systems for beef cattle, at least, the plasticity of the carcass tissues, particularly of muscle, allows animals that are growth-retarded early in life to attain normal or near normal carcass composition at equivalent weights in the long-term, albeit at older ages. This may well relate to regulation of nutrient intake to a level appropriate for the size and lean tissue growth capacity of the animal, coupled with the capacity of the myosatellite cell population to generate myonuclei in support of muscle growth over a prolonged recovery period, as discussed previously for sheep [49–51]. However, the availability of feed and quality of nutrition during recovery from severe growth retardation early in life may be important in determining the subsequent composition of young, lightweight ruminants relative to their heavier counterparts.

It also needs to be emphasised that further research on long-term consequences of more specific, acute environmental and exogenous influences during specific stages of embryonic, fetal and neonatal development is required. This need for further research extends to consequences of nutrition and growth early in life for subsequent reproductive and lactational performance. There is also a need for further research on interactions between prenatal and pre-weaning growth and nutrition for subsequent growth, efficiency, carcass, yield and meat quality characteristics, although these were not evident in cattle within our pasture-based production systems [48, 54]. Furthermore interactions between genotype and nutrition early in life require further research, despite our studies using offspring of Piedmontese and Wagyu sires mated to Hereford cows failing to identify interactions for growth, efficiency, carcass, yield and beef quality parameters [48, 54].

Finally, in managing the prenatal environment to enhance livestock productivity the importance of economic [3, 122, 138] and social factors, and of the contribution of survival and health of offspring and the capacity of dams to re-breed to the profitability of production systems, cannot be overemphasised.

References

1. Alexander, G. 1974. Birth weight of lambs: influences and consequences, pp. 213–239. K. Elliott, and J. Knight (eds.), Size at Birth, Elsevier, Amsterdam.
2. Alexander, G. 1984. Constraints to lamb survival, pp. 199–209. K. Elliott, and J. Knight (eds.) Reproduction in Sheep, Australian Academy of Science and Australian Wool Corporation, Canberra.
3. Alford, A.R., L.M. Cafe, P.L. Greenwood, and G.R. Griffith. 2007. The economic effects of early-life nutritional constraints in crossbred cattle bred on the NSW North Coast. *Economic Research Report* No. 33, NSW Department of Primary Industries, Armidale.
4. Allden, W.G. 1970. The effects of nutritional deprivation on the subsequent productivity of sheep and cattle. *Nutr. Abstr. Rev.* **40:**1167–1184.
5. Allden, W.G. 1979. Undernutrition of the Merino sheep and its sequelae. 5. The influence of severe growth retardation during early postnatal life on reproduction and growth in later life. *Aust. J. Agric. Res.* **30:**939–948.
6. Ashworth, C.J., T.G. McEvoy, J.A. Rooke, and J.J. Robinson. 2005. Nutritional programming of physiological systems throughout development. *Trends Dev. Biol.* **1:**117–229.
7. Barker, D.J.P. 1998. Mothers, Babies and Health in Later Life, 2nd ed. Churchill Livingstone, Edinburgh.
8. Barker, D.J.P. 2004. The developmental origins of chronic adult disease. *Acta Paed. Suppl.* **446:**26–33.
9. Bawden, C.S., D.O. Kleemann, C.J. McLaughlan, G.S. Nattrass, and S.M. Dunn. 2009. Mechanistic aspects of foetal development relating to postnatal fibre production and follicle

development in ruminants, pp. 121–159. P.L. Greenwood, A.W. Bell, P.E. Vercoe, and G.J. Viljoen (eds.), Managing Prenatal Development to Enhance Livestock Productivity, IAEA, Vienna.
10. Bell, A.W. 1992. Foetal growth and its influence on postnatal growth and development, p. 111–127. P.J. Buttery, K.N. Boorman, and D.B. Lindsay (eds.), The Control of Fat and Lean Deposition, Butterworth-Heinemann, Oxford.
11. Bell, A.W. 2006. Prenatal programming of postnatal productivity and health in livestock: a brief review. *Aust. J. Exp. Agric.* **46:**725–732.
12. Bell, A.W., P.L. Greenwood, and R.A. Ehrhardt. 2005. Regulation of metabolism and growth during prenatal life, pp. 3–34. D.G. Burrin, and H.J. Mersmann (eds.), Biology of Metabolism in Growing Animals, Elsevier, Amsterdam.
13. Bielli, A., R. Perez, G. Pedrana, J.T.B. Milton, A. Lopez, M.A. Blackberry, G. Duncombe, H. Rodriguez-Martinez, and G.B. Martin. 2002. Low maternal nutrition during pregnancy reduces the number of Sertoli cells in the newborn lamb. *Reprod. Fertil. Dev.* **14:**333–337.
14. Boyd, G.W., T.E. Kiser, and R.S. Lowrey. 1987. Effects of prepartum energy intake on steroids during late gestation and on cow and calf performance. *J. Anim. Sci.* **64:** 1703–1709.
15. Burt, B.E., B.W. Hess, P.W Nathanielz, and S.P. Ford. 2007. Flock differences in the impact of maternal dietary restriction on offspring growth and glucose tolerance in female offspring, pp. 411–424. J.L. Juengel, J.F. Murray, and M.F. Smith (eds.), Reproduction in Domestic Ruminants VI, Nottingham University Press, Nottingham.
16. Cafe, L.M., H. Hearnshaw, D.W. Hennessy, and P.L. Greenwood. 2006. Growth and carcass characteristics at heavy market weights of Wagyu-sired steers following slow or rapid growth to weaning. *Aust. J. Exp. Agric.* **46:**951–955
17. Cafe L.M., D.W. Hennessy, H. Hearnshaw, S.G. Morris, and P.L.Greenwood. 2006. Influences of nutrition during pregnancy and lactation on birth weights and growth to weaning of calves sired by Piedmontese or Wagyu bulls. *Aust. J. Exp. Agric.* **46:**245–255
18. Capuco, A.V., and M. Akers. 2009. Management and environmental influences on mammary gland development and milk production, pp. 259–292. P.L. Greenwood, A.W. Bell, P.E. Vercoe, and G.J. Viljoen (eds.), Managing Prenatal Development to Enhance Livestock Productivity, IAEA, Vienna.
19. Clarke, A.J., L.J. Cummins L.J., J.F. Wilkins, D.W. Hennessy, C.M. Andrews and B.J. Makings. 1994. Post weaning growth of twin cattle born at Hamilton and Grafton. *Proc. Aust. Soc. Anim. Prod.* **20:**34–35.
20. Corah, L.R., T.G. Dunn, and C.C. Kaltenbach. 1975. Influence of prepartum nutrition on the reproductive performance of beef females and the performance of their progeny. *J. Anim. Sci.* **41:**819–824.
21. Costine, B.A., E.K. Inskeep, and M.E. Wilson. 2005. Growth hormone at breeding modifies conceptus development and postnatal growth in sheep. *J. Anim. Sci.* **83:**810–815.
22. Cronje, P. 2003. Implications of nutritional programming of gene expression during early development for livestock production, pp. 321–332. L. 't Mannetje, L. Ramirez-Aviles, C.A. Sandoval Castro, and J.C. Ku-Vera (eds.), Proceedings of the 6th International Symposium on the Nutrition of Herbivores, Merida, Yucatan.
23. Cronje, P.B., and N.R. Adams. 2002. The effect of *in utero* nutrient restriction on glucose and insulin metabolism in merino sheep. *Anim. Prod. Aust.* **24:**41–44.
24. Daniel, Z.C.T.R., J.M. Brameld, J. Craigon, N.D. Scollan, and P.J. Buttery. 2007. Effect of maternal dietary restriction on lamb carcass characteristics and muscle fibre composition. *J. Anim. Sci.* **85:**1565–1576.
25. Da Silva, P., R.P. Aitken, S.M. Rhind, P.A. Racey, and J.M. Wallace. 2001. Influence of placentally mediated fetal growth restriction on the onset of puberty in male and female lambs. *Reproduction* **122:**375–383.
26. De Blasio, M.J., K.L. Gatford, J.S. Robinson, and J.S. Owens. 2007. Placental restriction of fetal growth reduces size at birth and alters postnatal growth, feeding activity, and adiposity in the young lamb. *Am. J. Physiol.* **292:**R875–R886.

27. Denney, G.D. 1990. Effects of preweaning farm environment on adult wool production of Merino sheep. *Aust. J. Exp. Agric.* **30:**17–25.
28. Dwyer, C.M., A.B. Lawrence, S.C. Bishop, and M. Lewis. 2003. Ewe-lamb bonding behaviours at birth are effected by maternal under nutrition in pregnancy. *Br. J. Nutr.* **89:** 123–136.
29. Everitt, G.C. 1965. Factors affecting foetal growth and development of merino sheep with particular reference to maternal nutrition. PhD Thesis, University of Adelaide.
30. Everitt, G.C. 1967. Residual effects of prenatal nutrition on the postnatal performance of merino sheep. *Proc. N.Z. Soc. Anim. Prod.* **27:**52–68.
31. Everitt, G.C. 1968. Prenatal development of uniparous animals, with particular reference to the influence of maternal nutrition in sheep, pp. 131–157. G.A. Lodge, and G.E. Lamming (eds.), Growth and Development of Mammals, Butterworths, London, UK.
32. Ferguson, M., D. Gordon, B. Paganoni, T. Plaisted, and G. Kearney. 2004. Lifetime wool. 6. Progeny birth weights and survival. *Anim Prod. Aust.* **25:**243.
33. Ferrell, C.L. 1991. Maternal and foetal influences on uterine and conceptus development in the cow: I. Growth of the tissues of the gravid uterus. *J. Anim. Sci.* **69:**1945–1953.
34. Ford, S.P., B.W. Hess, M.M. Schwope, M.J. Nijland, J.S. Gilbert, K.A. Vonnahme, W.J. Means, H. Han, and P.W. Nathanielz. 2007. Maternal undernutrition during early-mid-gestation in the ewe results in altered growth, adiposity, and glucose tolerance in male off-spring. *J. Anim. Sci.* **85:**1285–1294.
35. Freetly, H.C., C.L. Ferrell, and T.G. Jenkins. 2000. Timing of realimentation of mature cows that were feed-restricted during pregnancy influences calf birth weights and growth rates. *J. Anim. Sci.* **78:**2790–2796.
36. Gabbedy, B.J. (1974). A growth comparison of single and twin lambs during prenatal and early postnatal life. M. Agr. Sc. Thesis, University of Melbourne.
37. Gardner, D.S., K. Tingey, B.W.M. Van Bon, S.E. Ozanne, V. Wilson, J. Dandrea, D.H. Keisler, T. Stephenson, and M.E. Symonds 2005. Programming of glucose-insulin metabolism in adult sheep after maternal undernutrition. *Am. J. Physiol.* **289:**R947–R954.
38. Gardner, D.S., P.J. Butterty, Z. Daniel, and M.E. Symonds. 2007. Factors affecting birth weight in sheep: maternal environment. *Reproduction* **133:**297–307.
39. Geraseev, L.C., J.R.O. Perez, P.A. Carvalho, R.P. de Oliviera, F.A. Quintao, and A.L. Lima. 2006. Effects of pre and postnatal feed restriction on growth and production of Santa Ines lambs from birth to weaning. *Revista Brasilia de Zootecnia* **35:**245–251.
40. Geraseev, L.C., J.R.O. Perez, P.A. Carvalho, B.C. Pedriera, and T.R.V. Almeida. 2006. Effects of pre and postnatal feed restriction on growth and production of Santa Ines lambs from weaning to slaughter. *Revista Brasilia de Zootecnia* **35:**237–244.
41. Geraseev, L.C., J.R.O. Perez, F.A. Quintao, B.C. Pedriera, and P.A. Carvalho. 2007. Effects of pre and postnatal nutritional restriction on internal fat growth of Santa Ines lambs. *Arquivo Brasiliero de Medicina Veterinia e Zootecnia* **59:**782–788.
42. Gnanalingham, M.G., A. Mostyn, M.E. Symonds, and Stephenson, T. 2005. Ontogeny and nutritional programming of adiposity in sheep: potential role of glucocorticoid action and uncoupling protein-2. *Am. J. Physiol.* **289:**R1407–R1415.
43. Gootwine, E., T.E. Spencer, and F.W. Brazer. 2007. Litter-size-dependent intrauterine growth restriction in sheep. *Animal* **1:**547–564.
44. Gopalakrishnan, G.S., D.S. Gardner, S.M. Rhind, M.T. Rae, C.E. Kyle, A.N. Brooks, R.M. Walker, M.M. Ramsay, D.H. Keisler, T. Stephenson, and M.E. Symonds. 2004. Programming of adult cardiovascular function after early maternal undernutrition of sheep. *Am. J. Physiol.* **287:**R12–R20.
45. Gopalakrishnan G.S., D.S. Gardner, J. Dandrea, S.C. Langley-Evans, S. Pearce, L.O. Kurlak, R.M. Walker, I.W. Seetho, D.H. Keisler, M.M. Ramsay, T. Stephenson, and M.E. Symonds. 2005. Influence of maternal pre-pregancy body composition and diet during early-mid pregnancy on cardiovascular function and nephron number in juvenile sheep. *Br. J. Nutr.* **94:** 938–947.

46. Greenwood, P.L., and A.W. Bell. 2003. Consequences of intra-uterine growth retardation for postnatal growth, metabolism and pathophysiology. *Reprod. Suppl.* **61:**195–206.
47. Greenwood, P.L., and A.W. Bell. 2003. Prenatal nutritional influences on growth and development of ruminants. *Rec. Adv. Anim. Nutr. Aust.* **14:**57–73.
48. Greenwood, P.L., and L.M. Cafe. 2007. Prenatal and pre-weaning growth and nutrition of cattle: long-term consequences for beef production. *Animal* **1:**1283–1296.
49. Greenwood, P.L., A.S. Hunt, J.W. Hermanson, and A.W. Bell. 1998. Effects of birth weight and postnatal nutrition on neonatal sheep: I. Body growth and composition, and some aspects of energetic efficiency. *J. Anim. Sci.* **76:**2354–2367.
50. Greenwood, P.L., R.M. Slepetis, J.W. Hermanson, and A.W. Bell. 1999. Intrauterine growth retardation is associated with reduced cell cycle activity, but not myofibre number, in ovine foetal muscle. *Reprod. Fertil. Dev.* **11:**281–291.
51. Greenwood, P.L., A.S. Hunt, J.W. Hermanson, and A.W. Bell. 2000. Effects of birth weight and postnatal nutrition on neonatal sheep: II. Skeletal muscle growth and development. *J. Anim. Sci.* **78:**50–61.
52. Greenwood, P.L., R.M. Slepetis, and A.W. Bell. 2000. Influences on foetal and placental weights during mid and late gestation in prolific ewes well nourished throughout pregnancy. *Reprod. Fertil. Dev.* **12:**149–156.
53. Greenwood, P.L., L.M. Cafe, H. Hearnshaw, and D.W. Hennessy. 2005. Consequences of nutrition and growth retardation early in life for growth and composition of cattle and eating quality of beef. *Rec. Adv. Anim. Nutr. Aust.* **15:**183–195.
54. Greenwood, P.L., L.M. Cafe, H. Hearnshaw, D.W. Hennessy, J.M. Thompson and S.G. Morris. 2006. Long-term consequences of birth weight and growth to weaning on carcass, yield and beef quality characteristics of Piedmontese- and Wagyu-sired cattle. *Aust. J. Exp. Agric.* **46:**257–269.
55. Greenwood, P.L., S. Harden, and D.L. Hopkins. 2007. Myofibre characteristics of ovine *longissimus* and *semitendinosus* muscles are influenced by sire breed, gender, rearing type, age, and carcass weight. *Aust. J. Exp. Agric.* **47:**1137–1146.
56. Gregory, K.E., S.E. Echternkamp, and L.V. Cundiff. 1996. Effects of twinning on dystocia, calf survival, calf growth, carcass traits and cow productivity. *J. Anim. Sci.* **74:**1223–1233.
57. Guilloteau, P., T. Corring, , R. Toullec, Y. Villette, and J. Robelin. 1985. Abomasal and pancreas enzymes in the newborn ruminant: Effects of species, breed, sex and weight. *Nutr. Rep. Int.* **31:**1231–1236.
58. Gunn, R.G. 1977. The effects of two nutritional environments from 6 weeks pre partum to 12 months of age on lifetime performance and reproductive potential of Scottish Blackface ewes in two adult environments. *Anim. Prod.* **25:**155–164.
59. Gunn, R.G., D.A. Sim, and E.A. Hunter. 1995. Effects of nutrition *in utero* and in early life on the subsequent lifetime reproductive performance of Scottish Blackface ewes in two management systems. *Anim. Sci.* **60:**223–230.
60. Hales, N.C., and D.J.P. Barker. 2001. The thrifty phenotype hypothesis. *Br. Med. Bull.* **60:** 5–20.
61. Hardy, M.H., and A.G. Lyne. 1956. The pre-natal development of wool follicles in Merino sheep. *Aust. J. Biol. Sci.* **9:**423–441.
62. Hennessy, D.W., and J.F. Wilkins. 1997. The nutrition of single and twin suckled calves and their growth-12 months post weaning, pp. 45–47. D.W. Hennessy, S.R. McLennan, and V.H. Oddy (eds.), Growth and Development of Cattle, *Proceedings of the Growth and Development Workshop*, Cooperative Research Centre for Cattle and Beef Quality, Armidale.
63. Hight, G.K. 1966. The effects of undernutrition in late pregnancy on beef cattle production. *N.Z. J. Agric. Res.* **9:**479–490.
64. Hight, G.K. 1968. Plane of nutrition effects in late pregnancy and lactation on beef cows and their calves to weaning. *N.Z. J. Agric. Res.* **11:**71–84.
65. Hight, G.K. 1968. A comparison of the effects of three nutritional levels in late pregnancy on beef cows and their calves. *N.Z. J. Agric. Res.* **11:**477–486.

66. Houssin, Y., and M.J. Davico. 1979. Influence of birthweight on the digestibility of a milk-replacer in newborn lambs. *Ann. Rech. Vet.* **10:**419–421.
67. Hocking-Edwards, J.E., M.J. Birtles, P.M. Harris, A. Parry, E. Paterson, G.A. Wickham, and S.N. McCutcheon. 1994. Prenatal follicle development in Romney, Merino and Merino-Romney cross sheep. *Proc. N.Z. Soc. Anim. Prod.* **54:**131–134.
68. Holland, M.D., and K.G. Odde. 1992. Factors affecting calf birth weight: A review. *Theriogenology* **38:**769–798.
69. Hunnicutt, L.K., R.H. Stobart, D.W. Sanson, M.L. Riley, W.R. Taliaferro, and G.E. Moss. 1993. Acute nutritional stress during early pregnancy in ewes: Effect on lambs at birth and weaning. *Sheep Res. J.* **9:**109–114.
70. Hutchinson, G., and D.J. Mellor. 1983. Effects of maternal nutrition on the initiation of secondary wool follicles in fetal sheep. *J. Comp. Path.* **93:**577–583.
71. Jackson, N., T. Nay, and H.N. Turner. 1975. Response to selection in Australian Merino sheep. VII. Phenotypic and genetic parameters for some wool follicle characteristics and their correlation with wool and body traits. *Aust. J. Agric. Res.* **26:**937–957.
72. Jopson, N.B., G.H. Davis, P.A. Farquar, and W.E. Bain. 2002. Effects of mid-pregnancy nutrition and shearing on ewe body reserves and foetal growth. *Proc. N.Z. Soc. Anim. Prod.* **62:**49–52.
73. Kelly, R.W., I. Macleod, P. Hynd, and J.C. Greef. 1996. Nutrition during fetal life alters annual wool production and quality in young Merino sheep. *Aust. J. Exp. Agric.* **36:** 259–267.
74. Kelly, R.W., J.C. Greef, and I. Macleod. 2006. Lifetime changes in wool production of Merino sheep following differential feeding in fetal and early life. *Aust. J. Agric. Res.* **57:**867–876.
75. Kenyon, P.R., S.T. Morris, and S.N. McCutcheon. 2002. Does an increase in lamb birth weight through mid-pregnancy shearing necessarily mean an increase in lamb survival rates to weaning? *Proc. N.Z. Soc. Anim. Prod.* **62:**53–56.
76. Kenyon, P.R., S.T. Morris, D.K. Revell, and S.N. McCutcheon. 2002. Maternal constraint and birth weight response to mid-pregnancy shearing. *Aust. J. Agric. Res.* **53:**511–517.
77. Kenyon, P.R., D.K. Revell, and S.T. Morris. 2006. Mid-pregnancy shearing can increase birthweight and survival to weaning of multiple-born lambs under commercial conditions. *Aust. J. Exp. Agric.* **46:**821–825.
78. Kenyon, P.R., R.G. Sherlock, S.T. Morris, and P.C.H. Morel. 2006. The effect of mid- and late-pregnancy shearing of hoggets on lamb birthweight, weaning weight, survival rate and wool follicle and fibre characteristics. *Aust. J. Exp. Agric.* **46:**877–882.
79. Knight, C.H., and A.S. Sorenson, 2001. Windows in early mammary development: critical or not? *Reproduction* **122:**337–345.
80. Knight, T.W., P.R. Lynch, D.R. Hall, and H.P. Hockey. 1988. Identification of factors contributing to the lamb survival in Marshal Romney sheep. *N.Z. J. Agric. Res.* **31:**259–71.
81. Krausgrill, D.I., N.M. Tulloh, W.R. Shorthose, and K. Sharpe. 1999. Effects of weight loss in ewes in early pregnancy on muscles and meat quality of lambs. *J. Agric. Sci., Camb.* **132:**103–116.
82. Langlands, J.P., G.E. Donald, and D.R. Paull. 1984. Effects of different stocking intensities in early life on the productivity of Merino ewes grazed as adults at two stocking rates. 2. Reproductive performance. *Aust. J. Exp. Agric.* **24:**47–56.
83. Lea, R.G., L.P. Andrade, M.T., Rae, L.T. Hannah, C.E. Kyle, J.F. Murray, S.M. Rhind, and D.W. Miller. 2006. Effects of maternal undernutrition during early pregnancy on apoptosis regulators in the ovine fetal ovary. *Reproduction* **131:**113–124.
84. Lewer, R.P., R.R. Woolaston, and R.R. Howe. 1992. Studies of Western Australian Merino sheep. 1. Stud, strain and environmental effects on hogget performance. *Aust. J. Agric. Res.* **43:**1381–98.
85. Louey, S., M.L. Cock, K.M. Stevenson, and R. Harding. 2000. Placental insufficiency and fetal growth restriction lead to postnatal hypotension and altered postnatal growth in sheep. *Pediatric Res.* **48:**808–814.

86. Louey, S., M.L. Cock, and R. Harding. 2005. Long-term consequences of low birthweight on postnatal growth, adiposity and brain weight at maturity in sheep. *J. Reprod. Devel.* **51:** 59–68.
87. McMillen, I.C., C.L. Adams, and B.S. Muhlhausler. 2005. Early origins of obesity: programming the appetite regulatory system. *J. Physiol.* **565:**9–17.
88. Maier, A., J.C. McEwan, K.G. Dodds, D.A. Fischman, R.B. Fitzimons, and A.J. Harris. 1992. Myosin heavy chain composition of single fibres and their origins and distribution in developing fascicles of sheep tibialis cranialis muscle. *J. Muscle Res. Cell Motil.* **13:** 551–572.
89. Martin J.L., K.A. Vonnahme, D.C. Adams, G.P. Lardy, and R.N. Funston. 2007. Effects of dam nutrition on growth and reproductive performance of heifer calves. *J. Anim. Sci.* **85:**841–847.
90. Mellor, D.J. 1983. Nutritional and placental determinants of fetal growth rate in sheep and consequences for the new born lamb. *Br. Vet J.* **139:**307–324.
91. Mellor, D.J. 1988. Integration of perinatal events, pathophysiological changes and consequences for the newborn lamb. *Br. Vet J.* **144:**552–569.
92. Mexia, A.A, F. de A.F. de Macedo, R.M.G. de Macedo, E.S. Sakaguti, G.A. Santello, L.C.T. Capovilla, M. Zundt, and A. Sasa. 2006. Performance and skeletal muscular fiber characteristics of lambs born from ewes supplemented at different stages of pregnancy. *Revista Brasileira de Zootecnia* **35:**1780–1787.
93. Moore, R.W., C.M. Millar, and P.R. Lynch. 1986. The effect of prenatal nutrition and type of birth and rearing of lambs on vigour, temperature and weight at birth, and weight and survival at weaning. *Proc. N.Z. Soc. Anim. Prod.* **46:**259–262.
94. Morris, S.T., S.N. McCutcheon, and D.R. Revell. 2000. Birth weight responses to shearing ewes in early to mid gestation. *Anim. Sci.* **70:**363–369.
95. Morris, S.T., P.R. Kenyon, and D.M. West. 2005. Effect of hogget nutrition in pregnancy on lamb birthweight and survival to weaning. *N.Z. J. Agric. Res.* **48:** 165–175.
96. Munoz, C., A.F. Carson, M.A. McCoy, and L.E.R. Dawson. 2006. Effects of nutrition of ewes during early and mid pregnancy on lamb growth and carcass characteristics, p. 56. *Proceedings of the Agricultural Research Forum*, Tullamore, Ireland.
97. Nordby, D.J., R.A. Field, M.L. Riley, C.L. Johnson, and C.J. Kercher. 1986. Effects of maternal undernutrition during early pregnancy on postnatal growth in lambs. *Proc. Western Section Am. Soc. Anim. Sci.* **37:**92–95.
98. Nordby, D.J., R.A. Field, M.L. Riley, and C.J. Kercher. 1987. Effects of maternal undernutrition during early pregnancy on growth, muscle cellularity, fiber type and carcass composition in lambs. *J. Anim. Sci.* **64:**1419–1427.
99. Oliver, M.H., J.E. Harding, and P.D. Gluckman. 2001. Duration of maternal undernutrition during late gestation determines the reversibility of intrauterine growth restriction in sheep. *Prenatal Neonatal Med.* **6:**271–279.
100. Oliver, M.H., B.H. Breier, P.D. Gluckman, and J.E. Harding. 2002. Birth weight rather than maternal nutrition influences glucose tolerance, blood pressure, and IGF-I levels in sheep. *Paediatric Res.* **52:**516–524.
101. Paganoni, B.L., R. Banks, C.M. Oldham, and A.M. Thompson. 2004. Lifetime wool. 9. Progeny back fat and eye muscle depth. *Anim. Prod. Aust.* **25:**296.
102. Parr, R.A., A.H. Williams, I.P. Campbell, G.F. Witcombe, and A.M. Roberts. 1986. Low nutrition of ewes in early pregnancy and the residual effect on the offspring. *J. Agric. Sci., Camb.* **106:**81–87.
103. Penning, P.D., P. Corcuera, and T.T. Treacher. 1980. Effect of dry-matter concentration of milk substitute and method of feeding on intake and performance by lambs. *Anim. Feed Sci. Technol.* **5:**321–336.
104. Quigley, S.P., D.O. Kleeman, M.A. Kaker, J.A. Owens, G.S. Nattrass, S. Maddocks, and S.K. Walker. 2005. Myogenesis in sheep is altered by maternal feed intake during the periconception period. *Anim. Reprod. Sci.* **87:**241–251.

105. Rae, M.T., C.E. Kyle, D.W. Miller, A.J. Hammond, A.N. Brooks, and S.M. Rhind. 2002. The effects of undernutrition, in utero, on reproductive function in adult male and female sheep. *Anim. Reprod. Sci.* **72:**63–71.
106. Redmer, D.A., J.M. Wallace, and L.P. Reynolds. 2004. Effect of nutrient intake during pregnancy on fetal and placental growth and vascular development. *Domest. Anim. Endocrinol.* **27:**199–217.
107. Revell, D.K., S.F. Main, B.H. Breier, Y.H. Cottam, M. Hennies, and S.N. McCutcheon. 2000. Metabolic responses to mid-pregnancy shearing that are associated with a selective increase in the birth weight of twin lambs. *Domest. Anim. Endocrinol.* **18:**409–422.
108. Revell, D.K., S.T. Morris, Y.H. Cottam, J.E. Hanna, D.G. Thomas, S. Brown, and S.N. McCutcheon. 2002. Shearing ewes at mid pregnancy is associated with changes in fetal growth and development. *Aust. J. Agric. Res.* **53:**697–705.
109. Rhind, S.M. 2004. Effects of maternal nutrition on fetal and neonatal reproductive development and function. *Anim. Reprod. Sci.* **82–83:**169–181.
110. Rhind, S.M., D.A. Elston, J.R. Jones, M.E. Rees, S.R McMillen, and R.G. Gunn. 1998. Effects of restriction in growth and development of Brecon Cheviot ewe lambs on subsequent lifetime reproductive performance. *Small Ruminant Res.* **30:**121–126.
111. Rhind, S.M., M.T. Rae, and A.N. Brooks. 2003. Environmental influences on the fetus and neonate – timing, mechanisms of action and effects on subsequent adult function. *Domest. Anim. Endocrinol.* **25:**3–11.
112. de Rose, E.P., and J.W. Wilton. 1991. Productivity and profitability of twin births in beef cattle. *J. Anim. Sci.* **69:**3085–3093.
113. Roussel, S., and P. Hemsworth. 2002. The effect of prenatal stress on the stress physiology and liveweight of lambs. *Anim. Prod. Aust.* **24:**346.
114. Ryley, J.W., and R.J.W. Gartner. 1962. Drought feeding studies with cattle. 7. The use of sorghum grain as a drought fodder for cattle in late pregnancy and early lactation. *Qld. J. Agric. Sci.* **19:**309–330.
115. Schinckel, P.G., and B.F. Short. 1961. The influence of nutritional level during prenatal and early postnatal life on adult fleece and body characteristics. *Aust. J. Agric. Res.* **12:**176–202.
116. Short, B.F. (1955). Developmental modification of fleece structure by adverse maternal nutrition. *Aust. J. Agric. Res.* **6:**863–872.
117. Sibbald, A.M., and G.C. Davidson. 1998. The effect of nutrition during early life on voluntary food intake by lambs between weaning and 2 years of age. *Anim. Sci.* **66:**697–703.
118. Swali, A., and D.C. Wathes. 2006. Influence of the dam and sire on size at birth and subsequent growth, milk production and fertility in dairy heifers. *Theriogenology* **66:**1173–1184.
119. Swatland, H.J., and R.G. Cassens. 1973. Inhibition of muscle growth in foetal sheep. *J. Agric. Sci., Camb.* **80:**503–509.
120. Taplin, D.E., and G.C. Everitt. 1964. The influence of prenatal nutrition on postnatal performance of merino lambs. *Proc. Aust. Soc. Anim. Prod.* **5:**72–81.
121. Thompson, A.N., and C.M. Oldham. 2004. Lifetime Wool. 1. Project overview. *Anim. Prod. Aust.* **25:**326.
122. Thompson, A.N., and J.M. Young. 2002. A comparison of the profitability of farming systems using wool and wool/meat sheep genotypes in south west Victoria. *Wool Technol. Sheep Breeding* **50:**502–509.
123. Thompson, A.N., K. Webb, G. Kearney, and B. Leury. 2006. Poor nutrition in utero and preweaning reduced lean tissue and increased fat tissue mass in adult merino sheep. *Aust. Soc. Anim. Prod. 26th Biennial Conf.*, Short Communication 88.
124. Tudor, G.D. 1972. The effect of pre- and post-natal nutrition on the growth of beef cattle. I. The effect of nutrition and parity of the dam on calf birth weight. *Aust. J. Agric. Res.* **23:**389–395.
125. Tudor, G.D., and P.K. O'Rourke. 1980. The effect of pre- and post-natal nutrition on the growth of beef cattle. II. The effect of severe restriction in early postnatal life on growth and feed efficiency during recovery. *Aust. J. Agric. Res.* **31:**179–189.

126. Tudor, G.D., D.W. Utting, and P.K. O'Rourke. 1980. The effect of pre- and post-natal nutrition on the growth of beef cattle. III. The effect of severe restriction in early postnatal life on the development of the body components and chemical composition. *Aust. J. Agric. Res.* **31**:191–204.
127. Tygesen, M.P. 2005. *The effect of, and interactions between, maternal nutrient restriction in late gestation and paternal genetics on ovine productivity.* PhD Thesis, The Royal Veterinary and Agricultural University, Copenhagen.
128. Tygesen, M.P., A.P. Harrison, and M. Therkildsen. 2007. The effect of maternal nutrient restriction during late gestation on muscle, bone and meat parameters in five month old lambs. *Livestock Sci.* **110**:230–241.
129. Villette, Y., and M. Theriez. 1981. Influence of birth weight on lamb performances. I. Level of feed intake and growth. *Ann. Zootech.* **30**:151–168.
130. Villette, Y., and M. Theriez. 1981. Influence of birth weight on lamb performances. II. Carcass and chemical composition of lambs slaughtered at the same weight. *Ann. Zootech.* **30**:169–182.
131. Villette, Y., and M. Theriez. 1983. Milk intake in lambs suckled by their dams during the first week of life. *Ann. Zootech.* **32**:427–440.
132. Wallace, J. M., R.P. Aitken, and M.A. Cheyne. 1996. Nutrient partitioning and fetal growth in rapidly growing adolescent ewes. *J. Reprod. Fertil.* **107**:183–190.
133. Wallace, J. M., D.A. Bourke, and R.P. Aitken. 1999. Nutrition and fetal growth: paradoxical effects in the overnourished adolescent sheep. *J. Reprod. Fertil.* Supplement **54**:385–399.
134. Wilkins J.F., D.W. Hennessy, and R.J. Farquharson. 1994. Twinning in Beef Cattle. Roles of nutrition and early weaning in herds of high calving rate. Final Report to Meat Research Corporation, NSW Agriculture, Grafton.
135. Wilkins, J.F., R.C. Fry, H. Hearnshaw, L.M. Cafe, and P.L. Greenwood. 2006. Ovarian activity in heifers at 30 months of age following high or low growth *in utero* or from birth to weaning. *Aust. Soc. Anim. Prod. 26th Biennial Conf.* Short Communication 17.
136. Wilson, S.J., J.C. McEwan, P.W. Sheard, and A.J. Harris. 1992. Early stages of myogenesis in a large mammal: formation of successive generations of myotubes in sheep tibialis cranialis muscle. *J. Muscle Res. Cell Motility* **13**:534–550.
137. Winks, L., P.K. O'Rourke, P.C. Venamore, and R. Tyler. 1978. Factors affecting birth weight of beef calves in the dry tropics of north Queensland. *Aust. J. Exp. Agric. Anim. Husb.* **18**:494–499.
138. Young, J.M., A.N. Thompson, and C.M. Oldham. 2004. Lifetime wool. 15. Whole-farm benefits from optimising lifetime wool production. *Anim. Prod. Aust.* **25**:338.
139. Wu, G., F.W. Bazer, J.M. Wallace, and T.E. Spencer. 2006. Board-invited review: Intrauterine growth retardation: Implications for the animal sciences. *J. Anim. Sci.* **84**:2316–2337.
140. Zhu, M.J., S.P. Ford, P.W. Nathanielz, and M. Du. 2004. Effect of maternal nutrient restriction in sheep on the development of fetal skeletal muscle. *Biol. Reprod.* **71**:1968–1973.
141. Zhu, M.J., S.P. Ford, W.J. Means, B.T Hess, P.W. Nathanielsz, and M. Du. 2006. Maternal nutrient restriction affects properties of skeletal muscle fibres in offspring. *J. Physiol.* **575**:241–250.
142. Zundt, M., F.D.F. de Macedo, J.L. de Lima Astolphi, A.A. Mexia, and E.S. Sakaguti. 2006. Production and carcass characteristics of confined lambs born from Santa Ines ewes supplemented in different stages of pregnancy. *Revista Brasilia de Zootecnia* **35**:928–935.

Chapter 2
Quantification of Prenatal Effects on Productivity in Pigs

Pia M. Nissen and Niels Oksbjerg

Introduction

Pigs are a litter bearing species that give birth to an average between 9 and 13 pigs per litter depending on breed and country. The gestation length is 113–115 days. Early gestation ends at around day 40, and mid gestation at about day 80. In conventional, indoor pig production the lactation period is 3–4 weeks, thus pigs are weaned at around days 21–28 at a mean live body weight between 6 and 8 kg. In most countries, pigs are slaughtered when they reach a live body weight between 90 and 120 kg, although production of heavier pigs is common in some counties.

Sow performance and daily gain, feed efficiency and carcass composition of the offspring are important economic traits in pig production. Factors that affect these traits are of high interest to pig producers. Birth weight is a good indicator of fetal development, and is also highly related to postnatal growth performance, especially in the early postnatal growth phases [26, 31, 54, 73]. Differences in fetal muscle fibre development are also of great importance for postnatal growth performance, as the number of muscle fibres is positively correlated to average daily gain [13, 47]. In recent years, as average litter size has increased, the intra-litter variation in birth weight and postnatal growth performance has also increased. Consequently, more pigs of low birth weight and reduced viability are born. This presents a great challenge to the pig producer of today, and efforts to decrease this variation are required.

Muscle is the most economically valuable tissue in the pig, and therefore the development of muscle during fetal development has been the focus of many studies. In the pig, two major populations of muscle fibres develop during fetal life. The first population, primary fibres, is formed from days 25 to 55 of gestation, and the second population, secondary fibres, is formed from days 55 to 80–90 after which the number of fibres is constant (for review see [44]). In several studies where the

P.M. Nissen (✉)
Department of Food Science, Faculty of Agricultural Sciences, Aarhus University, DK-8830 Tjele, Denmark
e-mail: piam.nissen@agrsci.dk

P.L. Greenwood et al. (eds.), *Managing the Prenatal Environment to Enhance Livestock Productivity*, DOI 10.1007/978-90-481-3135-8_2,

main focus has been to support fetal muscle development by nutrient supplementation via sow feeding, the periods during which primary and secondary muscle fibres are formed provide the framework within which it is possible to manipulate muscle fibre formation.

Fetal growth is dependent on nutrient transfer from the mother across the placenta to the developing fetuses, and restricting sow nutrient intake may lead to permanent changes in postnatal growth and growth efficiency in pigs [15]. It is generally accepted that glucose is the major energy substrate for fetal growth. Fetal uptake of glucose is a consequence of the glucose gradient between maternal and fetal blood. In many countries gestating sows are fed restrictively (2–3 kg/day with a diet containing 16–18% crude protein) or according to their condition, at 30–40% of their voluntary feed intake. This low feed intake during gestation has led several research groups to question whether this allowance is adequate to support fetal muscle development in particular. However, increased maternal glucose can also be obtained by treating gestating sows with porcine growth hormone (pGH), and furthermore, dietary L-carnitine may be essential in the transportation of fatty acids across the mitochondrial membrane for β-oxidation. Thus, various methods have been evaluated in an attempt to support larger litters during gestation, and thereby decrease the variation in birth weight, survival, muscle development and postnatal growth performance.

In this chapter we describe quantitative aspects of the performance of sows and their offspring in relation to (i) litter variability, (ii) increased sow feed intake, (iii) sow protein intake, (iv) sow dietary supplementation with L-carnitine, and (v) sow treatment with pGH.

2.1 Litter Variability

There is a large variation in birth weight within a litter of pigs. This variation is believed to be partly caused by undernutrition in utero of the low birth weight pigs [14], but also the influence of genetics on the growth of the fetuses in utero. Generally, pigs with a low birth weight grow more slowly during postnatal life than their heavier littermates and therefore have a lower body weight when slaughtered at the same age [29].

Within some litters of pigs, one or two pigs are observed to be much smaller at birth than the rest of the litter. These pigs are called runts and are characterised by having a birth weight 2.5 standard deviations below the mean of the litter (i.e. < 0.8 kg). Runt pigs are thought to be a subpopulation within a litter, which has suffered from extreme undernutrition during fetal development [28, 30, 53]. They have a lower average postnatal daily gain, and when fed ad libitum their feed intake is less than normal-sized pigs [10, 62]. Reference to 'low birth weight' pigs in this chapter does not include runt pigs, unless otherwise stated.

2.1.1 Litter Size and Birth Weight

During the last 10–20 years genetic selection for increased litter size has increased the number of pigs born by 0.3 pigs/litter/year under Danish conditions resulting in an average litter size of approximately 13 in 2006 [68]. This increase in litter size has caused the variation in pig body weight at birth to increase, with the consequence of more pigs born with low birth weight. The negative relationship between litter size and pig birth weights has been studied by Milligan et al. [38] and Quiniou et al. [54]. In the research by Milligan et al. [38], 52 sows were studied through eight parities, and the litters were categorised into those below 9, between 9 and 11, and above 12 pigs per litter. The results from their study showed that the mean birth weight of the smallest pigs decreased from 1.05 to 0.90 kg in litters with below 9 to those above 12. The mean birth weight of the largest pigs decreased from 1.57 to 1.38 kg in the same litters. Mean weaning weights also decreased when litter size increased. Survival rate between birth and weaning decreased in low birth weight pigs from 83.6 to 63.1% and in high birth weight pigs from 92.3 to 91.9% in litters of less than 9 and above 12, respectively. In the study by Quiniou et al. [54], the litters were divided into four litter size classes: below 11, between 12 and 13, between 14 and 15, and above 16. In this study, the mean birth weight decreased from 1.59 to 1.26 kg in litters with less than 11 and above 16 pigs, respectively. The within litter variation in birth weight increased with increasing litter size, and the percentage of small pigs at birth increased from 7 to 23% in litters with less than 11 and above 16 pigs, respectively.

Birth weights of the lightest and the heaviest pigs within litters are presented for several studies in Table 2.1. There are significant differences between the littermate birth weights in all studies. The lowest birth weight was found to be 1.05 kg by Gondret et al. [26] and the heaviest birth weight to be 1.9 kg by Poore and Fowden [52]. Generally, it is difficult to compare birth weights between studies, as the definitions of birth weight classes differ. Thus, Gondret et al. [26] grouped their pigs in two groups, where the low birth weight pigs were between 0.75 and 1.25 kg, and pigs within the heavy birth weight group were between 1.75 and 2.05 kg. Poore and Fowden [52] defined their low birth weight class as being below 1.47 kg, and the heavy birth weight class as being above 1.53 kg. There are also large differences in litter size between countries and breeds, and consequently between studies, which influence birth weights as referred to previously.

Variation in litter size and birth weight among breeds has been the focus of several studies. Comparison of Chinese Meishan pigs with commercial European or U.S. pig breeds has revealed large differences in the efficiency of the placenta to support fetal growth with subsequent impacts on litter size and birth weights. Meishan sows farrow three to four more live pigs per litter than European and U.S. sows [22]. The higher number of fetuses in the Meishan sows has consequences for fetal weight, with the average weight at day 110 of gestation being 1.4 kg in Yorkshire and 0.9 kg in Meishan breeds. In the European breeds the embryos develop asynchronically during the early implantation period, which causes a high loss of

Table 2.1 Intra-litter variation in birth weight and postnatal performance

Body weight class[1]	Birth weight, kg	Period	Feeding strategy	Daily gain, g/day	Feed intake, kg/day	Feed: Gain, kg/kg	Meat percentage	References
LW	1.05^{a}	Birth to 100 kg	ad libitum	650^{a}	2.28	2.69^{a}	61.1^{a}	[26]
HW	1.89^{b}			690^{b}	2.25	2.49^{b}	63.0^{b}	
LW	1.47^{a}	30 kg to 104 kg	ad libitum	810^{a}			60.0	[42]
MW	1.50^{a}			935^{b}			60.1	
HW	1.71^{b}			1,038^{b}			59.5	
LW	1.13^{a}	Birth to 3 months	ad libitum	249^{a}				[52]
HW	1.90^{b}			392^{b}				
LW	1.27^{a}	25 kg to 105 kg	Restrictive	800	1.87	2.33	59.8	[3]
HW	1.76^{b}			810	1.87	2.33	59.7	
LW	1.39^{a}	30 kg to 110 kg	Restrictive	813^{a}			57.9^{a}	[31]
MW	1.55^{b}			865^{b}			58.8^{b}	
HW	1.73^{c}			869^{b}			59.5^{b}	
LW	1.32^{a}	6 kg to 110 kg	ad libitum	796^{a}	1.78^{a}	2.24	55.1	[73]
HW	1.83^{b}			851^{b}	1.87^{b}	2.19	54.9	

[1]LW = low weight; MW = middle weight; HW = heavy weight. Classes are birth weight classes within litter in all papers except for [42] where the classes refer to slaughter weights within litter at the same age.
[a, b, c]Within study and column, numbers with different superscripts differ significantly ($P < 0.05$).

embryos at this stage, whereas in the Meishan breed embryos develop more synchronously and less embryos are lost [22, 72]. This may be due to changes in the uterine environment in the highly selected U.S. and European breeds, which may have negative consequences for less developed fetuses. As the fetuses in the U.S. and European breeds have higher fetal growth rates, continued placental growth to support fetal growth is required, whereas the Meishan breed increases the density of placental blood vessels and, thereby, the efficiency of nutrient transfer across the placenta without increasing placental growth. Thus, placental growth and efficiency are very important for litter size and birth weight and, therefore, for the overall outcome of pregnancy.

Litter size in early gestation also affects the birth weight of the pigs. A larger number of fetuses per litter in early gestation cause lighter pigs at birth, even if the litter size is the same at birth as in litters with fewer fetuses in early gestation [48]. This may be due to a higher competition among fetuses during the implantation period in large litters, which has a negative effect on early placental development and growth, and thereby on placental capacity during later gestation. In support of this notion, Père et al. [48] found the weight of placenta at day 112 of gestation to be less for fetuses originating from litters with a larger number of fetuses in early gestation. The position of the fetus within the uterus has also been suggested to affect fetal growth and therefore birth weight [46], as fetuses positioned near the ovaries are heavier than other fetuses. This may only be apparent in late gestation, where the competition for nutrients becomes more important, as other studies have not been able to show an association between fetal growth and uterine position in mid gestation [2, 43].

The extent to which growth and development of individual fetuses is similar between sexes has also been studied. Parfet et al. [46] found the average birth weight of male pigs to be 5.5% higher than that of female pigs and the survival rate to be equivalent. In contrast, Bee [3] did not find any significant difference in birth weight between sexes. Furthermore, the sex of adjacent fetuses does not seem to influence the birth weight of either males or females [46].

2.1.2 Postnatal Performance

Positive relationships between birth weight and postnatal growth rate have been observed in several studies (Table 2.1). Generally, daily gain over different time periods from birth to slaughter shows that low birth weight pigs grow more slowly than high birth weight pigs. When dividing the postnatal growth into growth phases, there are contradictory results on the impact of birth weight on the growth rate within these phases. The average daily gain during suckling has been found to be significantly different between low and high birth weight pigs in most studies [3, 26–27, 31, 52, 54], whereas Wolter et al. [73] did not find this relationship to be significant. In the growth phase between weaning and approximately 30 kg live body weight, the average daily gain was found by several authors [26–27, 73] to be significantly lower for pigs of low compared to high birth weight, although Bee [3] did not find this relationship. In the final growth phase from approximately 30 kg until slaughter at 100–110 kg, Bee [3] and Gondret et al. [27] did not find a significant difference in average daily gain between low and high birth weight pigs, whereas Heyer et al. [31] and Gondret et al. [26] did find a significant difference. Wolter et al. [73] divided this latter growth phase in two and found a significant difference in daily gain from 65 to 110 kg between low and high birth weight pigs, but not between 25 and 65 kg. Taking the overall average daily gain from birth until slaughter at about 100–110 kg into consideration, the results from most studies show a significant positive relationship between birth weight and daily gain. Poore and Fowden [52] also studied growth until the pigs reached 12 months of age. At 12 months the low birth weight pigs weighed 152.8 kg, whereas the high birth weight pigs weighed 169.4 kg. In this study the growth rate was found to be significantly different for low and high birth weight pigs between birth and 1 month as well as to 3 months of age, whereas it was not significant over the entire postnatal period from birth to 12 months of age. Taken together, there appears to be a close relationship between birth weight and postnatal growth during the early growth phases, whereas the relationship becomes weaker as the pigs age.

When pigs with a large difference in birth weight compete for milk, those with the highest weight have an initial advantage. The largest pigs generally find the best teats and are more efficient in activating the teat to produce more milk. During suckling it is therefore likely that the feed intake of the low birth weight pigs is less than that of the high birth weight pigs, and this difference in feed intake will affect growth performance. In accordance with this, Campbell and Dunkin [7] showed that high birth weight pigs consume more milk per suckle than low birth weight pigs, although the relative milk consumption (kg milk/kg body weight) did not differ between birth weight classes. When supplementing pigs with milk replacer during suckling, a significant difference in milk replacer intake between low and high birth weight littermates was found, with the high birth weight pig having the highest intake [73]. Thus, even when pigs have the opportunity to increase their feed intake by supplementation, low birth weight pigs consume less than their heavier littermates during the suckling period.

The average daily feed intake from weaning until slaughter was found by Wolter et al. [73] to be significantly higher for high than for low birth weight pigs when fed ad libitum. When fed restrictively, the potential of the high birth weight pigs for feed intake, feed utilisation and growth may not be reached, and therefore possible differences in growth potential between low and high birth weight pigs may be levelled out. Thus, the difference in growth potential between the low and high birth weight pigs may depend on the feeding strategy. The ability of the pigs to utilise the feed for growth, i.e. the feed-to-gain ratio, has been measured in some studies. Bee [3] and Wolter et al. [73] did not find any difference in the feed-to-gain ratio of low and high birth weight pigs from day 30 until slaughter or from weaning to slaughter. Even though Gondret et al. [26] did not find a difference in feed intake, they did find a significantly lower feed-to-gain ratio in high birth weight pigs than in low birth weight pigs, which showed that pigs of higher birth weight grew more efficiently.

Variation in growth between the low and high birth weight littermates may also be related to the number of muscle fibres developed during fetal life. Accordingly, fewer fibres in various muscles were found in the low compared to the high birth weight littermates selected at birth [26–28, 60] or at slaughter [42]. Significant correlations have been found between the number of muscle fibres and growth performance, with the number of muscle fibres formed during fetal development directly related to postnatal daily gain ($r = 0.42$) [13, 47] and inversely related with feed-to-gain ratio ($r = -0.41$) [13].

It is well known that gender affects postnatal growth in pigs. Poore and Fowden [52] also found an interaction between birth weight and gender in relation to postnatal growth rate. In this study, low birth weight entire male pigs exhibited catch-up growth during the first 3 months of age, and this was not evident to the same extent in low birth weight female pigs. This catch-up growth resulted in the weight of low and high birth weight male pigs being equal at both 3 and 12 months of age, whereas low birth weight female pigs continued to be smaller than high birth weight female pigs at both 3 and 12 months. The gender of adjacent fetuses during development may affect the postnatal growth rate of male but not female pigs, depending on the feeding strategy. When pigs are fed a restricted diet during growth and are kept in groups, male pigs that were positioned between two males in utero exhibit significantly higher growth rates postnatally than males positioned between females or between a male and a female [46]. The authors explain this as either a direct effect of the environment in utero, or an indirect effect of the behaviour of the males postnatally dependent on the position in utero, as males positioned between males generally behave more aggressively than males positioned next to a female. This may favour the feed intake of the more aggressive males compared to the other littermates under restrictive feeding conditions, and thereby enhance the postnatal growth rate of these aggressive males. However, this difference in growth rate between aggressive males positioned between two males in utero and other littermates is not seen under ad libitum feeding postnatally, as the other littermates have free access to feeding and therefore can feed at other times than the aggressive males [46].

2.1.3 Carcass Composition

Factors such as genetics, nutrition and hormones that affect fetal growth may also influence carcass composition at slaughter, as the development of different tissues may not be equally affected. In pigs, the distribution of lean and fat tissue is of great importance for the meat industry, and is a basis for payment to producers. Intra-litter variation in carcass composition may therefore be of considerable importance for both producers and for the meat and processing industry. In most countries a lean meat percentage is estimated at the slaughterhouses, and this estimate provides a general indicator of the distribution of meat and fat content in the carcass. There are contradictory results regarding differences in lean meat percentage between low and high birth weight pigs (Table 2.1). When the pigs are slaughtered at the same weight, both Heyer et al. [31] and Gondret et al. [26] found a significant difference in lean meat percentage, which was higher in high than in low birth weight pigs. This is in contrast to Wolter et al. [73], Bee [3] and Gondret et al. [27], who did not find any difference in lean meat percentage between birth weight categories. Also, a comparison of the slowest and the fastest growing pig within a litter did not reveal any differences in lean meat percentage at the same age [42]. The feeding strategy used differed among these studies but did not explain the differing results.

Backfat thickness can be measured at different anatomical positions and may therefore not be directly comparable between studies. Generally, backfat thickness is measured between the third and fourth lumbar vertebra or just behind the last rib and approximately 5 cm from the mid-line. When pigs within a litter are slaughtered at the same weight, Gondret et al. [26] found a significantly higher backfat thickness in low than in high birth weight pigs. When performing a carcass dissection into meat and fat, Bee [3] also found a significantly higher adipose tissue yield in low than in high birth weight pigs. In contrast to this, others have not found this difference in backfat or adipose tissue yield between low and high birth weight pigs when slaughtered at the same weight [27, 31, 73].

The different muscles in the body have their own specific function, and some muscles are more vital than others. In the case where the low birth weight pigs have suffered from undernutrition during fetal development, it may well be that more vital muscles have developed normally, while other less vital muscles have been more influenced by undernutrition. In accordance with this, Heyer et al. [31] found that the percentage of *M. semimembranosus et adductor*, *M. gluteus* and *M. biceps femoris* in the ham was higher for high than for low birth weight pigs, whereas *M. psoas major*, *M. semitendinosus* and *M. quadriceps* did not differ. Also, Bee [3] and Gondret et al. [27] did not find the weight of *M. semitendinosus* differed between low and high birth weight littermates. The percentage of ham in the carcass was also found to be higher in high than low birth weight pigs [26, 31]. The percentage of loin in the carcass was found by Gondret et al. [26] to be higher in high birth weight pigs, while Heyer et al. [31] did not find any difference between different birth weight classes.

Gender is also known to affect the carcass composition in pigs and, as for postnatal growth rate, Poore and Fowden [52] found an interaction between gender and birth weight in relation to backfat thickness at 5 different anatomical locations along the back. This interaction showed that low birth weight of female pigs resulted in increased fat depth at 12 months, whereas fat depth was not related to birth weight in male pigs at 12 months of age.

2.2 Feeding of the Sow During Gestation

Nutrient supply reaching the fetus is critical for fetal growth, and is largely affected by the nutrients fed to the sow during gestation. Both the amount and the composition of the feed fed to the sow during pregnancy affect fetal growth as, to some extent, the nutrients regulate nutrient uptake and utilisation by the placenta and fetuses either directly or indirectly through hormonal mechanisms. The nutrient supply to the fetuses during late gestation, where most of the fetal growth takes place, is of great importance, although the nutrient supply during early and mid gestation may also affect fetal growth. Proliferation of progenitor cells and differentiation of cells into specific tissues occur in early gestation, and differences in these processes may have effects on later development [5, 63].

Requirements for energy during gestation vary to a certain degree during the gestation period. Generally, pregnant sows are restrictively fed, and energy requirements during gestation are between 20 and 35 MJ ME per day (2–3 kg/day) [24, 41].

In this section, differences in sow feed intake, protein intake and supplementation with dietary L-carnitine in relation to sow and offspring performance will be discussed. Reference to increased or decreased feed intake in the following section equates with animals fed above or below requirements.

2.2.1 Sow Performance

2.2.1.1 Feed Intake

Several authors have investigated the influence of increased or decreased sow feed intake on sow and litter performance (Table 2.2). However, the extent and timing of

Table 2.2 The influence of sow feed intake during gestation on sow and litter performance

Treatment window during gestation	Feed intake, kg/day	Fetal/pig age	Litter size	Litter weight, kg	Average fetal/pig weight, g	Sow weight gain, kg	References
Throughout	0.79	Birth			760		[6]
	2.35				1,100		
Throughout	2.5	Birth	11.4	15.6	1,570		[15]
Day 25–50	5.0		14.6	17.4	1,410		
Day 50–80	5.0		11.2	15.4	1,500		
Day 25–80	5.0		11.8	15.9	1,510		
Throughout	2.0	Birth	12.1		1,500	50^a	[41]
Day 25–50	ad lib		12.5		1,500	85^b	
Day 25–70	ad lib		12.9		1,500	104^c	
Throughout	1.81	Day 46	9.83		21.1	4.32	[39]
Day 30–45	7.0		11.33		18.4	34.0	
Throughout	2.11	Day 56	11.67		96.6	2.11	
Day 30–55	7.0		10.50		97.6	41.2	
Throughout	2.3	Birth	11.5		1,540		[31]
Day 25–85	3.11		12.1		1,640		
Day 25–85	3.90		13.0		1,480		
Day 25–85	4.60		12.4		1,600		
Throughout	2.8	Birth			1,580	20.4	[3]
Day 0–50	4.0				1,420	24.3	
Day 0–50	1.7				1,540	18.6	
Throughout	2.0	Day 50	14.0	0.594	45		[43]
Day 25–50	ad lib		16.3	0.641	44		
Throughout	2.0	Day 70	13.4	3.322	262		
Day 25–70	ad lib		12.8	3.390	290		

[a,b,c] Within study and column, numbers with different superscripts differ significantly ($P < 0.05$).

the increase or decrease in feed intake during gestation differs greatly among studies. Provision of sows with ad libitum access to feed during early to mid gestation increases the feed intake during the period of ad libitum feeding from two to four times compared with sows fed according to requirements [41, 39]. The weight gain of these sows during gestation is significantly increased by ad libitum compared with restrictive feeding. In other studies, the feed intake during gestation has been doubled [15], increased by 30% [3, 24], or increased by 35, 70 or 100% [31] during early to mid gestation. In most studies the increased feed intake in early to mid gestation caused a significant increase in sow body weight gain. A 30% increase in feed intake during early to mid gestation did not increase the backfat thickness significantly although the body weight increased [3, 24]. However, a 40% reduction in feed intake during early to mid gestation compared with feeding to requirements significantly reduced the sow body weight gain and backfat thickness [3].

Litter size measured as the total number of pigs born or the number of pigs born alive was not affected by increased sow feed intake in most studies [3, 15, 24, 41]. In the study by Heyer et al. [31] a 70 or 100% increase in feed intake caused a tendency towards increased total number of pigs born in the second, but not the first parity, but the number of live-born pigs did not differ significantly despite a numerical increase (from 11 in control sows, to 13.8 and 13.0 in sows offered 70 or 100% additional feed, respectively). More pigs were weaned per litter when feed intake was increased by 70 or 100% compared with control sows [31]. Taken together, increased feed intake from around day 25 until mid gestation does not seem to significantly affect the number of pigs born per litter. Even though sows are already fed restrictively when fed to requirements, Bee [3] found no effect on litter size at birth or at weaning after a 40% reduction in feed intake during early to mid gestation.

Several authors have found that increased feed intake throughout or during stages of gestation has a negative effect on sow feed intake during lactation [9, 12, 71], which may influence milk yield. In the study by Nissen et al. [41] no effect of ad libitum feed intake from day 25 to day 50 or 70 of gestation was found on estimated milk yield over a 3-week lactation period. In this study, the feed intake during lactation was also not affected by gestational feed intake, consistent with the findings of Mahan [37].

2.2.1.2 Protein Intake

Besides overall feed intake, protein intake during gestation is also important for fetal growth and development. To some extent, maternal tissues can buffer the fetal demand for protein through tissue mobilisation, but in cases where the protein restriction is very severe or prolonged, this may be inadequate to support fetal growth and development. In these latter cases, birth weights, postnatal performance and carcass composition will be affected. A high degree of mobilisation of maternal tissues during severe and prolonged protein restriction will affect maternal weight and fat distribution, which may have consequences for milk production and the success of the subsequent gestation.

Feeding protein-free diets (< 0.5% protein) either throughout or in specific periods of gestation has far-reaching consequences for sow body weight gain and backfat thickness during gestation. A protein-free diet throughout or within specific periods of gestation reduces the body weight of the sow from breeding to parturition compared with sows fed a control diet [49–50]. This means that despite being pregnant the sow actually loses weight during gestation, and that the weight loss is influenced by duration during which the protein-free diet is fed to the sow. In a more recent study the dietary intake of protein was either reduced or increased relative to requirements from day 25 of gestation until parturition [35]. In this study, the body weight gain of the sow was significantly influenced by protein level during gestation. A protein-free diet throughout gestation also reduced backfat thickness of the sow [50]. Reduced or increased protein intake from day 25 of gestation to parturition does not have any significant effect on backfat thickness compared with sows fed according to requirements, but the backfat thickness was significantly reduced in sows fed a low compared with a high protein diet [35]. Feeding a protein-free diet over 2 successive parities showed that both body weight and backfat thickness mainly decreases during the first parity, whereas in the second parity there is little change [50]. Both body weight and backfat thickness are lower from the start of the second parity on a protein-free diet compared with control fed sows.

The total number of pigs born and the number of pigs born alive are not affected by protein level during gestation. Several authors have investigated the effect of feeding either a protein-free diet throughout or during specific periods of gestation, and of increased or decreased protein intake at different stages of gestation, and there was no effect on litter size at birth found in any of these studies [23, 49–50, 65].

The established negative effects of sow protein intake on body weight and backfat gain during gestation may have serious consequences for mammary gland development, lactational performance and therefore on pig milk intake, survival and weight gain during suckling. Unfortunately, this has not been the focus of most studies on protein intake during gestation. The authors are only aware of one study in which the milk yield has been measured after differences in protein intake during gestation. In this study, protein level from day 25 of gestation to parturition was either decreased or increased compared with requirements [35]. The milk yield increased significantly with increased protein intake, and this was evident at days 8 and 18 of lactation.

2.2.1.3 Supplementation with L-Carnitine

The primary role of L-carnitine is to transport long and medium chain fatty acids across the mitochondrial membrane for β-oxidation, although it is also involved in protein synthesis and glucose homeostasis [11, 45]. Feeding L-carnitine to pigs during growth enhances protein accretion and decreases backfat thickness, suggesting an increased ability of the pigs to utilise fat as an energy source, and to use carbon for amino acid synthesis and branched-chain amino acids for protein synthesis [45]. This change in metabolism upon supplementation with L-carnitine may be beneficial for sow performance, and for fetal development and growth when supplemented to sows during gestation. In most studies, L-carnitine is supplemented at 125 mg/day per sow during gestation and at 250 mg/day per sow during lactation. Unless stated otherwise, these levels of supplementation apply within the sections below concerning L-carnitine. Results from research on L-carnitine supplementation of sows are summarised in Table 2.4.

There seems to be controversy regarding the weight gain of sows during gestation when L-carnitine is supplemented from mating until parturition. Ramanau et al. [55, 57] found no effect of L-carnitine supplementation during gestation on sow body weight gain, whereas Eder et al. [16] and Ramanau et al. [56] found an increase in body weight gain from mating until day 85 of gestation, and Musser et al. [40] found an increase in body weight from breeding until day 112 of gestation. In this latter study, the supplementation during gestation was 100 mg/day per sow. Also, changes in backfat thickness with L-carnitine supplementation are reported in some studies but not in others. Ramanau et al. [55, 57] found no effect of L-carnitine supplementation during gestation on backfat thickness at parturition, whereas there was an increase in backfat thickness from breeding until day 112 of gestation after L-carnitine supplementation in the study of Musser et al. [40]. The reasons for these differences are not obvious.

Litter size is related to ovulation rate and survival rate of embryos especially in early period gestation. Changes in sow metabolism during gestation when supplemented with L-carnitine may well affect embryo survival and therefore the number of pigs born. In support of this hypothesis, fetal litter size at approximately day 57 of gestation was found to be increased from 10.8 in control sows to 15.5 in sows supplemented with 100 mg/day of L-carnitine from the day of breeding until day 57 of gestation [70]. Some authors have also found an increase in the total number of pigs and number of live pigs born to sows supplemented with L-carnitine during gestation [4, 57]. In contrast, several other studies have not found any difference in total number of pigs and number of live pigs born to control or supplemented sows [16, 40, 55–56].

Supplementation with L-carnitine during both gestation and lactation has a large impact on milk yield. In two studies, milk production was measured at days 11 and 18 of lactation, and in both studies milk yield was increased significantly in sows that were supplemented with L-carnitine [55, 57]. Concentrations of fat, protein, lactose and gross energy did not differ between milk from supplemented and non-supplemented sows in either of these two studies.

Thus, there are very conflicting results regarding sow performance when L-carnitine is supplemented throughout gestation, although the positive effect on milk yield of supplementation during both gestation and lactation are very convincing.

2.2.2 *Fetal Growth and Birth Weight*

2.2.2.1 Feed Intake

An increased sow feed intake may increase the supply of nutrients to the fetuses, and thereby influence fetal development and growth. A summary of the influence of sow feed intake on fetal and birth weight is presented in Table 2.2. In two studies, sows were fed either restrictively according to requirements or ad libitum during early (from approximately day 25) to mid-gestation [39, 43]. After the period of ad libitum feeding the sows were slaughtered, and the fetal weights recorded. There was no difference in average fetal weight, average litter weight or number of fetuses per litter between ad libitum and restrictively fed sows in either study. In the study by Musser et al. [39] the relationship between number of fetuses per sow and average fetal weight was examined separately for ad libitum- and control-fed sows. This revealed that a negative relationship existed in control-fed sows, whereby average fetal weight decreased with increasing litter size, whereas in ad libitum-fed sows no significant relationship was evident. Thus, variation in fetal weight seems to be less in litters from ad libitum-fed than control-fed sows.

In most studies the birth weight of pigs has not been found to be greater as a result of increased sow feed intake during gestation. In one study, Buitrago et al. [6] found a difference of 30% in birth weight of pigs from dams fed either a low or high feed intake throughout gestation, although this difference was not significant. This is consistent with several other studies where increased feeding level was in early to mid gestation only and no differences in birth weight were found among treatments [3, 15, 24, 31, 41]. An increase in sow feed intake throughout or in specific periods of gestation does not seem to have an effect on body weight of individual fetuses nor on litter weight. Most studies have been conducted during the first two thirds of gestation, when fetal growth rate is low compared with the last third. The nutrient requirements to support fetal growth may well be met when sows are fed according to requirements during this part of gestation, and therefore no beneficial effect of increased nutrient supply on fetal growth occurs. Whether it has an effect on development of the individual tissues however is discussed later.

2.2.2.2 Protein Intake

Results from some studies where sows have been fed different amounts of protein during gestation on fetal growth and birth weights are summarised in Table 2.3.

Severe protein restriction during early gestation (days 0–63) tended to decrease the fetal body weight at day 63, but there was no effect on crown-rump length and heart girth [65]. In this same study, placental weight was significantly reduced when sow protein intake was restricted, suggesting that the capacity of the placenta to support continuous fetal growth during late gestation may be impaired. This is

Table 2.3 The influence of sow protein intake during gestation on sow and litter performance

Treatment window during gestation	Protein content of feed	Fetal/pig age	Number of fetuses/pigs	Litter weight, kg	Average fetal/pig weight, g	References
Throughout	Control	Birth	10.7		1,120[a]	[49]
Day 0–16, 21 to term	0		11.0		880[b]	
Day 25 to term	0		9.2		1,090[a]	
Throughout	0		10.0		750[c]	
Throughout	13%	Birth	11.1	14.4[a]	1,365[a]	[50]
Throughout	0.5%		8.4	8.3[b]	1,000[b]	
Day 0–63	13%	Day 63	10.5		155	[65]
Day 0–63	0.5%		12.2		145	

[a, b, c] Within study and column, numbers with different superscripts differ significantly ($P < 0.05$).

supported by several other studies in which birth weight was significantly reduced when the feed intake of sows was severely restricted or they were fed protein-free diets either throughout or for periods of gestation [49–50, 66]. Less severe protein restriction during gestation does not seem to have an effect on pig birth weight or measurements of anatomical proportions such as crown-rump length, abdominal circumference and skull width[23, 32].

2.2.2.3 Supplementation with L-Carnitine

Effects of L-carnitine supplementation on fetal growth and birth weights from different studies are shown in Table 2.4.

L-carnitine supplementation from the day of mating until approximately day 57 of gestation did not have any significant effect on individual fetal weight or crown-rump length at day 57 of gestation compared with non-supplemented sows [70]. Although not significant ($P = 0.07$), total litter weight was numerically higher for supplemented compared to non-supplemented sows (1,450 versus 989 g). This could be explained by a significantly higher number of fetuses in supplemented sows, as discussed earlier.

At birth, there are again conflicting results regarding the effect of L-carnitine supplementation on individual and litter weights. Two studies showed that average pig birth weight increased when sows were supplemented with L-carnitine during gestation [40, 56], while another study found that birth weight decreased with supplementation [57]. Total litter weight at birth was found by Ramanau et al. [56–57] to be higher after supplementation. In contrast, neither Eder et al. [16], Birkenfeld et al. [4] nor Ramanau et al. [55] found any significant effect of L-carnitine supplementation on individual or litter birth weight, although a tendency ($P = 0.055$) for an increase in birth weight of individuals after supplementation of sows was found by Eder et al. [16].

The reason for the conflicting results regarding birth weight and litter size is not obvious, but in some experiments the number of animals studied has been low and may therefore not be suitable for investigations of litter traits. More and possibly larger studies are needed to elucidate the effects of L-carnitine on litter performance.

2.2.3 Postnatal Growth

2.2.3.1 Feed Intake

The influence of increased sow feed intake during gestation on postnatal growth performance of the offspring has been the focus of several studies, and some of the results from these studies are summarised in Table 2.5.

Weaning weights are closely linked to birth weights and to the milk yield of the sow from parturition until weaning. As discussed in the previous sections, birth weights and milk yield were not found to be influenced by increased sow feed intake throughout or in specific periods during early to mid gestation. In most of the studies reviewed for this chapter, weaning weights were not affected by sow feed

Table 2.4 The influence of sow supplementation with L-carnitine during gestation on sow and litter performance

L-Carnitine in gestation	Number of fetuses/pigs born alive		Litter weight, kg		Fetal/pig weight, g		Sow performance during gestation		References
	At birth	At weaning	At birth	At weaning	At birth	At weaning	Weight change, kg	Fat thickness change, mm	
0	10.33	8.91	14.65	41.14	1,480^{a}	4,680^{a}	+46.5^{a}	+1.6^{a}	[40]
100 mg/day	10.43	9.02	15.45	45.63	1,580^{b}	4,960^{b}	+55.4^{b}	+2.6^{b}	
0	10.0	8.1	13.6	66.8^{a}	1,370	8,240	+32.7		[16]
125 mg/day	9.90	8.5	14.5	74.0^{b}	1,470	8,710	+39.2		
0	9.95^{a}		15.8^{b}		1,620^{a}	9,210^{a}	+30.5	+5.0	[57]
125 mg/day	12.8^{b}		18.2^{b}		1,460^{b}	9,770^{b}	+30.0	+5.0	
0	12.1		17.9	92^{a}	1,550		+55	+5.5	[55]
125 mg/day	12.3		20.1	108^{b}	1,650		+49	+6.7	
0	10.8^{a}		0.989		91.4				[70][1]
100 mg/day	15.5^{b}		1.450		92.4				

[a, b, c] Within study and column, numbers with different superscripts differ significantly ($P < 0.05$).
[1] Sows were fed L-carnitine during gestation until approximately day 57 of gestation where sows were slaughtered and fetal measurements made.

Table 2.5 The influence of sow feed intake during gestation on postnatal growth performance of the offspring

Treatment window during gestation	Sow feed intake, kg/d	Postnatal performance of offspring					References
		Feed intake, kg/day	Daily gain, g/day	Feed:Gain	Meat percentage		
Throughout	2 kg	ad libitum	924		59.9		[41]
Day 25–50	ad libitum		893		59.9		
Day 25–70	ad libitum		967		59.9		
Throughout	2.8	1.86	810[a]	2.27[a]	59.9		[3]
Day 0–50	4.0	1.88	790[b]	2.38[b]	59.7		
Day 0–50	1.7	1.88	810[a]	2.33[c]	59.7		
	1st parity						[31]
Throughout	2.3	Pigs were	830[a]		58.6		
Day 25–85	3.11	scale fed	872[b]		59.1		
Day 25–85	3.90	according	819[a]		59.1		
Day 25–85	4.60	to Swedish	874[b]		58.2		
	2nd parity	standards					
Throughout	2.3		880[a]		57.4		
Day 25–85	3.11		848[ab]		58.4		
Day 25–85	3.90		814[b]		58.9		
Day 25–85	4.60		815[b]		58.4		
Throughout	2.5	2.09	840[a]	2.49[a]			[15]
Day 25–80	5.0	2.13	924[b]	2.31[b]			
Throughout	3.0	1.82	799	2.30	54.2		[8]
Day 45–85	+50% or +75%[1]	1.82	786	2.33	53.0		

[1] +50% for 1st parity and +75% for 2nd parity.
[a, b, c] Within study and column, numbers with different superscripts differ significantly ($P < 0.05$).

intake during gestation [3, 6, 15, 31, 41]. In contrast, Gatford et al. [24] did find an increased weaning weight of pigs from sows with high feed intake from days 25 to 50 of gestation. As the birth weight of these pigs did not differ between sow feeding strategies during gestation, the growth rate of pigs from sows fed a higher allowance was increased from birth to weaning compared with pigs from sows fed according to requirements. Whether this increase in growth rate during suckling was due to greater milk production of the sow, better feed conversion efficiency of the pigs, or whether the offspring were better at stimulating milk production, is not known.

With regard to the growth rate from weaning until slaughter of offspring from sows fed different amounts of feed during specific periods of gestation, the results are more contradictory. In two studies, where the sows were fed either 35, 70 or 100% above requirements from days 25 to 85 of gestation [31] or 30% above requirements from days 0 to 50 of gestation [3], average daily gain was found to be lower in offspring from sows fed above requirements except from first parity sows.

For first parity sows the results were less clear, as the growth rate was increased in offspring from sows fed 35 or 100% above requirements, whereas it was not different from controls in offspring from sows fed at 70% above requirements [31]. In contrast, studies by Nissen et al. [41] and Gatford et al. [24] showed no positive effect of increasing the feed intake of sows from days 25 to 50 or 70 of gestation on average daily gain of the offspring. A non-significant negative effect of higher feed intake by sows from days 25 to 50 of gestation was found on average daily gain, carcass weight and muscle deposition rate of the offspring. In this same study an interaction showed that it was mainly the slow-growing pigs that were negatively affected by increased sow feed intake during gestation. Thus, the slow-growing pigs within a litter exhibited even slower growth postnatally following increased sow feed intake compared to slow-growing pigs from control fed sows, whereas faster-growing littermates were not affected by sow feed intake [41]. Cerisuelo et al. [8] increased the feed intake of sows during mid gestation only (days 45–85). In one of two replicates average daily gain was increased from weaning to day 62 of age in offspring from supplemented sows, whereas no effect of increased feed intake was found between days 63 and 184. When pigs were divided into weight groups at weaning, only the lightest pigs showed an increase in growth rate [8]. In an earlier study where sows were fed at double the requirement from days 25 to 80 of gestation, the average daily gain of the offspring from weaning to day 70 of age was not different between treatments, whereas offspring from sows fed double the requirement had significantly faster growth from days70 to 130 [15].

Feed efficiency (feed-to-gain ratio) has only been measured in a few of the above mentioned studies. In that of Bee [3], the feed-to-gain ratio was increased in offspring from sows fed increased amounts of feed during early gestation, meaning that the amount of feed per weight gain was higher in these pigs. In contrast, Dwyer et al. [15] found feed-to-gain ratio decreased in offspring from sows fed double the amount normally fed in early to mid gestation. Accordingly, pigs from sows fed increased amounts during early gestation had a slower rate of postnatal growth in the study by Bee [3], whereas the growth rate was increased in the study by Dwyer et al. [15].

There are inconsistent results with respect to the number of muscle fibres in the offspring in response to increased sow nutrition in early to mid gestation. Dwyer et al. [15] found that the smaller pigs within litters from sows with high intakes in early and mid gestation had increased muscle fibre number. Also, Gatford et al. [24] suggested that increased sow feed intake during early gestation increases muscle fibre number. In contrast, pigs in the study by Nissen et al. [41] did not have an increase in muscle fibre number in response to increased sow nutrition in early to mid gestation.

In summary, the results regarding the effect of increased sow feed intake during early to mid gestation on postnatal growth, feed efficiency and muscle fibre number of the offspring are inconsistent and no clear conclusions can be drawn. However, it should be stressed that interactions between feeding of the sow during gestation and weight and growth rate of the offspring may be important for inclusion in future studies.

2.2.3.2 Protein Intake

Severe protein restriction has a considerable impact on growth performance of the offspring from birth to slaughter, and results from various studies are summarised in Table 2.6. Feeding a protein-free diet throughout gestation decreases the average daily gain of the offspring from birth until weaning, resulting in a significantly lower weaning weight [49–50, 66]. As the milk yield is negatively affected by decreased protein intake during gestation [35], this difference in average daily gain during suckling may well be related to the milk supply and intake of the pigs as well as the protein supply to the fetuses during gestation. Unfortunately, in most studies it is not possible to distinguish between the direct effect of decreased protein supply during fetal life and the indirect effect on milk yield and therefore milk intake during suckling on the postnatal growth performance of the offspring, as cross fostering to sows with a different protein supply during gestation has not taken place. In the study by Schoknecht et al. [66] this effect of cross fostering on weaning weights was examined, and pigs that were bred and suckled by a protein-restricted sow tended to weigh less than pigs bred and suckled by sows fed an adequate protein level and pigs cross fostered between restricted and adequately fed sows.

This suggests that the decreased growth performance of offspring from protein restricted sows during suckling is largely related to a decrease in milk yield of the sow and to a lesser degree on fetal protein supply. Even in studies where the protein restriction during gestation has been less severe, pig weight gain during suckling remains affected, and an increase in protein intake causes increased weight gain [32, 35]. In the study by Kusina et al. [35], high or low sow protein intake during both gestation and lactation was studied, and protein intake during lactation also had a significant positive impact on pig weight gain during suckling.

Table 2.6 The influence of sow protein intake during gestation on postnatal growth performance of the offspring

		Postnatal performance of offspring			
Treatment window during gestation	Protein content of feed	Feed intake, kg/day	Daily gain, g/day	Feed:Gain	References
Throughout	Control	ad libitum	705[a]		[49]
Day 0–16, 21 to term	0		657[b]		
Day 25 to parturition	0		658[b]		
Throughout	0		568[b]		
Throughout	13%	2.78[a]	1,216[a]	2.29[a]	[50]
Throughout	0.5%	2.43[b]	948[b]	2.56[b]	
Throughout	13%	ad libitum	560[ab]		[66]
Throughout	0.5%		530[b]		
Day 1–44	0.5%		580[b]		
Day 82 to term	0.5%		610[a]		

[a, b]Within study and column, numbers with different superscripts differ significantly ($P < 0.05$).

Fewer studies have examined the effect of protein intake during gestation on postnatal growth from weaning to slaughter of the offspring. In studies in which a severe protein restriction occurs throughout gestation, daily gain from weaning to slaughter or from birth to slaughter is significantly decreased [49–50, 66]. Protein restriction during various stages of gestation, but not throughout, either shows intermediate weight gain between pigs from control sows and those restricted throughout [49], or does not differ compared with controls [66]. The results of both these studies imply that the sow protein intake during early gestation, when embryo implantation occurs, is of great importance for later development and growth. In one study, the mature size of young gilts following protein restriction during gestation was analysed, and revealed that both adult weight and length were decreased due to sow protein restriction [51]. Also, backfat thickness was lower in mature gilts from protein-restricted sows. Taken together, these results suggest that protein supply during fetal life has long-lasting and probably permanent effects on weight and size in adulthood.

2.2.3.3 Supplementation with L-Carnitine

As earlier discussed, milk yield seems to be positively affected by supplementation with L-carnitine during gestation and lactation. Weight gain during suckling and hence weaning weight are highly influenced by milk yield, and positive effects of L-carnitine supplementation should be expected on litter performance during suckling. This was also the case in some studies in which increased weight gain and weaning weight of individual pigs have been found when sows were supplemented with L-carnitine during gestation and lactation [40, 57]. In contrast, Ramanau et al. [56] did not find any effect on individual pig weight gain during suckling, but increased litter weight gain was evident upon supplementation. Higher litter weight gain was also found by Eder et al. [16] and Ramanau et al. [55], whereas no effect was found by Musser et al. [40].

Although the results for individual and litter weight gain during suckling provide no clear conclusions, overall increases in either individual or litter weight gain upon L-carnitine supplementation have been found in all studies.

The authors are only aware of one study in which offspring from sows supplemented with L-carnitine have been examined for postnatal growth performance beyond weaning. In this study pigs from supplemented sows were followed until day 35 post-weaning, reaching an approximate live body weight of 25 kg [4]. At weaning, pigs from sows treated with L-carnitine were heavier than pigs from control sows, but from weaning to the end of the experiment no effects of supplementation with L-carnitine during gestation on postnatal growth performance of the offspring were observed, including feed intake and feed-to-gain ratio. The experimental design of this study, where pigs from sows with and without supplementation were allotted into groups with an initial equal weigh at weaning, makes it difficult to compare the results of post-weaning growth performance between offspring from supplemented and non-supplemented sows, as these pigs do not represent all pigs from the two experimental groups.

2.2.4 Carcass Composition

Differences in sow feed intake may well affect the development of different tissues of the fetuses, as feed intake influences the hormone levels and the metabolism of the sow and thereby may influence fetal nutrient supply and/or fetal hormonal levels. Dependent on the timing and the magnitude of the difference in sow feed supply during gestation, fetal tissues which vary in the timing of their development may be affected differently.

2.2.4.1 Feed Intake

With regard to the estimated relationship between lean and fat tissue in the carcass at slaughter, neither Nissen et al. [41], Bee [3] nor Heyer et al. [31] found any difference in lean meat percentage or lean tissue yield between offspring from sows fed either according to or above energy requirements during early to mid gestation (summarised in Table 2.5). Adipose tissue yield was found to be higher in offspring from sows fed more during early gestation [3]. In contrast, Heyer et al. [31] found no difference in backfat thickness in offspring from sows fed more during early to mid gestation, either in the first or second parity.

Several authors have studied the effects of sow feed intake on the cross-sectional area and/or weight of several muscles in the offspring at slaughter. The *M. semitendinosus* has been most studied, as this muscle is commonly used when studying muscle fibre development and growth. When sows were supplemented with feed above requirements from days 0 to 50 [3], days 25 to 50 [24, 41] or from days 25 to 70 of gestation [41] no effect on the cross-sectional area of *M. semitendinosus* of the offspring was found. When examining the weight of *M. semitendinosus* the results are less clear, as Bee [3] and Nissen et al. [41] did not find a positive effect of increased feed intake from days 0 to 50 or days 25 to 70. In contrast, increased feed intake from days 25 to 50 revealed a significant decrease in the weight of *M. semitendinosus* of the offspring [41]. Similarly, Heyer et al. [31] observed that the weight of several different muscles (*M. psoas major*, *M. semimembranosus et adductor*, *M. semitendinosus*, *M. quadriceps*, *M. gluteus* and *M. biceps femoris*) did not differ as a percent of carcass weight among offspring from sows fed greater amounts from days 45 to 85 of gestation. The weight of *M. longissimus* and *M. semimembranosus* was also measured by Cerisuelo et al. [8], and in one of two experiments they found a significant increase in the weight of *M. semimembranosus* but not of *M. longissimus*.

Again, the results regarding effects of sow feed intake on carcass composition are inconsistent although again it is important to be mindful that the time periods and amounts of extra feed offered differ among these studies, making direct comparisons difficult. Generally, increased sow feed supply during early to mid gestation does not appear to have a large influence on offspring muscle weights at slaughter, although supplementation within more specific time periods may cause differences.

2.2.4.2 Protein Intake

There are only a few studies in which the consequences of protein restriction during gestation on carcass composition of offspring have been examined. In two studies where carcass composition was examined at cold carcass weights of 60–80 kg, backfat thickness of the offspring was not influenced by severe protein restriction throughout gestation or during specific periods of gestation [50, 66]. The cross-sectional area of *M. longissimus dorsi* was found to be significantly smaller in offspring from dams restricted throughout gestation compared with control animals in one study [66], whereas this was not evident in another study [50]. Surprisingly, protein restriction in specific periods of gestation actually caused the cross-sectional area of *M. longissimus dorsi* to increase compared with control animals [66]. The weight of trimmed ham and loin were also measured in these two studies, and both cuts were significantly lighter in offspring from dams restricted in protein throughout gestation [50, 66]. The weight of the cuts in offspring from sows that had been protein-restricted during specific periods of gestation did not differ from controls. The explanation for the greater weight of muscles and cuts of offspring from sows that have been protein-restricted in specific time periods is not obvious, although this may be a consequence of compensatory growth.

In mature female offspring (12 months-old) from sows that had been protein-restricted throughout gestation, backfat thickness was significantly lower than in offspring from sows fed adequate amounts of protein [51]. The cross-sectional area of *M. longissimus dorsi* was also significantly smaller in offspring from protein-restricted sows. These reductions in backfat thickness and muscle cross-sectional area appear to be proportional to the reduction in mature body weight and size.

2.3 Sow Porcine Growth Hormone Treatment During Gestation

A vast number of experiments have demonstrated that daily injections with porcine growth hormone (pGH) increases muscle tissue growth rate and reduces fat accretion rate in grower-finishing pigs in a dose-dependent manner [17]. This results in increased lean and decreased fat content of the carcass [69]. The increase in muscle growth is related to muscle fibre hypertrophy supported by increased satellite cell proliferation. The response in muscle tissue growth rate varies among experiments, and it was suggested that this variation was related to the cross-sectional area of muscle fibres. Therefore Sørensen et al. [69] suggested an inverse relation between the size of the response and the cross-sectional area of the muscle fibres. Thus, pig populations with small muscle fibres would respond more than pig populations with large muscle fibres. Growth hormone is produced in the pituitary gland and is regulated mainly by two peptides in the hypothalamus; i.e., somatostatin, which inhibits GH, and Growth Hormone-Releasing Hormone (GHRH), which stimulates the production of GH. Besides its growth-promoting effects, GH also has a diabetogenic effect, which results in hyperglycaemia and hyperinsulineamia [1, 19].

Similar changes in blood metabolites were found in pregnant sows [24, 36, 61, 64]. Because glucose is the major energy substrate for the fetus, hyperglycaemia following pGH may increase the maternal/fetal blood glucose gradient resulting in higher fetal uptake of glucose and hence increased fetal growth. This hypothesis has been tested in some studies, and the results on the influence of sow treatment with pGH on sow and litter performance have been summarised in Table 2.7.

2.3.1 Sow Performance

In a study by Gatford et al. [25] pregnant sows were treated in the second quarter of gestation with 0, 2 or 4 mg pGH/day and slaughtered at day 52. In this study the sows gained weight in a dose-dependent manner with increasing levels of pGH. At the same time the backfat thickness was reduced by pGH treatment. These data indicate that muscle growth is responsible for the increase in weight gain. In agreement, Rehfeldt et al. [59] found increased lean meat percentage of up to 4 percentage units and unchanged body weight when sows were treated with 6 mg pGH/day in the first quarter of gestation and slaughtered at days 28, 37 or 62 of gestation, again indicating increased muscle growth as a result of pGH treatment. However, when gestating sows were treated with 2 or 4 mg pGH/day in the second quarter of gestation, body gain was unaffected by pGH treatment at parturition, although sows treated with 2 mg pGH/day had increased backfat thickness compared with control and sows treated with 4 mg pGH/day [23]. This indicates that compensation in sow adipose tissue growth may occur following treatment with pGH early in gestation dependent on the dose of pGH.

Long-term (days 25–100) treatment of gestating sows with 2 mg pGH/day was found to affect sow body weight and backfat thickness both at day 100 and one day after parturition, depending on the dietary protein content [23]. Porcine growth hormone increased body weight and reduced backfat thickness when the diet contained 22.2% protein compared with control sows and with sows fed a diet containing 16.6% protein and treated with 0 or 2 mg pGH/day. This indicates that pGH increases muscle growth and decreases fat growth at high dietary protein concentration. Etienne et al. [18] and Farmer et al. [20] treated pregnant sows late in gestation with GHRH and found a numeric increase in sow body weight of 4 and 7 kg at parturition, and GHRH injection increased the plasma concentration of GH [18].

In conclusion, it is most likely that treatment of pregnant sows in gestation with pGH or GHRH during specific periods of gestation results in unchanged or increased body weight and carcasses with a higher lean meat percentage at the end of the treatment period. However, the influence of treatment with pGH in early gestation on carcass composition of sows is not maintained at parturition. In contrast, long-term treatment with pGH increases body weight and lean meat percentage at parturition.

Table 2.7 The influence of sow pGH or GHRH treatment during gestation on sow and litter performance

Treatment window during gestation	Dose of pGH/GHRH	Feed level kg/day	Fetal/pig age	Number of fetuses/pigs	Litter weight, kg	Average fetal/pig weight, g	Sow weight gain, kg	Fat thickness change, mm	References
Day 92–113	0	2.0	Birth	10.6		1,371			[36]
	10 mg/d			11.3		1,408			
Throughout	0		Birth	10.8	15.0	1,392[a]			[58]
Day 10–24	6 mg/d			9.10	12.4	1,364[a]			
Day 50–64	6 mg/d			10.3	14.0	1,356[a]			
Day 80–94	6 mg/d			9.5	13.9	1,459[b]			
Day 28–40	0	3.0	d 41	10.0		9.0			[33]
Day 28–40	30 μg/kg/d			12.3		10.2			
Day 28–40	0		Birth	7.6	13.3				
Day 28–40	30 μg/kg/d			7.6	14.1				
Day 30–43	0		d 44	9.92		16.44[a]			[67]
Day 30–43	5 mg/d			9.50		18.06[b]			
Day 25–51	0	1.8	d 51	unchanged		46[a]	+9	–2	[25]
	13.3 μg/kg/d					52[a]	+12	–5	
	25.6 μg/kg/d					54[b]	+17	–5	

Table 2.7 (continued)

Treatment window during gestation	Dose of pGH/GHRH	Feed level kg/day	Fetal/pig age	Number of fetuses/pigs	Litter weight, kg	Average fetal/pig weight, g	Sow weight gain, kg	Fat thickness change, mm	References
Day 10–27	0		d28	13.4		1.19	pGH	pGH	[59]
	6 mg/d			12.4		1.14	treatment did	treatment	+
	0		d37	14.8		6.11	not affect sow	increased	[60]
	6 mg/d			16.7		6.53	body weight	percentage of	
	0		d62	11.7		156.3		lean by 4%	
	6 mg/d			11.7		164.8			
	0		Birth	10.8	16.4	1,410			
	6 mg/d			10.1	16.5	1,480			
Day 102–112	0	2.5	Birth	10.8		1,240	+33.3		[18]
	50 μg/kg/d GHRH			10.2		1,190	+37.1		
Day 100 to term	0	2.2	Birth	9.9		1,520	+39		[20]
	12 mg of GHRH			10.8		1,570	+46		

[a, b] Within study and column, numbers with different superscripts differ significantly ($P < 0.05$).

2.3.2 Litter Size

Embryo survival and number of live pigs at birth following treatment of sows with pGH have been measured in several studies (Table 2.7). Kelley et al. [33] found an increased embryo survival at day 41 of gestation following treatment of sows with 30 μg pGH/day/kg body weight (approximately 3.9 mg pGH/day at breeding) during the second quarter of gestation. In contrast, Sterle et al. [67] did not find altered embryo survival at day 44 of gestation following treatment of sows with 5 mg of pGH/day in the second quarter of gestation. Furthermore, Sterle et al. [67] found there was no effect of treatment with pGH (5 mg/day) from either days 0 to 30 or from days 30 to 64 in gestation on embryo survival at day 64 of gestation. Similarly, Rehfeldt et al. [59] found no change in embryo or fetal survival at days 28, 37 and 62 of gestation following treatment with 5 mg pGH/day of sows during the first quarter of gestation.

Although inconsistent results on embryo survival have been reported, no effects of pGH treatment of the pregnant sow during specific periods of early gestation have been found on the number of viable embryos/fetuses in the period from days 29 to 64. Likewise, total number of pigs and the number of live pigs per litter at birth were unaffected by pGH treatment during specific periods of gestation [23, 36, 58–59]. Long-term treatment with pGH from days 25 to 100 also failed to alter the number of live born pigs [23], and GHRH did not affect number of live pigs at birth [18, 20].

In conclusion, data from these experiments indicate that treatment of the pregnant sow with either pGH or GHRH does not influence litter size. However, the studies carried out so far have only been conducted on relatively few sows, and larger studies are needed to establish the relationship between pGH treatment of sows and litter size.

2.3.3 Fetal Growth

Sow treatment with pGH may influence fetal growth and birth weight depending on the treatment windows during gestation and on the dietary protein content fed to sows. Sterle et al. [67] and Gatford et al. [25] found fetal weight increased approximately 10% at days 44 and 51 following treatment of sows with pGH during the second quarter of gestation. Also, Kelley et al. [33] found a 10% increase (non significant) in fetal body weight at day 41 of gestation following treatment with pGH during the second quarter and found a significant increase in the crown-rump length. It seems that this increase in fetal body weight is not necessarily maintained at birth as Rehfeldt et al. [58, 59], Kelley et al. [33] and Gatford et al. [23–24] found neither increased litter weight nor increased average birth weight when sows were treated with pGH in early gestation. The smallest pigs at birth within a litter may benefit most from treatment of the sows with pGH. Rehfeldt et al. [59] reported that the increase in pig birth weight (70 g) when dams had been treated with pGH in early gestation, although non-significant, was more pronounced in small littermates (+242 g) than in medium (+73 g) and larger littermates (+43 g).

When sows are treated with pGH in late gestation birth weight may increase [36, 58]. Long-term treatment of gestating sows from days 25 to 100 increased mean litter birth weight [23]. These researchers found that birth weight was increased in the second, third and fourth quartiles of the birth weight groups within their litter after pGH treatment regardless of the protein content of the sow diet. Pigs in the first quartile also had increased birth weight after sows were treated with pGH, but only when the sow diet contained a high level of protein (22.2%).

Treatment of sows with pGH during the final 21 days of gestation did not show any effect on carcass weight of offspring at birth, or on dry matter and protein concentration, while a small but significant increase in lipid concentration of 10% was evident [36]. Kelley et al. [33] showed that the cross-sectional areas of the *M. longissimus dorsi* muscle in neonatal offspring from sows treated with pGH were 23% larger and the thickness of backfat at the 10th rib was 22% lower. Moreover, when sows were treated with pGH in early gestation, the total protein, fat and ash content in the empty body of the pigs increased compared with pigs born to control sows, although on a body weight-specific basis the composition in these components did not differ due to treatment [60]. In the same study there were no effects on the weights of the *M. psoas major* and *M. semitendinosus* at day 62 of gestation and at birth, although at birth the weight of the muscles of offspring from sows treated with pGH were numerically larger (16–30%). Treatment of sows late in gestation with GHRH did not affect the birth weight [18, 20].

In summary, treatment with pGH in early gestation leads to increased fetal weight, an effect which does not persist to increase average birth weight. On the other hand, treatment of pregnant sows late in gestation or long-term treatment with pGH increases birth weights.

2.3.4 Postnatal Performance and Carcass Composition

The influence of treating sows with pGH during gestation on postnatal growth traits of the offspring has also been investigated. Kelley et al. [33] did not find any effects of treatment of sows with pGH in the second quarter of gestation on the body weight of 7-day old pigs or on weaning weight of the offspring compared with those of control sows. In contrast, slaughter pigs tended to have *M. longissimus dorsi* muscles with cross-sectional areas that were 8% larger due to pGH treatment. Kveragas et al. [36] did not find any effect of sow treatment with pGH during late gestation on the body weight of the offspring at day 14 and at weaning. Treatment of gestating sows in the second quarter with pGH did not alter postnatal growth of the offspring until day 61 of age [24], and finally, Kuhn et al. [34] found no difference in growth performance from birth to slaughter due to pGH treatment of the sows in the first quarter of gestation.

Possible effects on postnatal growth and carcass composition due to sow pGH treatment may be related to changes in fetal muscle development. Rehfeldt et al. [60] showed an increased number of muscle fibres in various muscles in the small and medium body weight littermates, but not in the heaviest littermates, following treatment with pGH during early gestation compared with pigs born to control sows. In contrast, Gatford et al. [24] did not find any change in muscle fibre number, but did observe an increase in muscle fibre cross-sectional area in offspring from sows treated with pGH in the second quarter of gestation. These two studies differed with respect to the treatment window: Rehfeldt [60] treated the sows prior to formation of primary fibres, whereas Gatford et al. [24] treated sows during the period where primary fibres are formed and proliferation of secondary myoblasts prior to secondary fibre formation occurs.

When sows were treated with GHRH in early gestation, the offspring had reduced daily gain and required 6.6 days more to reach slaughter weight. This was associated with unaltered muscle fibre number after treatment with GHRH [21]. Similarly, Etienne et al. [18] did not find any effect on daily gain from 24 to 100 kg body weight or on carcass composition in pigs born to sows treated with GHRH during late gestation. However, the feed-to-gain ratio was lower in offspring from sows treated with GHRH. More studies are needed to clarify the effect of sow pGH and GHRH treatment on postnatal growth potential of the offspring.

In summary, there is no convincing evidence that sow treatment with pGH in early gestation leads to better performance of the offspring. However, the long-term effects on postnatal growth need clarification.

2.4 Summary

The large variation in birth weight within litters and the increase in small pigs at birth, which is a consequence of genetic selection, provides pig producers with a great challenge. Pigs that are small at birth are weaker than large pigs, and therefore have greater environmental requirements during early life in order to survive. These requirements involve stable, high temperatures, feed supply, competition for feed, and avoidance of parasites, bacteria and viruses that can cause illness. Results from most studies show that pigs with low birth weights grow more slowly during the postnatal period than high birth weight pigs, especially at time points closer to birth. Also, carcass composition may differ depending on birth weight, although there are contradictory results regarding lean meat percentage and backfat thickness. It is possible that pigs that have developed differently during fetal life may need different postnatal strategies to achieve the most efficient growth and develop optimal carcass composition.

The gestating sow is normally fed a restricted diet in order to obtain nutrients in accordance with requirements. Different alternative feeding strategies have been studied to reduce litter variation in fetal development and thereby increase postnatal performance of the offspring without compromising sow performance. The following is a summary of results from studies where gestating sows have been fed either increased or decreased total feed or protein, supplemented with dietary L-carnitine, or treated with pGH.

Higher feed or protein intake will cause an increase in sow body weight and possibly backfat thickness, depending on the magnitude of the increase. In contrast, protein restriction will cause a decrease in body weight and backfat thickness, depending on the degree and period of restriction. The results from studies in which sows have been supplemented with L-carnitine in their diet during gestation vary: some authors report an increase in sow body weight and backfat thickness, whereas others do not. Litter size does not seem to be affected by increased or decreased feed or protein intake during gestation, whereas supplementation of L-carnitine has been reported by some authors to increase the number of total and live born pigs. Milk production does not seem to be affected by an increased feed intake during early to mid gestation, whereas it increases with increasing amount of protein in the gestational diet. Supplementation with dietary L-carnitine during both gestation and lactation has a positive impact on milk yield.

An increase in feed supply during gestation does not seem to affect average fetal weight, average fetal litter weight, or birth weight. Severe protein restriction causes a decrease in fetal as well as birth weights of pigs, whereas less severe protein restriction has no effect. Studies which examined the effects of supplementing sows with L-carnitine during gestation on offspring birth and litter weights are more confusing, with unchanged, positive and negative effects all being reported.

Average daily gain during suckling is highly correlated with birth weight and milk yield. In most studies, weaning weights are not affected by gestational feed intake although there was a positive effect reported in one study. Protein restriction reduces weight gain during suckling, possibly due to adverse effects of decreased protein supply on both fetal development and milk yield of the sow. Although results from sows supplemented with L-carnitine are not conclusive, a general increase in average pig and/or litter weight after supplementation are evident in most studies.

Effects on growth rate and feed efficiency from weaning until slaughter of offspring from sows allowed increased amounts of feed during early to mid gestation are not clear. Results differ with studies showing negative, positive or no effects. Also, the influence of increased feed intake on muscle fibre formation in offspring is variable among studies. Protein restriction throughout gestation seems to cause reduced daily gain of the offspring, and the effects of protein restriction during specific time periods vary. An adequate protein supply during embryonic implantation in the uterus seems to be of greatest importance. In one study, the effect of sow supplementation with L-carnitine on performance of offspring up to 25 kg was investigated, and L-carnitine did not influence daily gain, feed intake or feed-to-gain ratio.

Increased feed intake during gestation does not seem to affect the lean meat percentage of the offspring. The cross-sectional areas and weights of muscles in offspring from sows with an increased feed intake have been measured in several studies, but in most cases there were no differences from control offspring. Restricting the sow's intake of protein in early to mid gestation does not influence backfat thickness of the offspring, although restriction throughout gestation causes muscle weight and muscle cross-sectional area of several muscles to decrease. If the restriction is only applied in more specific periods of gestation, there do not seem to be effects on muscle weights or cross-sectional areas.

It is most likely that treatment of pregnant sows in gestation with GHRH and pGH within specific periods of gestation results in unchanged or increased body weight and altered carcass composition in sows towards a higher lean meat percentage at the end of the treatment period. However, the influence of treatment with pGH in early gestation on carcass composition of sows is not maintained at parturition. In contrast, long-term treatment with pGH increases sow body weight and lean meat percentage at parturition. Treating the pregnant sow with either pGH or GHRH does not influence the litter size. Treatment with pGH in early gestation leads to increased fetal weight, an effect which is not maintained at birth. On the other hand, treatment of pregnant sows late in gestation or long-term treatment of sows with pGH increases birth weights. From the data available, there is no evidence of a beneficial effect of treating sows during early pregnancy with pGH, or in early or mid gestation with GHRH, on growth performance or carcass quality of the offspring. However, the effect of long-term treatment of gestating sows with pGH on the growth performance of the offspring needs further research.

2.5 Future Perspectives

In this chapter we have reviewed studies in which the effects of sow feed and protein intake, supplementation with dietary L-carnitine and treatment with pGH on sow and litter performance have been examined. From the published work there seems no obvious new strategy that could improve the existing strategy for sow management during gestation, with the aim of reducing within-litter variation in birth weight and muscle development, and ultimately of increasing postnatal growth performance of the offspring. L-carnitine seems to have a positive effect on milk yield, but how supplementation affects sow performance, litter variation and postnatal growth performance of the offspring needs further investigation. A global increase in sow feed intake during early to mid gestation does not seem to be of great benefit to the offspring. A better approach in the future may be to study the effects of increasing single nutrients within the sow diet, with the intention of increasing the availability of glucose to the fetuses without a marked increase in the growth of sow tissues such as fat. An approach to decrease the litter variability in birth weight and thereby the survival rate of all pigs within a litter would be of great economic benefit to the pig farmer. As for the use of long-term pGH treatment of gestational sows, more work may also be needed to clarify the effects on birth weight and offspring performance. The use of hormone treatment in animal production is currently a much discussed topic, and the possibility exists that bans may be imposed in the future in other parts of the world beyond Europe where they are presently banned. This is an important consideration before starting large and expensive experiments within this field of study.

Acknowledgements The authors wish to thank trilingual secretary Aase K. Sørensen for linguistic comments to this chapter.

References

1. Agergaard, N., N. Oksbjerg, and M.T. Sørensen. 1991. Influence of exogenous growth hormone on homeostasis in finishing pigs. *Proceedings of the 6th International Symposium on Protein Metabolism and Nutrition,* EAAP-publication No 59. **2**:194–197.
2. Ashworth, C.J., A.M. Finch, K.R. Page, M.O. Nwagwu, and H.J. McArdle. 2001. Causes and consequences of fetal growth retardation in pigs. *Reprod. Suppl.* **58**:233–246.
3. Bee, G. 2004. Effect of early gestation feeding, birth weight, and gender of progeny on muscle fiber characteristic of pigs at slaughter. *J. Anim. Sci.* **82**:826–836.
4. Birkenfeld, C., A. Ramanau, H. Kluge, J. Spilke, and K. Eder. 2005. Effect of dietary L-carnitine supplementation on growth performance of piglets from control sows or sows treated with L-carnitine during pregnancy and lactation. *J. Anim. Physiol. Anim. Nutr.* **89**:277–283.
5. Brameld, J.M., P.J. Buttery, J.M. Dawson, and J.M.M. Harper. 1998. Nutritional and hormonal control of skeletal-muscle cell growth and differentiation. *Proc. Nutr. Soc.* **57**:207–217.
6. Buitrago, J.A., E.F. Walker, W.I. Snyder, and W.G. Pond. 1974. Blood and tissue traits in pigs at birth and at 3 weeks from gilts fed low or high energy diets during gestation. *J. Anim. Sci.* **38**:766–771.
7. Campbell, R.G. and A.C. Dunkin. 1982. The effect of birth weight on the estimated milk intake, growth and body composition of sow-reared piglets. *Anim. Prod.* **35**:193–197.
8. Cerisuelo, A., R. Sala, J. Coma, D. Carrión, J. Gasa, and M.D. Baucells. 2006. Effect of maternal feed intake during mid-gestation on pig performance and meat quality at slaughter. *Arch. Anim. Breed.* **49**:57–61.
9. Close, W.H. and D.J.A. Cole. 2000. Nutritional history, pp. 184–188. Nutrition of sows and boars 1st ed., Nottingham University Press, Nottingham, U.K.
10. Dauncey, M.J., K.A. Burton, and D.R. Tivey. 1994. Nutritional modulation of insulin-like growth factor-I expression in early postnatal piglets. *Pediat. Res.* **36**:77–84.
11. De Gaetano, A., G. Mingrone, M. Castagneto, and M. Calvani. 1999. Carnitine increases glucose disposal in humans. *J. Am. Coll. Nutr.* **18**:289–295.
12. Dourmad, J.Y. 1991. Effect of feeding level in the gilt during pregnancy on voluntary feed intake during lactation and changes in body composition during gestation and lactation. *Livest. Prod. Sci.* **27**:309–319.
13. Dwyer, C. M., J.M. Fletcher, and N.C. Stickland. 1993. Muscle cellularity and postnatal growth in the pig. *J. Anim. Sci.* **71**:3339–3343.
14. Dwyer, C.M. and N.C. Stickland. 1991. Sources of variation in myofibre number within and between litters of pigs. *Anim. Prod.* **52**:527–533.
15. Dwyer, C.M., N.C. Stickland, and J.M. Fletcher. 1994. The influence of maternal nutrition on muscle fiber number development in the porcine fetus and on subsequent postnatal growth. *J. Anim. Sci.* **72**:911–917.
16. Eder, K., A. Ramanau, and H. Kluge. 2001. Effect of L-carnitine supplementation on performance parameters in gilts and sows. *J. Anim. Physiol. Anim. Nutr.* **85**:73–80.
17. Etherton, T.D., J.P. Wiggins, C.S. Chung, C.M. Evock, J.F. Rebhun, and P.E. Walton. 1986. Stimulation of pig growth performance by porcine growth hormone and growth hormone-releasing factor. *J. Anim. Sci.* **63**:1389–1399.
18. Etienne, M., M. Bonneau, G. Kann, and F. Deletang. 1992. Effects of administration of growth hormone-releasing factor to sows during late gestation on growth hormone secretion, reproductive traits, and performance of progeny from birth to 100 kg live weight. *J. Anim. Sci.* **70**:2212–2220.
19. Evock-Clover, C.M., N.C. Steele, T.J. Caperna, and M.B. Solomon. 1992. Effects of frequency of recombinant porcine somatotropin administration on growth performance, tissue accretion rates, and hormone and metabolite concentrations in pigs. *J. Anim. Sci.* **70**:3709–3720.
20. Farmer, C., D. Petitclerc, G. Pelletier, and P. Brazeau. 1992. Lactation performance of sows injected with growth hormone-releasing factor during gestation and (or) lactation. *J. Anim. Sci.* **70**:2636–2642.

21. Faucitano, L., C. Pomar, C. Gariépy, and C. Farmer. 2005. Growth-hormone-releasing factor given to early-pregnant Genex-Meishan and large White gilts: Effects on growth, carcass, meat quality and histochemical traits of the progeny. *Can. J. Anim. Sci.* **85**:37–46.
22. Ford, S.P. 1997. Embryonic and fetal development in different genotypes in pigs. *J. Reprod. Fert. Suppl.* **52**:165–176.
23. Gatford, K.L., J.M. Boyce, K. Blackmore, R.J. Smits, R.G. Campbell, and P.C. Owens. 2004. Long-term, but not short-term, treatment with somatotropin during pregnancy in underfed pigs increases the nody size of progeny at birth. *J. Anim. Sci.* **82**:93–101.
24. Gatford, K.L., J.E. Ekert, K. Blackmore, M.J. De Blasio, J.M. Boyce, J.A. Owens, R.G. Campbell, and P.C. Owens. 2003. Variable maternal nutrition and growth hormone treatment in the second quarter of pregnancy in pigs alter semitendinosus muscle in adolescent progeny. *Brit. J. Nutr.* **90**:283–293.
25. Gatford, K.L., J.A. Owens, R.G. Campbell, J.M. Boyce, P.A. Grant, M.J. De Blasio, and P.C. Owens. 2000. Treatment of underfed pigs with GH throughout the second quarter of pregnancy increases fetal growth. *J. Endocrinol.* **166**:227–234.
26. Gondret, F., L. Lefaucheur, H. Juin, I. Louveau, and B. Lebret. 2006. Low birth weight is associated with enlarged muscle fiber area and impaired meat tendesness of the longissimus muscle in pigs. *J. Anim. Sci.* **84**:93–103.
27. Gondret, F., L. Lefaucheur, I. Louveau, B. Lebret, X. Pichodo, and Y. Le Cozler. 2005. Influence of piglet birth weight on postnatal growth performance, tissue lipogenic capacity and muscle histological traits at market weight. *Livest. Prod. Sci.* **93**:137–146.
28. Handel, S.E. and N.C. Stickland. 1987. Muscle cellularity and birth weight. *Anim. Prod.* **44**:311–317.
29. Handel, S.E. and N.C. Stickland. 1988. Catch-up growth in pigs: A relationship with muscle cellularity. *Anim. Prod.* **47**:291–295.
30. Hegarty, P.V.J. and C.E. Allen. 1978. Effect of pre-natal runting on the post-natal development of skeletal muscles in swine and rats. *J. Anim. Sci.* **46**:1634–1640.
31. Heyer, A., H.K. Andersson, J.E. Lindberg, and K. Lundström. 2004. Effect of extra maternal feed supply in early gestation on sow and piglet performance and production and meat quality of growing/finishing pigs. *Acta Agric. Scand., Sect. A, Anim. Sci.* **54**:44–55.
32. Johnston, L.J., M. Ellis, G.W. Libal, V.B. Mayrose, and W.C. Weldon. 1999. Effect of room temperature and dietary amino acid concentration on performance of lactating sows. NCR-89 Committee on Swine Management. *J. Anim. Sci.* **77**:1638–1644.
33. Kelley, R.L., S.B. Jungst, T.E. Spencer, W.F. Owsley, C.H. Rahe, and D.R. Mulvaney. 1995. Maternal treatment with somatotropin alters embryonic development and early postnatal growth of pigs. *Domest. Anim. Endocrinol.* **12**:83–94.
34. Kuhn, G., E. Kanitz, M. Tuchscherer, G. Nürnberg, M. Hartung, K. Ender, and C. Rehfeldt. 2004. Growth and carcass quality of offspring in response to porcine somatotropin (pST) treatment of sows during early pregnancy. *Livest. Prod. Sci.* **85**:103–112.
35. Kusina, J., J.E. Pettigrew, A.F. Sower, M.E. White, B.A. Crooker, and M.R. Hathaway. 1999. Effect of protein intake during gestation and lactation on the lactational performance of primiparous sows. *J. Anim. Sci.* **77**:931–941.
36. Kveragas, C.L., R.W. Seerley, R.J. Martin, and W.L. Vandergrift. 1986. Influence of exogenous growth hormone and gestational diet on sow blood and milk characteristics and on baby pig blood, body composition and performance. *J. Anim. Sci.* **63**:1877–1887.
37. Mahan, D.C. 1998. Relationship of gestation protein and feed intake level over a five-parity period using a high-producing sow genotype. *J. Anim. Sci.* **76**:533–541.
38. Milligan, B.N., D. Fraser, and D.L. Kramer. 2002. Within-litter birth weight variation in the domestic pig and its relation to pre-weaning survival, weight gain, and variation in weaning weights. *Livest. Prod. Sci.* **76**:181–191.
39. Musser, R.E., D.L. Davis, S.S. Dritz, M.D. Tokach, J.L. Nelssen, J.E. Minton, and R.D. Goodband. 2004. Conceptus and maternal responses to increased feed intake during early gestation in pigs. *J. Anim. Sci.* **82**:3154–3161.

40. Musser, R.E., R.D. Goodband, M.D. Tokach, K.Q. Owen, J.L. Nelssen, S.A. Blum, S.S. Dritz, and C.A. Cicis. 1999. Effects of L-carnitine fed during gestation and lactation on sow and litter performance. *J. Anim. Sci.* **77**:3289–3295.
41. Nissen, P.M., V.O. Danielsen, P.F. Jorgensen, and N. Oksbjerg. 2003. Increased maternal nutrition of sows has no beneficial effects on muscle fiber number or postnatal growth and has no impact on the meat quality of the offspring. *J. Anim. Sci.* **81**:3018–3027.
42. Nissen, P.M., P.F. Jorgensen, and N. Oksbjerg. 2004. Within-litter variation in muscle fiber characteristics, pig performance, and meat quality traits. *J. Anim. Sci.* **82**:414–421.
43. Nissen, P.M., I.L. Sørensen, M. Vestergaard, and N. Oksbjerg. 2005. Effects of sow nutrition on maternal and fetal serum growth factors and on fetal myogenesis. *Anim. Sci.* **80**:299–306.
44. Oksbjerg, N., F. Gondret, and M. Vestergaard. 2004. Basic principles of muscle development and growth in meat-producing mammals as affected by the insulin-like growth factor (IGF) system. *Domest. Anim. Endocrinol.* **27**:219–240.
45. Owen, K.Q., H. Ji, C.V. Maxwell, J.L. Nelssen, R.D. Goodband, M.D. Tokach, G.C. Tremblay, and S.I. Koo. 2001. Dietary L-carnitine suppresses mitochondrial branched-chain keto acid dehydrogenase activity and enhances protein accretion and carcass characteristics of swine. *J. Anim. Sci.* **79**:3104–3112.
46. Parfet, K.A.R., W.R. Lamberson, A.R. Rieke, T.C. Cantley, V.K. Ganjam, F.S. vom Saal, and B.N. Day. 1990. Intrauterine position effects in male and female swine: Subsequent survivability, growth rate, morphology and semen characteristics. *J. Anim. Sci.* **68**:179–185.
47. Pedersen, P.H., N. Oksbjerg, A.H. Karlsson, H. Busk, E. Bendixen, and P. Henckel. 2001. A within litter comparison of muscle fibre characteristics and growth of halothane carrier and halothane free crossbreed pigs. *Livest. Prod. Sci.* **73**:15–24.
48. Père, M.C., J.Y. Dourmad, and M. Etienne. 1997. Effect of number of pig embryos in the uterus on their survival and development, and on maternal metabolism. *J. Anim. Sci.* **75**:1337–1342.
49. Pond, W.G., D.N. Strachan, Y,N. Sinha, E.F. Walker, J.A. Dunn, and R.H. Barnes 1969. Effect of protein deprivation of swine during all or part of gestation on birth weight, postnatal growth rate and nucleic acid content of brain and muscle of progeny. *J. Nutr.* **99**:61–67.
50. Pond, W.G., J. Yen, and H.J. Mersmann. 1987. Effect of severe dietary protein, nonprotein calories or feed restriction during gestation on postnatal growth of progeny in swine. *Growth* **51**:355–371.
51. Pond, W.G., J. Yen, H.J. Mersmann, and R.R. Maurer. 1990. Reduced mature size in progeny of swine severely restricted in protein intake during pregnancy. *Growth, Dev. Aging* **54**:77–84.
52. Poore, K.R. and A.L. Fowden. 2004. The effects of birth weight and postnatal growth patterns on fat depth and plasma leptin concentrations in juvenile and adult pigs. *J. Physiol.* **558**:295–304.
53. Powell, T.G. and E.D. Aberle. 1981. Skeletal muscle and adipose tissue cellularity in runt and normal birth weight swine. *J. Anim. Sci.* **52**:748–756.
54. Quiniou, N., J. Dagorn, and D. Gaudré. 2002. Variation of piglets' birth weight and consequences on subsequent performance. *Livest. Prod. Sci.* **78**:63–70.
55. Ramanau, A., H. Kluge, and K. Eder. 2005. Effects of L-carnitine supplementation on milk production, litter gains and back-fat thickness in sows with a low energy and protein intake during lactation. *Brit. J. Nutr.* **93**:717–721.
56. Ramanau, A., H. Kluge, J. Spilke, and K. Eder. 2002. Reproductive performance of sows supplemented with dietary L-carnitine over three reproductive cycles. *Arch. Anim. Nutr.* **56**: 287–296.
57. Ramanau, A., H. Kluge, J. Spilke, and K. Eder. 2004. Supplementation of sows with L-carnitine during pregnancy and lactation improves growth of the piglets during the suckling period through increased milk production. *J. Nutr.* **134**:86–92.
58. Rehfeldt, C., I. Fiedler, R. Weikard, E. Kanitz, and K. Ender. 1993. It is possible to increase skeletal muscle fibre number *in utero*. *Biosci. Rep.* **13**:213–220.

59. Rehfeldt, C., G. Kuhn, G. Nürnberg, E. Kanitz, F. Schneider, M. Beyer, K. Nürnberg, and K. Ender. 2001. Effects of exogenous somatotropin during early gestation on maternal performance, fetal growth, and compositional traits in pigs. *J. Anim. Sci.* **79**:1789–1799.
60. Rehfeldt, C., G. Kuhn, J. Vanselow, R. Fübass, I. Fiedler, G. Nürnberg, A.K. Clelland, N.C. Stickland, and K. Ender. 2001. Maternal treatment with somatotropin during early gestation affects basic events of myogenesis in pigs. *Cell Tissue. Res.* **306**:429–440.
61. Rehfeldt, C., P.M. Nissen, G. Kuhn, M. Vestergaard, K. Ender, and N. Oksbjerg. 2004. Effects of maternal nutrition and porcine growth hormone (pGH) treatment during gestation on endocrine and metabolic factors in sows, fetuses and pigs, skeletal muscle development, and postnatal growth. *Domest. Anim. Endocrinol.* **27**:267–285.
62. Ritacco, G., S.V. Radecki, and P.A. Schoknecht. 1997. Compensatory growth in runt pigs is not mediated by insulin-like growth factor I. *J. Anim. Sci.* **75**:1237–1243.
63. Robinson, J.J., K.D. Sinclair, and T.G. McEvoy. 1999. Nutritional effects on fetal growth. *Anim. Sci.* **68**:315–331.
64. Schneider, F., E. Kanitz, D.E. Gerrard, G. Kuhn, K.P. Brüssow, K. Nürnberg, I. Fiedler, G. Nürnberg, K. Ender, and C. Rehfeldt, C. 2002. Administration of recombinant porcine somatotropin (rpST) changes hormone and metabolic status during early pregnancy. *Domest. Anim. Endocrinol.* **23**:455–474.
65. Schoknecht, P.A., G.R. Newton, D.E. Weise, and W.G. Pond. 1994. Protein restriction in early pregnancy alters fetal and placental growth and allantoic fluid proteins in swine. *Theriogenol.* **42**:217–226.
66. Schoknecht, P.A., W.G. Pond, H.J. Mersmann, and R.R. Maurer. 1993. Protein restriction during pregnancy affects postnatal growth in swine progeny. *J. Nutr.* **123**:1818–1825.
67. Steerle, J.A., T.C. Cantley, W.R. Kanberson, M.C. Kucy, D.E. Gerrard, R.L. Matteri, and B.N. Day. 1995. Effects of recombinant somatotropin on placental size, fetal growth, and IGF-I and IGF-II concentration in pigs. *J. Anim. Sci.* **73**:2980–2985.
68. Sørensen, G. 2006. Personal communication. Danish Meat Association – Axelborg, Department of Nutrition and Reproduction, Copenhagen, Denmark.
69. Sørensen, M.T., N. Oksbjerg, N. Agergaard, and J.S. Petersen. 1996. Tissue deposition rates in relation to muscle fibre and fat cell characteristics in lean female pigs (Sus Scofa) following treatment with porcine growth hormone (pGH). *Comp. Biochem. Physiol.* **113A**:91–96.
70. Waylan, A.T., J.P. Kayser, D.P. Gnad, J.J. Higgins, J.D. Starkey, E.K. Sissom, J.C. Woodworth, and B.J. Johnson. 2005. Effects of L-carnitine on fetal growth and the IGF system in pigs. *J. Anim. Sci.* **83**:1824–1831.
71. Weldon, W.C., A.J. Lewis, G.F. Louis, J.L. Kovar, M.A. Giesemann, and P.S. Miller. 1994. Postpartum hypophagia in primiparous sows: Effects of gestation feeding level on feeding level on feed intake, feeding behaviour, and plasma metabolite concentrations during lactation. *J. Anim. Sci.* **72**:387–394.
72. Wilson, M.E., N.J. Biensen, and S.P. Ford. 1999. Novel insight into the control of litter size in pigs, using placental efficiency s a selection tool. *J. Anim. Sci.* **77**:1654–1658.
73. Wolter, B.F., M. Ellis, B.P. Corrigan, and J.M. De Decker. 2002. The effect of birth weight and feeding of supplemental milk replacer to piglets during lactation on preweaning and postweaning growth performance and carcass characteristics. *J. Anim. Sci.* **80**:301–308.

Chapter 3
Managing Prenatal Development of Broiler Chickens to Improve Productivity and Thermotolerance

Zehava Uni and Shlomo Yahav

Introduction

In recent decades, the rate of weight gain of meat-type broiler chickens has increased significantly [25, 26]. In the 1960s, a commercial broiler chicken reached the market weight of 2.2 kg by 12 weeks of age; today, broilers attain this same market weight at less than 6 weeks. This impressive improvement is due to the knowledge accumulated by industry poultry managers and researchers from different disciplines – geneticists, nutritionists, physiologists and veterinarians – who have contributed to the optimisation of broiler performance.

Since the embryonic period (21 days in a hatchery) now represents half of the broiler's post-hatch life span, there is a need for research focused on the embryonic phase. The rationale is that anything that hinders or promotes development during this period will have a marked effect on poultry's overall performance, resistance and health.

Hence, this chapter focuses on:

a. The impact of pre-hatch nutritional manipulations by *in-ovo* feeding on gastrointestinal tract (GIT) development.
b. The effect of thermal manipulations during critical phases of the perinatal period on improved acquisition of thermotolerance.
c. The improvement of broiler chicken performance resulting from different manipulations during the prenatal or postnatal periods.

Z. Uni (✉)
The Faculty of Agricultural, Food and Environmental Quality Sciences, The Hebrew University of Jerusalem, Rehovot 76100, Israel
e-mail: uni@agri.huji.ac.il

P.L. Greenwood et al. (eds.), *Managing the Prenatal Environment to Enhance Livestock Productivity*, DOI 10.1007/978-90-481-3135-8_3,

3.1 The Complexity of the Transition from Embryo to Independent Chick

The transition from embryo to an independent chick, which can optimise the bird's potential for rapid growth, is mediated by processes that occur during the critical period from a few days pre- to a few days post-hatch. During this period, chicks undergo the metabolic and physiological transition from reliance on egg-based nutrients in the yolk sac and amniotic fluid to reliance on exogenous feed.

In birds, one of the major physiological processes during prenatal embryonic development is the maintenance of glucose homeostasis. This is dependent upon the amount of glucose held in reserves, primarily as glycogen, in the liver and the glycolytic muscles [25, 26, 35], and upon the degree of glucose generated by gluconeogenesis from glucogenic amino acids in protein [12–14].

Glycogen reserves are utilised as embryos progress through the hatching process [6] and therefore, the late-term embryo (prenatal chick) depends on gluconeogenesis from amino acids, first mobilised from the amnion albumen, then probably from muscle [10, 22, 36]. Vieira and Moran [80] suggested that the gluconeogenesis occurring in the pre- to post-hatch period leads to the depletion of muscle protein reserves, thereby limiting early growth and development [80, 81].

Although the pattern of glycogen utilisation as hatch approaches is similar in embryos from breeding flocks of different ages, there are differences in glycogen concentration: embryos from mature breeding flocks have greater glycogen reserves in the embryonic liver than embryos from young breeding flocks (Fig. 3.1). These glycogen reserves begin to be replenished when the newly hatched chick has full access to feed and oxygen [57, 58].

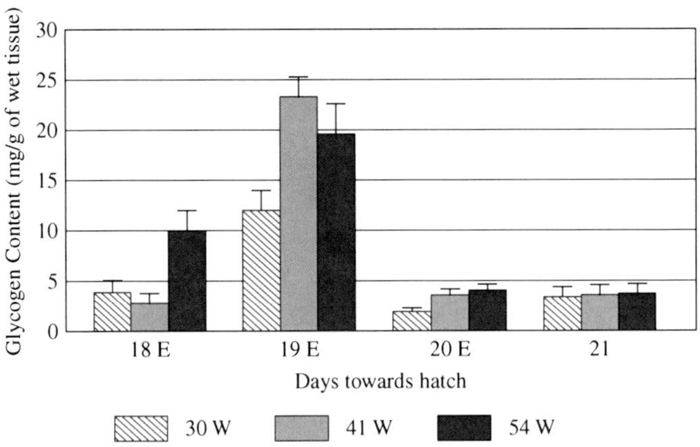

Fig. 3.1 Glycogen levels in the liver of late-term embryos from different ages of breeding flocks (Cobb strain): 30 weeks (30 W), 41 weeks (41 W) and 54 weeks (54 W). Values are average ± SD of 20 embryos for each breeding flock at each sampling day

Immediately post-hatch, the chick draws from its limited body reserves and undergoes rapid physical and functional development of the GIT in order to digest feed and assimilate nutrients. Therefore, the sooner the GIT achieves full functionality, the sooner the young bird can utilise dietary nutrients and efficiently approach its genetic growth potential, while resisting infectious and metabolic diseases.

An exploration of intestinal development shows that the GIT develops throughout incubation, but the functional abilities of the small intestine only begin to develop during the last quarter of incubation. Towards the end of incubation, extensive morphological, cellular and molecular changes occur in the small intestine. Research in broiler embryos has shown that during the last days of incubation there is a significant increase in the weight of the intestine relative to embryonic weight: from 1.4% at 17 days of incubation to 3.4% at hatch [79]. Activity and mRNA expression of brush-border enzymes responsible for digestion of disaccharides (sucrase-isomaltase) and small peptides (aminopeptidase), and of major transport proteins (sodium-glucose transporter and ATPase), begin to increase a few days before hatch and continue increasing on the day of hatch [79].

In the hatching chick, as in neonatal mammals, the small intestinal mucosa appears immature. At birth, mammals exhibit proliferating enterocytes along the villi and have a few well-defined crypts. Our studies with birds have indicated a similar situation in the young chick. On day 4 post-hatch, the percentage of proliferating-cell nuclear antigen (PCNA)-positive cells in the three regions of the crypt-villus axis was 55% in the crypt, 32% in the mid-section and 8% in the upper region [77]. In the post-hatch period, the intestinal mucosa exhibits organisation and establishment of the crypt region, a several-fold increase in villus height and area, an increase in the number and polarity of enterocytes, and maturation of the goblet cells which are capable of producing both acidic and neutral mucins.

The immediate post-hatch period seems to be critical for intestinal development. Intestinal development was retarded when chickens were fasted for 36–48 h post-hatch. This "fasting" condition is common in the poultry industry. Since chicken embryos have a wide "hatching window", commercial hatcheries do not remove birds until the maximum number of eggs have hatched; thus, chick age at exit from the hatchery averages over 1 day-old [46]. Hatchery treatments such as sexing, vaccination and transport to farms result in an additional time lag before birds receive first access to food and water. Thus, most chicks are fasted for 48 h or more before first access to feed. The rationale underlying this procedure was the concept, adopted by farmers and industry poultry managers alike, that the yolk sac can maintain the hatching bird during the initial post-hatch period until stable feed intake is available.

However, it has been shown that fasting for 36–48 h immediately post-hatch decreases enterocyte number, crypt size, the number of crypts per villus, crypt proliferation, villus area, rate of enterocyte migration, goblet-cell size and mucin dynamics [18, 78]. This withholding of feed also results in a decrease in growth at an early age and reduced body weight and proportion of breast muscle at marketing [52].

It can thus be surmised that during the last quarter of incubation embryos suffer from low glycogen status. Insufficient glycogen and albumen forces the embryo to mobilise more muscle protein for gluconeogenesis, thus restricting early growth and impacting productivity at market age. Furthermore, late access to external nutrient sources, which delays development of small intestinal functionality, limits the capacity to digest and absorb at an early age and, therefore, inhibits maximal chicken growth.

3.2 "Feeding the Embryo" Before Hatch

In order to overcome the physiological limitations described above and to improve intestinal functionality and nutritional status of hatchlings, a method for "feeding the embryo before hatch" (*in-ovo* feeding) was developed. This method which involves inserting nutrient solutions into the embryonic amniotic fluid was created for poultry, and it has been patented [73]. The method makes use of the knowledge that neonatal birds naturally consume the amniotic fluids towards hatch (Fig. 3.2). Therefore, addition of a nutrient solution into the embryonic amniotic fluid delivers essential nutrients to the embryos intestine.

Many potential nutrient supplements can be included in the *in-ovo* feeding solution. Carbohydrates can be used as a source for glucose, which is crucial for the hatching process and hatchling development [45]. Na^+ and Cl^- ions play a major role in the activity of apical and basolateral transporters and in the absorption of glucose and amino acids. β-hydroxy-β-methylbutyrate (HMB), a leucine metabolite which increases carcass yield [51], is a good candidate for the *in-ovo* feeding solution, as are minerals, vitamins and enteric modulators which support the development of the skeletal, immune and digestive systems in chickens.

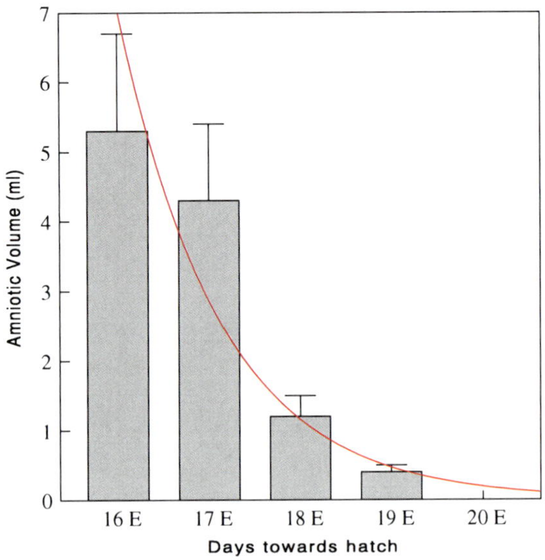

Fig. 3.2 Amniotic volume and rate of amniotic uptake during the last quarter of incubation in broiler (Cobb strain) embryos. The curve shows that the optimal period for the *in-ovo* feeding procedure (described in the text in section on "Feeding the Embryo" Before Hatch) is between days 17 E and 18 E (E= Embryonic days)

Studies have shown that the administration of 1 mL of *in-ovo* feeding solution including dextrin (as a source of carbohydrates), Na^+, Cl^-, zinc-methionine and HMB markedly enhances enteric development [64, 65]. Morphological evaluation of enteric sections from embryos and hatchlings has revealed a significant acceleration of development 48 h after *in-ovo* feeding relative to non-injected controls. The *in-ovo*-fed birds exhibited increased pancreatic capacity for carbohydrate digestion, increased villus dimensions, higher levels of mRNA expression and activity of brush-border digestive enzymes and transporters (leucine-aminopeptidase, sucrase-isomaltase, sodium-glucose co-transporter, Na^+K^+-ATPase, zinc transporter). The results indicate that provision of nutrients, including carbohydrates as an energy source, to late-term embryos have trophic effects on the small intestine and enhances goblet-cell development [62]. We have concluded that at the time of hatch, the small intestine of *in-ovo*-fed birds is at a functional stage similar to that of conventionally fed 2-day-old chicks.

Moreover, the hatchling's nutritional status is improved by provision of nutrients in the pre-hatch period by *in-ovo* feeding. Several experiments, in which glycogen levels in the liver were determined as a measure of the nutritional status of embryos and hatchlings, showed that administration of 1 mL of *in-ovo* feeding solution containing dextrin, maltose and sucrose into the amnion of the broiler embryo leads to increased total liver glycogen in the pre-hatch period (by 75% on the day before hatch, and by 47% on the day of hatch – Fig. 3.3) compared to controls.

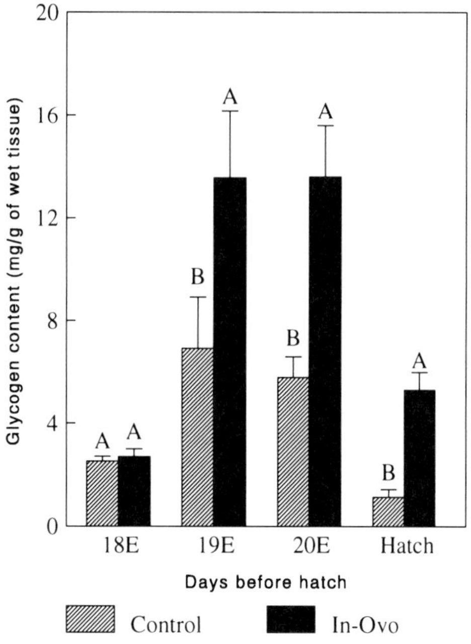

Fig. 3.3 Glycogen concentration in the liver of late-term embryos (Ross) from control and *in-ovo*-fed (*in-ovo*) treatments [74]. The *in-ovo* feeding solution contained 50 g/L maltose, 200 g/L Dextrin and 1 g/L β-hydroxy-β-methylbutyrate (HMB), all dissolved in 5 g/L NaCl. Within age, bars followed by different letters differ significantly ($P \leq 0.05$)

Table 3.1 Body weight and pectoral muscle weight of chickens (Ross strain) from control and *in-ovo* fed treatments on day of hatch and 10 and 25 days of age

	Age					
	Day of hatch		Day 10		Day 25	
	Control	*In ovo* fed	Control	*In ovo* fed	Control	*In ovo* fed
Body weight[1] (g)	45.3±0.3[b]	47.0±0.3[a]	243±4[b]	254±3[a]	943±21[b]	997±20[a]
Body weight difference (%)		+ 3.7		+ 4.2		+ 5.7
Pectoral weight[2] (g)	0.86±0.03[b]	0.95±0.06[a]	27.9±0.8[b]	30.3±0.8[a]	114±3[b]	130±6[a]
Pectoral weight/ body weight (%)	1.93±0.06[b]	2.05±0.05[a]	11.4±0.3[b]	12.3±0.3[a]	12.0±0.4[b]	13.0±0.2[a]
Difference in pectoral muscle weight (%)		+ 6.2		+ 5.2		+ 8.3

[1]Body weight values are means ± SEM of 80–100 birds (equal numbers of males and females). [2]Pectoral muscle weights are means ± SE of 10 birds for day of hatch, 10 birds for day 10, and 80 birds for day 25 (equal numbers of males and females). [a,b]Within age, values followed by different letters differ significantly ($P \leq 0.05$).

Results from several experiments demonstrated that *in-ovo* feeding of embryos from young breeding flocks on day 17 of incubation improves liver glycogen, increases hatching weights by 5% over controls and elevates breast-muscle size on a body weight-specific basis by 6%. These weight advantages (Table 3.1) were sustained throughout the experiments, to 25 days of age [74].

It is reasonable to assume (Fig. 3.4) that the elevated glycogen levels in the *in-ovo* treatment reduce the need to produce glucose via gluconeogenesis and therefore contribute to less utilisation of muscle protein and higher pectoral muscle weight on a body weight-specific basis in these birds. Furthermore, the chicks hatch with

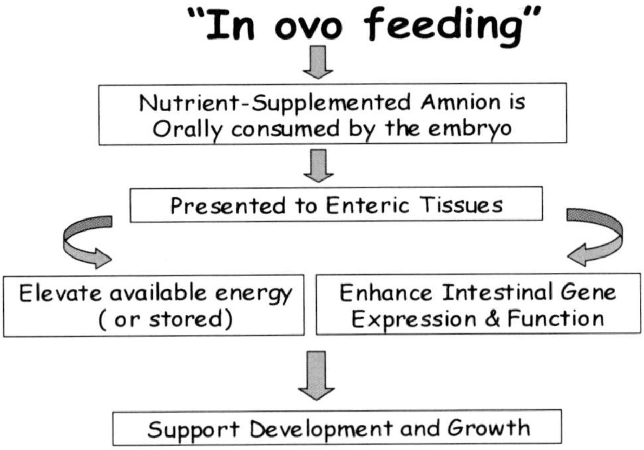

Fig. 3.4 The *in-ovo* feeding model

a more mature intestine which is beneficial during the transition from embryo to independent chick.

In summary, the improved nutritional status and intestinal function of *in-ovo*-fed birds support the embryo during its transition to independent chick. *In-ovo* feeding also enables growth at an early age to better approach genetic potential. *In-ovo* feeding is expected to yield several other advantages, among them: reduced post-hatch mortality and morbidity; greater efficiency of feed-nutrient utilisation at an early age; improved immune response to enteric antigens; reduced incidence of developmental skeletal disorders; and increased muscle development and breast-meat yield.

In-ovo feeding offers the promise of sustaining progress in production efficiency and welfare of commercial poultry. With *in-ovo* feeding, a new science of neonatal nutrition can be established for birds, which is expected to yield a better understanding of the developmental transition from embryo to chick. *In-ovo* feeding can be adapted for commercial practice by using existing automated systems for *in-ovo* vaccination.

3.3 Coping with Extreme Environmental Temperatures – Thermotolerance

In recent decades, significant strides have been made in the genetic selection of fast-growing meat-type broiler chickens [25–27]. However, the progress of genetic selection in broiler chickens, coupled with their high sensitivity to changes in environmental conditions, contribute to difficulties in coping with conditions of heat or cold stress.

Birds are homeotherms, that is, they are able to maintain their body temperature (T_b) within a narrow range. An increase in T_b above or below the regulated range, as a result of exposure to heat or cold stress, respectively, may initiate an irreversible cascade of thermoregulatory events that can be lethal. It has been well documented that the potential growth rate of broilers as a result of advances in genetic selection deteriorates during exposure to harsh environmental conditions [32, 92]. Furthermore, the late twentieth and twenty-first centuries were both characterised by increases in global mean surface temperature of 0.8–1.7°C (U.S. National Climatic Center, 2001). Scientists expect that the average global surface temperature will rise by 0.6–2.5°C over the next 50 years. This situation, in which growth rate (heat production) improves on a yearly basis while the global surface temperature conditions deteriorate, necessitates an efficient means of economically improving the acquisition of thermotolerance by broiler chickens in order to retain the performance potential obtained by the genetic-selection process.

In general, the assumption is that thermotolerance acquisition and performance improvement are two conflicting processes [15]. Therefore, simultaneously improving both, particularly in domestic animals characterised by a high rate of production, is considered difficult to impossible.

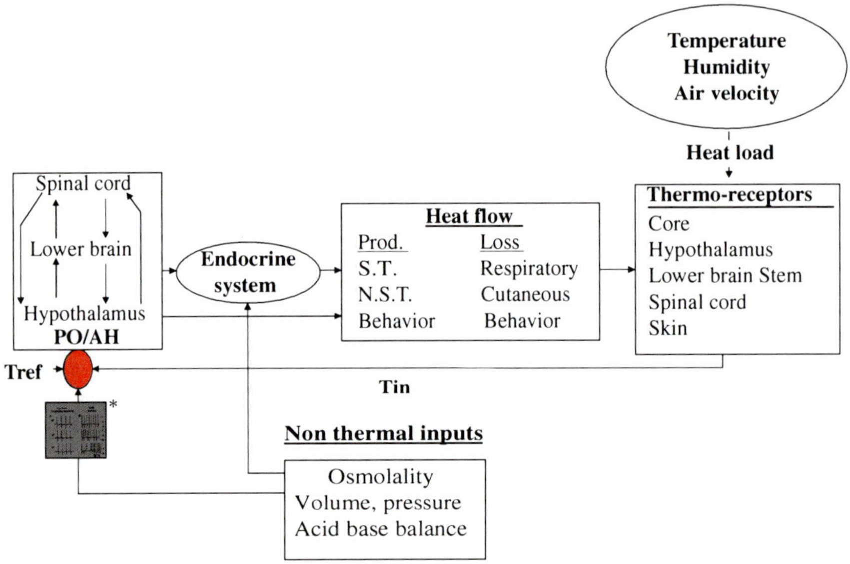

Fig. 3.5 Schematic flow chart describing how environmental heat load is transformed to thermal information by thermoreceptors and transferred (Tin) to the pre-optic anterior hypothalamus (PO/AH) to elicit thermoregulatory responses by the animal (heat production and heat loss) mediated by the endocrine system. Non-thermal inputs from the blood system can modify the response. Tref = the set point temperature in the PO/AH; S.T. = shivering thermogenesis; N.S.T. = non-shivering thermogenesis *Schematic figure of sensitive and non-sensitive neurons in the PO/AH [3]

3.4 Different Strategies for Improving Thermotolerance

To develop thermotolerance, three direct responses are employed by the bird: the rapid thermal-shock response [53]; acclimation [30, 31, 61]; and epigenetic adaptation [48, 70]. All three strategies are based mainly on the assumption that ambient temperature (T_a) is a major factor affecting thermotolerance. However, the last two strategies have a strong influence on the determination of the "set-point" for physiological control systems [11]. Thermosensitive neurons located in the pre-optic anterior hypothalamus (PO/AH) are able to sense core T_b and control physiological, endocrinological, and behavioural responses in order to maintain core T_b relatively constant [2, 3, 4] (Fig. 3.5). The thermoregulatory response is mediated mainly by the level of metabolism induced or permitted by the thyroid hormones (thyroxine, T_4; triiodothyronin, T_3) [28, 43] and the hydration status of the animal (non-thermal input), which is mediated by arginine vasotocin (AVT) [59, 91]. Changing the sensitivity of the warm- and/or cold-sensitive neurons located in the PO/AH may change the threshold for heat production and/or heat loss in the animal.

The rapid thermal or heat-shock response (HSR) is mediated by the production of heat-shock proteins (HSPs). The HSPs protect existing proteins and membranes against heat stress and facilitate repair or degradation of damaged proteins [53], as well as protecting against protein translocation [5]. This response is universal [40], and has been related to core T_b [89].

Acclimation has been defined as a physiological response that reduces the strain or enhances endurance to the strain caused by experimental climatic factors (acclimation) or ambient climatic factors (acclimatisation) [33]. Heat acclimation in broiler chickens is mediated by reduced metabolic rate, blood-volume expansion [92, 93], reduced T_b [86, 93], reduced heart rate, reduced threshold for heat dissipation, increased cardio-vascular reserves, and increased capacity for evaporative heat loss [30, 31, 60, 61]. This process requires from 4 to 7 days for completion in the domestic fowl [84]. Early marketing age coupled with: (a) the necessity to keep the environmental temperature controlled up to the age of 21 days for brooding; (b) the deleterious effect of heat acclimation on broiler chickens, and (c) the enormous cost of temperature-controlled poultry houses, makes this process relatively impractical.

3.5 The Epigenetic Response

3.5.1 Thermal Manipulations During the Postnatal Period

Epigenetic adaptation, which has been defined [50, 69, 71] as a lifelong adaptation occurring during prenatal (embryogenesis) or early postnatal ontogeny within critical developmental phases that affect gene expression, seems suited to achieving the goal of improved thermotolerance acquisition in broilers. During early development, most functional systems evolve from an open-loop system without feedback into a closed control system with feedback ("transformation rule", [11]). Thermal manipulations during the critical phases of development may induce alterations in the thermoregulatory control system.

The epigenetic response has been successfully modulated by thermal manipulation at an early age in postnatal chicks. Thermal manipulations of broilers on day 3 of life improved acquisition of thermotolerance, resulting in a significant reduction in heat production during exposure to acute thermal challenge at market age [87]. This coincided with an alteration in sensible heat loss by convection and radiation [90] and reduced stress levels in the thermally manipulated chickens, as evaluated by monitoring plasma corticosterone concentration. These changes facilitated lower T_b and thus dramatically reduced mortality. Thermal manipulation at day 3 of age also induced compensatory growth, leading to improvement of performance and muscle growth due to the proliferation of skeletal muscle satellite cells during the manipulation [19]. In terms of the GIT, this treatment was performed at a critical period of GIT development, when the main process in the developing tissue is hyperplasia and before the tissue reaches an equilibrium state of continuous cell proliferation, migration and differentiation [72, 76]. This led to a positive effect on GIT growth and function which modulated the GIT for compensatory growth [75]. Furthermore, in chick PO/AH, a significant increase in R-Ras3 [39] and brain-derived neurotrophic factor (BDNH) gene expression [38] was detected during thermal manipulation, suggesting involvement of these genes in thermal manipulation.

It is now well documented that despite the assumption of Emmans and Kyriazakis [15] that thermotolerance acquisition and performance improvement are two conflicting processes, it is possible to improve both in broilers simultaneously. Nevertheless, uniform post-hatch temperature manipulation for the improvement of thermotolerance and performance is difficult to achieve, while the use of such manipulations during incubation would likely be more efficient and uniform.

3.5.2 Thermal Manipulations During the Prenatal Period

Thermal manipulations during embryogenesis are based on the following assumptions:

a. During embryogenesis, it is possible to induce long-lasting physiological memory, based on epigenetic adaptation.
b. This long-lasting memory can be defined, most probably, as an alteration in the hypothalamic threshold response to changes in the environment.
c. Thermal manipulations during sensitive periods in embryogenesis, by means of a specific level and duration of heat exposure, will impart improved thermotolerance for the entire span of the bird's life.
d. Improving thermotolerance throughout the bird's life will reduce broiler morbidity and mortality and may even contribute to the improvement of performance, as was demonstrated in birds which were thermally manipulated post-hatch.

In general, the incubation temperature of domestic fowl is relatively constant. In contrast to the uniform temperature of commercial incubation, in nature incubation conditions are non-uniform as a result of the need to search for food and escape from predators, and due to uneven nest insulation. This may be one of the reasons why birds in the wild are better able to cope with extreme environmental temperatures.

The optimum incubation temperature for the improvement of long-term thermotolerance of broilers and its combination with timing (critical phases of embryogenesis) and duration are still unknown. Nevertheless, the exact timing for optimal thermal manipulation should logically occur during periods of greatest gene activity. According to Hamburger and Hamilton [21], chick embryogenesis can be divided into three major phases: two early phases in which the organs and systems of the body are formed; and the last phase, from 13 days of embryogenesis (E13) onward, during which growth and maturation occur. Changes in the duration of increased incubation T_a can affect embryos in different ways. A short-term increase in incubation T_a has been found to activate the heat-loss mechanism in chick embryos [29], whereas a long-term increase adversely affected embryo morphology [37], increased the incidence of malpositions, and decreased hatchability [17, 56].

The search for the optimal critical phase in embryogenesis for improving thermotolerance can be based on the development of two major axes related to thermoregulation; the hypothalamus-hypophysis-thyroid axis (HHTA), and the hypothalamus-hypophysis-adrenal axis (HHAA).

Thyroid hormones are well known to affect the development of chick embryos and to be responsible for the development of thermoregulation [43]. The thyroid gland possesses a limited ability to synthesise hormones [66] until mid-incubation. This period has been characterised by the synthesis of monoiodotyrosine on E8, of diiodotyrosine on E9, and of T_4 and thyroid-stimulating hormone (TSH) on E10 [54]. The hypothalamic-pituitary-thyroid axis is linked between E10.5 and E11.5 [67]. Levels of T_3 start increasing on E12 [36] and increase significantly prior to hatch in preparation for their role in the final maturation of many tissues and in the physiological integration of hatching [1, 8, 42]. Therefore, thermal manipulation during the sensitive development of this axis [55] may affect the "set-point" of the heat-production threshold.

Corticosterone modulates peripheral conversion of T_4 to T_3 prenatally [9, 47] and is the main stress hormone in the chick. Epple et al. [16] suggested that embryos are susceptible to stress. Increasing incubation T_a during and/or after the HHAA has been activated [82] may affect the stress response in the post-hatch chick, and it may be possible to beneficially modify the stress response by using prenatal thermal manipulation during this period.

Based on the development, maturation and function of the HHTA and HHAA axes, the period between E8 and E18 has been used to evaluate the critical phase for improvement of thermotolerance.

Thermal manipulation of embryos was performed during the critical period of E16–E18 (that is, during the functional phase of the two axes) by subjecting the embryos to 38.5°C for 3 h/day, and was shown to significantly affect heat production. Lower T_b coupled with lower plasma concentration of thyroid hormones (Table 3.2) lent supporting evidence for the positive effect of thermal manipulation on thermoregulation, which most probably resulted in reduced metabolic rate.

In a later study [85], E8–E10 (development phase) and E16–E18 (functioning phase) were chosen as critical phases, and two manipulated temperatures applied (39.5 and 41°C) for 3 h/day. This study demonstrated a pronounced effect on

Table 3.2 Effects of thermal conditioning (TC) of embryos at 38.5°C and 65% relative humidity for 3 h on E16, E17 and E18 of incubation on body temperature (T_b) and plasma thyroid hormone concentrations at hatch [88]

Variable	Control	TC
T_b (°C)	39.37±0.06[a]	38.94±0.05[b]
T_4 (ng/mL)	5.70±0.50[a]	4.41±0.37[b]
T_3 (ng/mL)	2.51±0.16[a]	2.03±0.14[b]

[a]Within rows, mean values ± SEM followed by different superscripts differ significantly ($P \leq 0.05$).
E – embryogenesis.

Table 3.3 The effect of thermal manipulation during early (E8–E10; EE) and late (E16–E18; LE) embryogenesis on hatchability, body weight and body temperature (T_b) of broiler chicks after hatch and feather drying [85]

Variable	Treatment control	EE (39.5°C)	EE (41.0°C)	LE (39.5°C)	LE (41.0°C)
Hatchability (%)	92.5^b	89.9^b	89.7^b	97.9^a	92.5^b
Body weight (g)	46.8±0.3	47.3±0.3	47.0±0.3	47.0±0.3	47±0.3
T_b (°C)	38.12±0.09^{ab}	37.75±0.09^b	37.97±0.09^{ab}	37.72±0.09^b	38.23±0.10^a

[a]Within rows, mean values ± SEM followed by different letters differ significantly ($P \leq 0.05$).
E – embryogenesis; EE – early embryogenesis; LE – late embryogenesis.
$n = 158$–183, depending on the treatment.

hatchability of thermal manipulation during the later embryogenic phase (LE) chicks treated at 39.5°C, which coincided with the lowest T_b (Table 3.3).

Challenging the chicks at 3 days of age by exposing them to 41°C for 6 h (Table 3.4) demonstrated a dramatic difference in development of hyperthermia. While the control chicks developed acute hyperthermia, those thermally manipulated developed only moderate hyperthermia, and those belonging to the LE 39.5°C group had the lowest T_b. This phenomenon was coupled with significantly lower plasma corticosterone concentration. Having both the lowest T_b and lowest plasma corticosterone concentration indicates that thermal manipulation during the later phase of embryogenesis when both the thyroid and adrenal axes are functional reduces metabolic rate and stress response, thus enabling the organism to better cope with heat stress.

Prolonged thermal manipulation (38.5°C) has also been applied to laying-strain chicken embryos from E18 until the end of incubation. On the final day of incubation, the thermally manipulated embryos had significantly higher heat production than controls [41]. Similar effects were found in Muscovy duck embryos subjected to thermal manipulation from E29 until hatch. Furthermore, prolonged exposure of Muscovy duck embryos to warm (38.5°C) or cold (34.5°C) conditions induced changes in the thermosensitivity of PO/AH neurons [41] which persisted until 10

Table 3.4 The effect of thermal manipulation during early (E8–E10; EE) and late (E16–E18; LE) embryogenesis on body temperature (T_b) and plasma concentrations of corticosterone (Cort) of thermally challenged chicks at the age of 3 days [85]

Variables	Treatment control	EE(39.5°C)	EE(41.0°C)	LE(39.5°C)	LE(41.0°C)
T_b (°C)	44.04±0.16^a	42.78±0.25^b	43.01±0.20^b	42.66±0.19^b	42.96±0.20^b
Cort (ng/mL)	34.3±3.4^a	23.8±3.8^{ab}	23.8±2.4^{ab}	18.6±3.7^b	35.7±5.0^a

[a]Within rows, mean values ± SEM followed by different superscripts differ significantly ($P \leq 0.05$).
E – embryogenesis; EE – early embryogenesis; LE – late embryogenesis.
$n = 20$ for T_b and $n = 10$ for corticosterone concentration.

days post-hatch [69]. During the first 10 days post-hatch, Muscovy ducklings and turkeys exposed to thermal manipulations during embryogenesis exhibited changes in heat production and in their preferred ambient temperatures[49].

Exposure of layer embryos to high or low incubation temperatures (38.5 or 34.5°C, respectively) on E18 and heat-stress on E20 resulted in altered expression of c-Fos in the hypothalamus on the last day of incubation [34]. This suggests the laying hens being under stress as a result of heat stress despite the previous thermal treatment.

The response of chicks exposed to thermal manipulation during embryogenesis and then raised to market age does not, however, reflect improved thermotolerance [7, 68]. In other words, thermal manipulations do not appear to create a long-lasting "physiological memory" for improved thermotolerance. Most embryonic thermal manipulation studies demonstrate an improvement in thermotolerance during the first 10 days post-hatch [44, 49, 69, 85, 88]. However, in the study of Collin et al. [7] presented in Fig. 3.6, immediately post-hatch the thermally treated chicks exhibited lower to significantly lower T_b than controls, indicating these chicks had a lower metabolic rate. However, chickens in the EL [early (E) E8–E10 and late (L) E16–E18], L and E treatment groups "lost" their advantage of significantly lower T_b after 28, 35 and 41 days of age, respectively (Fig. 3.6). A thermal challenge to these chickens at the age of 42 days by exposure to 35°C for 6 h did not result in differences in development of hyperthermia or in mortality rate.

These results raise the question of whether a long-lasting thermal "memory" can be imparted by thermal manipulation during embryogenesis at all, or whether it is just a question of selecting the correct critical period. In mammals, it appears that for different control systems, as well as for specific functions of a particular physiological system (as exhibited, for example, in the development of the mammalian visual system [24]), several different, partially overlapping "critical phases" can be identified. Furthermore, species-specific differences have to be taken into consideration. Therefore, in broilers, it may indeed be only a question of finding the correct "critical phase".

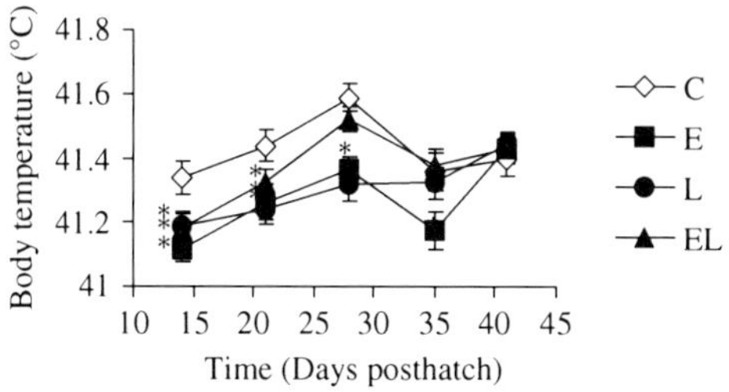

Fig. 3.6 Effect of thermal manipulation during incubation [C = control; thermal manipulation during early embryogenesis (E8–E10) = E, and during late embryogenesis (E16–E18) = L, or during both periods = EL [7]. All manipulations were at 39.5°C for 3 h/day. Body temperature was recorded between days 14 and 41 post-hatch. *Significant difference compared to control chickens of the same age ($P < 0.05$). E – embryogenesis

Previous studies on thermal manipulation at 3 days of age have shown induction of compensatory growth, leading to improved growth performance and muscle growth due to enhancement of proliferation and differentiation of skeletal muscle satellite cells during and immediately after the thermal manipulation [19].

In the chick, embryonic myoblasts are most abundant on E5, whereas fetal myoblasts are most abundant between E8 and E12 [63]. Individual myofibers become encased in a basal lamina during late embryogenesis (E15 onwards), and at this stage that it is possible to distinguish satellite cells by their morphology and location between the basal lamina and the sarcolemma [23, 83].

The induction of muscle growth due to enhancement of skeletal muscle satellite cell proliferation and differentiation in chicks, coupled with the fact that in embryos skeletal muscle satellite cells appear from E15 onwards, led Halevy et al. [20] to hypothesise that temperature manipulation at E16–E18 alters satellite cell proliferation and subsequently, enhances muscle growth in broiler chicks post-hatch.

In two separate experiments, body weight was significantly higher at 9 and 15 days of age in chicks thermally manipulated from E16 to E18 at 38.5°C (Fig. 3.7A) or at 9 days of age in chicks that had been manipulated at 39.5°C (Fig. 3.7B). Similar results were obtained for breast-muscle percentage of BW in chicks manipulated from E16 to E18 at 38.5°C (Fig. 3.8) and in females at market age [20]

These results suggest that thermal manipulations during the late stage of incubation influence muscle growth in post-hatch broilers and may therefore improve broiler performance.

In summary, there is accumulating evidence that the epigenetic adaptation approach, and its association with changes in the environment in mammals and birds, with an emphasis on fine tuning the level and duration of the stress to coincide with the "critical phase", can elicit an efficient epigenetic adaptive response. This

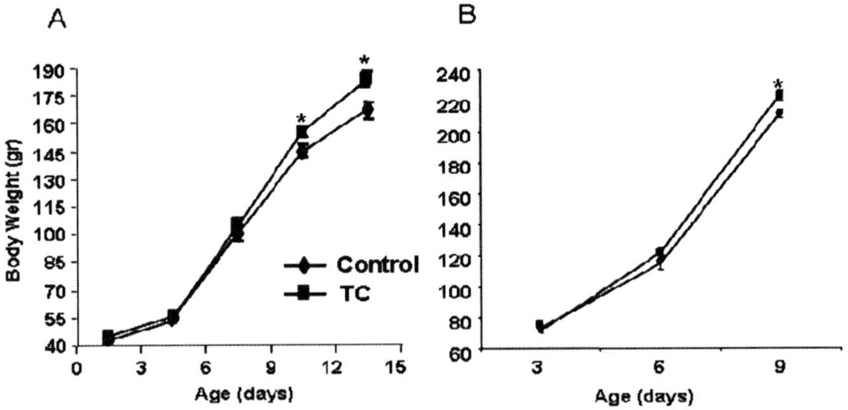

Fig. 3.7 Body weight (grams =g) of control and thermally manipulated (TC) male chicks at various days post-hatch. Thermal manipulation was conducted at 38.5°C (**a**) or 39.5°C (**b**) on embryonic days E16–E18 for 3 h/day. Results are means ± SE (n = 20 or n = 50 for **a** and **b**, respectively). $^{*}P < 0.05$ vs. control at the same age

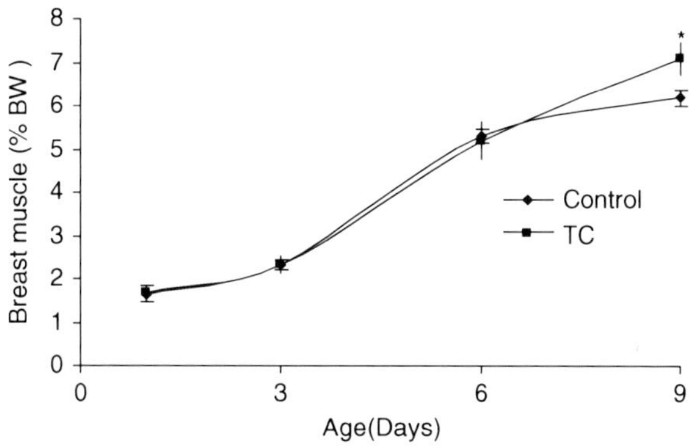

Fig. 3.8 Breast muscle as percentage of body weight (BW) in chicks thermally manipulated on embryonic days E16–E18 (TC) or untreated controls [20]. Results are means ± SD for $n = 6$ per group at each age. *$P < 0.05$ vs. control at the same age

complex issue needs intensive study in order to shed light on epigenetic adaptation in domestic fowl. It can be further concluded that thermal manipulations at different stages of embryogenesis may induce alterations in developmental processes.

3.6 Conclusion

The embryonic period represents half of the chicken's life span. This period is crucial for determination of post-hatch performance and resistance to stressors. Today's rapid-growing strains have not achieved their genetic potential for growth as a result of limitations in the pre- and post-hatch periods in available energy and intestinal functionality, and in their ability to cope with hot environments. This chapter describes two prenatal manipulations that enable broilers to overcome these limitations. The first, addition of nutrients to the developing embryo, leads to increased glycogen (energy) reserves and a more highly developed small intestine in the pre-hatch period, which contributes to heavier body weight and greater muscle mass post-hatch. The second, thermal manipulation, improves performance while increasing acquisition of thermotolerance, at least during the earlier post-hatch period.

References

1. Black, B.L. 1978. Morphological development of the epithelium of the embryonic chick intestine in culture: influence of thyroxine and hydrocortisone. *Am. J. Anat.* **153**:573–600.
2. Boulant, J.A. 1980. Hypothalamic control of thermoregulation: Neurophysiological basis, p. 1–82. *In* P.J. Morgane, and J. Panksepp (eds.), Handbook of Hypothalamus, vol. 3, part A, Marcel Dekker, New York.
3. Boulant, J.A. 1996. Hypothalamic neurons regulating body temperature, pp. 105–126. *In* M.J. Fregly, and C.M. Blatteis (eds.), APS Handbook of Physiology. Section 4: Environmental Physiology, Oxford Press, New York.
4. Boulant, J.A. and J.B. Dean. 1986. Temperature receptors in the central nervous system. *Ann. Rev. Physiol.* **48**:639–654.

5. Burel, C., V. Mezger, M. Rallu, S. Trigon, and M. Morange. 1992. Mammalian heat shock protein families. Expression and function. *Experientia* **48**:629–634.
6. Christensen, V.L., M.J. Wineland, G.M. Fasenko, and W.E. Donaldson. 2001. Egg storage effects on plasma glucose and supply and demand tissue glycogen concentrations of broiler embryos. *Poult. Sci.* **80**:1729–1735.
7. Collin, A., C. Berri, S. Tesseraud, F. Requena, S. Cassy, S. Crochet, M.J. Duclos, N. Rideau, K. Tona, J. Buyse, V. Bruggemann, E. Decuypere, M. Picard, and S. Yahav. 2007. Effects of thermal manipulation during early and late embryogenesis on thermotolerance and breast muscle characteristics in broiler chickens. *Poult. Sci.* **86**:795–800.
8. Decuypere, E., E. Dewil, and E.R. Kuhn. 1992. The hatching process and the role of hormones, p. 239–255. *In* S.G. Tullett (ed.), Avian Incubation, Butterworth-Heinemann, London.
9. Decuypere, E., C.G. Scanes, and E.R. Kühn. 1983. Effect of glucocorticoids on circulating concentrations of thyroxine (T4) and triiodothyronine (T3) and on the peripheral monodeiodination in pre and post-hatching chickens. *Horm. Metab. Res.* **15**:233–236.
10. Dickson, A.J. and D.R. Langslow. 1978. Hepatic gluconeogenesis in chickens. *Mol. Cell. Biochem.* **22**:167–181.
11. Dörner, G. 1974. Environment-dependent brain differentiation and fundamental process of life. *Acta Biol. Med. Germ.* **33**:129–148.
12. Elwyn, D.H. and S. Bursztein. 1993. Carbohydrate metabolism and requirements for nutritional support: Part I. *Nutrition* **9**:50–66.
13. Elwyn, D.H. and S. Bursztein. 1993. Carbohydrate metabolism and requirements for nutritional support: Part II. *Nutrition* **9**:164–177.
14. Elwyn, D.H. and S. Bursztein. 1993. Carbohydrate metabolism and requirements for nutritional support: Part III. *Nutrition* **9**:255–267.
15. Emmans, G.C. and I. Kyriazakis. 2000. Issues arising from genetic selection for growth and body composition characteristics in poultry and pigs, p. 39–53. *In* The Challenge of Genetic Changes in Animal Production, Occasional Publication No. 27. British Society of Animal Science, Edinburgh.
16. Epple, A., B. Gower, M.T. Busch, T. Gill, L. Milakofsky, R. Piechotta, B. Nibbio, T. Hare, and M.H. Stetson. 1997. Stress responses in avian embryos. *Am. Zool.* **37**:536–545.
17. French, N.A. 1994. Effect of incubation temperature on the gross pathology of turkey embryos. *Br. Poult. Sci.* **35**:363–371.
18. Geyra, A., Z. Uni, and D. Sklan. 2001. The effect of fasting at different ages on growth and tissue dynamics in the small intestine of the young chick. *Br. Poult. Sci.* **86**:53–61.
19. Halevy, O., A. Krispin, Y. Leshem, J.F. McMurtry, and S. Yahav. 2001. Early age heat stress accelerates skeletal muscle satellite cell proliferation and differentiation in chicks. *Am. J. Physiol.* **281**:R302–R317.
20. Halevy, O., M. Lavi, and S. Yahav. 2006. Enhancement of meat production by thermal manipulations during embryogenesis of broilers, pp. 77–88. *In* S. Yahav, and B. Tzschentke (eds.), New Insights into Fundamental Physiology and Perinatal Adaptation of Domestic Fowl, Nottingham University Press, UK.
21. Hamburger, V. and H.L. Hamilton. 1992. A series of normal stages in the development of the chick embryo. *Dev. Dyn.* **195**:232–272.
22. Hamer, M.J. and A.J. Dickson. 1989. Influence of developmental stage on glycogenolysis and glycolysis in hepatocytes isolated from chick embryos and neonates. *Biochem. Soc. Trans.* **17**:1107–1108.
23. Hartely, R., S. Bandman, and Z. Yablonka-Reuveni. 1992. Skeletal muscle satellite cells appear during late chicken embryogenesis. *Dev. Biol.* **153**:206–216.
24. Harwerth, R.S., L. Smith, G.C. Duncan, M.L. Crawford, and G.K. von Noorden. 1986. Multiple sensitive periods in the development of the primate visual system. *Science* **232**:235–238.
25. Havenstein, G.B., P.R. Ferket, and M.A. Qureshi. 2003. Growth, livability, and feed conversion of 1957 versus 2001 broilers when fed representative 1957 and 2001 broiler diets. *Poult. Sci.* **82**:1500–1508.

26. Havenstein, G.B., P.R. Ferket, and M.A. Qureshi. 2003. Carcass composition and yield of 1957 versus 2001 broilers when fed representative 1957 and 2001 broiler diets. *Poult. Sci.* **82**:1509–1518.
27. Havenstein, G.B., P.R. Ferket, S.E. Scheideler, and B.T. Larson. 1994. Growth, livability, and feed conversion of 1991 versus 1957 broilers when fed "typical" 1957 and 1991 broiler diets. *Poult. Sci.* **73**:1785–1794.
28. Hillman, P.E., N.R. Scott, and A. van Tienhoven. 1985. Physiological responses and adaptations to hot and cold environments, p. 27–71. *In* M.K. Yousef (ed.), Stress Physiology in Livestock, Vol. 3, Poultry. CRC Press, Inc., Boca Raton, FL.
29. Holland, S., M. Nichelmann, and J. Höchel. 1997. Development in heat loss mechanisms in avian embryos. *Verh. Dtsch. Zool. Ges.* **90**:105.
30. Horowitz, M. 1998. Do cellular heat acclimation responses modulate central thermoregulatory activity? *News Physiol. Sci.* **13**:218–225.
31. Horowitz, M. 2002. From molecular and cellular to integrative heat defense during exposure to chronic heat. *Comp. Biochem. Physiol.* **131A**:475–483.
32. Hurwitz, S., M. Weiselberg, U. Eisner, I. Bartov, G. Riesenfeld, M. Sharvit, A. Niv, and S. Bornstein. 1980. The energy requirements of growing chickens and turkeys as affected by environmental temperature. *Poult. Sci.* **59**:2290–2299.
33. IUPS Thermal Commission. 2001. Glossary of terms for thermal physiology. *Jap. J. Physiol.* **51**:245–280.
34. Janke, O. and B. Tzschentke. 2006. Hypothalamic c-fos expression of temperature experienced chick embryos after acute heat exposure, p. 109–115. *In* S. Yahav, and B. Tzschentke (eds.), New Insights into Fundamental Physiology and Perinatal Adaptation of Domestic Fowl, Nottingham University Press, UK.
35. John, T.M., J.C. George, and E.T. Moran, Jr. 1988. Metabolic changes in pectoral muscle and liver of turkey embryos in relation to hatching: influence of glucose and antibiotic-treatment of eggs. *Poult. Sci.* **67**:463–469.
36. Kameda, Y., K. Udatsu, M. Horino, and T. Tagawa. 1986. Localization and development of immunoreactive triiodothyronine in thyroid glands of dogs and chickens. *Anat. Res.* **214**: 168–176.
37. Kaplan, S., G. L. Kolesari, and J. P. Bahr. 1978. Temperature dynamics of the fertile chicken eggs. *Am. J. Physiol.* **234**:R183–187.
38. Katz, A. and N. Meiri. 2006. Brain-derived neurotrophic factor is critically involved in thermal-experience-dependent developmental plasticity. *J. Neurosci.* **12**:3899–3907.
39. Labunsky, G. and N. Meiri. 2006. R-Ras3/(M-Ras) is involved in thermal adaptation in the critical period of thermal control establishment. *J. Neurobiol.* **66**:56–70.
40. Lindquist, S. 1986. The heat shock response. *Ann. Rev. Biochem.* **55**:1151–1191.
41. Loh, B., I. Maier, A. Winar, O. Janke, and B. Tzschentke. 2004. Prenatal development of epigenetic adaptation processes in poultry: changes in metabolic and neuronal thermoregulatory mechanisms. *Avian Poultry Biol. Rev.* **15**:119–128.
42. Mallon, D.L. and T.W. Betz. 1982. The effects of hydrocortisone and thyroxine treatments on duodenal morphology, alkaline phosphatase and sugar transport in chicken (Gallus domesticus) embryos. *Can. J. Zool.* **60**:3447–3455.
43. McNabb, F.M.A. and D.B. King. 1993. Thyroid hormones effects on growth development and metabolism, p. 393–417. *In* M.P. Schreibman, C.G. Scanes, and P.K.T. Pang (eds.), The Endocrinology of Growth Development and Metabolism in Vertebrates, Academic Press, New York.
44. Moraes, V.M.B., D. Malehiros, V. Bruggemann, A. Collin, K. Tona, P. Van As, O.M. Onagbesan, J. Buyse, E. Decuypere, and M. Macari. 2004. Effect of thermal conditioning during embryonic development on aspects of physiological responses of broilers to heat stress. *J. Thermal Biol.* **28:**133–140.
45. Moran, E.T. 1985. Digestion and absorption in fowl and events through perinatal development. *J. Nutr.* **115**:665–674.

46. Moran, E.T. and B.S. Reinhart. 1980. Poult yolk amount and composition upon placement: effect of breeder age, egg weight, sex and subsequent change with feeding or fasting. *Poult. Sci.* **59**:1521–1528.
47. Neeuwis, R., R. Michielsen, and E. Decuypere. 1989. Thyrotrophic activity of the ovine corticotrophin-releasing factor in the chick embryo. *Gen. Comp. Endocrinol.* **76**:357–363.
48. Nichelmann, M., J. Höchel, and B. Tzschentke. 1999. Biological rhythms in birds—development, insights and perspectives. *Comp. Biochem. Physiol. A* **124**:437–439.
49. Nichelmann, M., B. Lange, R. Pirow, J. Langbein, and S. Herrmann. 1994. Avian thermoregulation during the perinatal period, p. 167–173. *In* E. Zeisberger, E. Schönbaum, and P. Lomax (eds.), Thermal Balance in Health and Disease. Advances in Pharmacological Science, Birkhäuser Verlag, Basel.
50. Nichelmann, M. and B. Tzschentke. 2002. Ontogeny of thermoregulation in precocial birds. Comp. *Biochem. Physiol. A* **131**:751–763.
51. Nissen, S., J.C. Fuller, Jr., J. Sell, P.R. Ferket, and D.V. Rives. 1994. The effect of beta-hydroxy-beta-methylbutyrate on growth, mortality, and carcass qualities of broiler chickens. *Poult. Sci.* **73**:137–155.
52. Noy, Y. and D. Sklan. 1998. Yolk utilisation in the newly hatched poult. *Br. Poult. Sci.* **39**: 446–451.
53. Parsell, D.A. and S. Lindquist 1994. Heat shock proteins and stress tolerance, p. 457–494. *In* R.I. Morimoto, A. Tissieres, and C. Georgopoulos (eds.), Biology of Heat Shock Proteins and Molecular Chaperones, Cold Spring Harbor Laboratory Press, New York.
54. Prati, M., R. Calvo, and G. Morreale de Escobar 1992. L-thyroxine and 3,5,3'-triiodothyronine concentrations in the chicken egg and in the embryo before and after the onset of thyroid function. *Endocrinology* **130**:2651–2659.
55. Reynes, G.E., K. Venken, G. Morreale de Escobar, E.R. Kühn, and V.M. Darras. 2003. Dynamics and regulation of intracellular thyroid hormone concentrations in embryonic chicken liver, kidney, brain and blood. *Gen. Comp. Physiol.* **134**:80–87.
56. Romanoff, A.L. 1972. Pathogenesis of the Avian Embryo. John Wiley and Sons, New York.
57. Rosebrough, R.W., E. Geis, K. Henderson, and L.T. Frobish. 1978. Glycogen depletion and repletion in the chick. *Poult. Sci.* **57**:1460–1462.
58. Rosebrough, R.W., E. Geis, K. Henderson, and L.T. Frobish. 1978. Glycogen metabolism in the turkey embryo and poult. *Poult. Sci.* **57**:747–751.
59. Saito, N. and R.Grossmann. 1998. Effect of short-term dehydration on plasma osmolality, levels of arginine vasotocin and its hypothalamic gene expression in the laying hen. Comp. *Biochem. Physiol. A* **121**:235–239.
60. Sawaka, M.N., C.B. Wenger, and K.B. Pandolf. 1996. Thermoregulatory responses to acute exercise-heat stress and heat acclimation, p. 157–187. *In* M.J. Fregly, and C.M. Blatteis (eds.), Handbook of Physiology, Section 4: Environmental Physiology, Vol. 1., Oxford University Press, Oxford.
61. Shido, O., Y. Yoneda, and T. Nagasaka. 1989. Changes of body temperatures in rats acclimated to heat with different acclimated schedule. *J. Appl. Physiol.* **67**:2154.
62. Smirnov, A., E. Tako, P.R. Ferket, and Z. Uni. 2006. Mucin gene expression and mucin content in the chicken intestinal goblet cells are affected by in ovo feeding of carbohydrates. *Poult. Sci.* **85**:669–673.
63. Stockdale, F.E. 1992. Myogenic cell lineages. *Dev. Biol.* **154**:284–298.
64. Tako, E., P.R. Ferket, and Z. Uni. 2004. The effects of In Ovo feeding of carbohydrates and beta-hydroxy-beta-methylbutyrate on the development of chicken intestine. *Poult. Sci.* **83**:2023–2028.
65. Tako, E., P.R. Ferket, and Z. Uni. 2005. Changes in chicken intestinal zinc exporter (ZnT1) mRNA expression and small intestine functionality following an intra amniotic zinc-methionine (ZnMet) administration. *J. Nutr. Biochem.* **16**:339–346.
66. Thommes, R.C. 1987. Ontogenesis of thyroid function and regulation in the developing chick embryo. *J. Exp. Zool. Suppl.* **1**:273–279.

67. Thommes, R.C., N.B. Clark, L.L.S. Mok, and S. Malone. 1984. Hypothalamo-adenohypophyseal-thyroid interrelationships in the chick embryo. V. The effects of thyroidectomy on T4 levels in blood plasma. *Gen. Comp. Endocrinol.* **54**:324–327.
68. Tona, K., O. Onagbesan, V. Bruggeman, A. Collin, C. Berri, M. Duclos, S. Tesseraud, J. Buyse, E. Decuypere, and S. Yahav. 2007. Effects of heat conditioning at d 16 to 18 of incubation or during early broiler rearing on embryo physiology, post-hatch growth performance and heat tolerance. *Br. Poult. Sci. Arch. Geflügelk* **72**(2):75–83.
69. Tzschentke, B. and D. Basta. 2002. Early development of neuronal hypothalamic sensitivity in birds: influence of epigenetic temperature adaptation. Comp. *Biochem. Physiol. A* **131**: 825–832.
70. Tzschentke, B., D. Basta, and M. Nichelmann. 2001. Epigenetic temperature adaptation in birds: peculiarities and similarities in comparison to acclimation. *News Biomed. Sci.* **1**:26–31.
71. Tzschentke, B., J. Basta, O. Janke, and I. Maier. 2004. Characteristics of early development of body functions and epigenetic adaptation to the environment in poultry: focused on development of central nervous mechanisms. *Avian Poult. Biol. Rev.* **15**:107–118.
72. Uni, Z. 1999. Functional development of the small intestine: cellular and molecular aspects. *Avian Poult. Biol. Rev.* **10**:167–179.
73. Uni, Z. and P.R. Ferket, inventors. 2003. Enhancement of development of oviparous species by in ovo feeding. US patent number 6,592,878. Issued: Jul 15, 2003.
74. Uni, Z., P.R. Ferket, E. Tako, and O. Kedar. 2005. In ovo feeding improves energy status of late term chicken embryos. *Poult. Sci.* **84**:764–770.
75. Uni, Z., O. Gal-Garber, A. Geyra, D. Sklan, and S. Yahav. 2001. Changes in growth and function of chick small intestine epithelium due to early thermal conditioning. *Poult. Sci.* **80**:438–445.
76. Uni, Z., Y. Noy, and D. Sklan. 1996. Developmental parameters of the small intestines in heavy and light strain chicks pre- and post-hatch. *Br. Poult. Sci.* **36**:63–71.
77. Uni, Z., R. Platin, and D. Sklan. 1998. Cell proliferation in chicken intestinal epithelium occurs both in the crypt and along the villus. *J. Comp. Physiol.* B **168**:241–247.
78. Uni, Z., A. Smirnov, and D. Sklan. 2003. Pre- and posthatch development of goblet cells in the broiler small intestine: effect of delayed access to feed. *Poult. Sci.* **82**:320–327.
79. Uni, Z., E. Tako, O. Gal-Garber, and D. Sklan. 2003. Morphological, molecular, and functional changes in the chicken small intestine of the late-term embryo. *Poult. Sci.* **82**: 1747–1754.
80. Vieira, S.L. and E.T. Moran. 1999. Effect of egg origin and chick post-hatch nutrition on broiler live performance and meat yields. *World's Poult. Sci. J.* **56**:125–142.
81. Vieira, S.L. and E.T. Moran. 1999. Effects of delayed placement and used litter on broiler yields. *J. Appl. Poult. Res.* **8**:75–81
82. Wise, P.M. and B.E. Frye. 1975. Functional development of the hypothalamo-hypophyseal-adrenal cortex axis in chick embryo, Gallus domesticus. *J. Exp. Zool.* **185**:277–292.
83. Yablonka-Reuveni, Z. 1995. Myogenesis in the chicken: the onset of differentiation of adult myoblasts is influenced by tissue factors. *Basic Appl. Myol.* **5**:33–41.
84. Yahav, S. 2000. Domestic fowl—strategies to confront environmental conditions. *Avian Poult. Biol. Rev.* **11**:81–95.
85. Yahav, S., A. Collin, D. Shinder, and M. Picard. 2004. Thermal manipulations during broiler chick's embryogenesis—the effect of timing and temperature. *Poult. Sci.* **83**:1959–1963.
86. Yahav, S., S. Goldfeld, I. Plavnik, and S. Hurwitz. 1995. Physiological responses of chickens and turkeys to relative humidity during exposure to high ambient temperature. *J. Thermal Biol.* **20**:245–253.
87. Yahav, S. and S. Hurwitz. 1996. Induction of thermotolerance in male broiler chickens by temperature conditioning at an early age. *Poult. Sci.* **75:**402–406.
88. Yahav, S., R. Sasson Rath, and D. Shinder. 2004. The effect of thermal manipulations during embryogenesis of broiler chicks *(Gallus domesticus)* on hatchability, performance and thermoregulation after hatch. *J. Thermal Biol.* **29**:245–250.

89. Yahav, S., A. Shamay, G. Horev, D. Bar-Ilan, O. Genina, and M. Friedman-Einat. 1997. Effect of acquisition of improved thermotolerance on the induction of heat shock proteins in broiler chickens. *Poult. Sci.* **76:**1428–1434.
90. Yahav, S., D. Shinder, J. Tanny, and S. Cohen. 2005. Sensible heat loss—the broiler's paradox. *World's Poult. Sci. J.* **61**:419–435.
91. Yahav, S., A. Straschnow, D. Luger, D. Shinder, J. Tanny, and S. Cohen. 2004. Ventilation, sensible heat loss, broiler energy and water balance under harsh environmental conditions. *Poult. Sci.* **83:**253–258.
92. Yahav, S., A. Straschnow, I. Plavnik, and S. Hurwitz. 1996. Effect of diurnal cyclic versus constant temperatures on chicken growth and food intake. *Br. Poult. Sci.* **37:**43–54.
93. Yahav, S., A. Straschnow, I. Plavnik, and S. Hurwitz. 1997. Blood system response of chickens to changes in environmental temperature. *Poult. Sci.* **76:**627–633.

Part II
Mechanistic Basis of Postnatal Consequences of Fetal Development

Chapter 4
Biological Mechanisms of Fetal Development Relating to Postnatal Growth, Efficiency and Carcass Characteristics in Ruminants

John M. Brameld, Paul L. Greenwood, and Alan W. Bell

Introduction

Over recent years there has been a lot of interest in the effects of prenatal environment on subsequent development of tissues and the postnatal consequences. In farm animal species this has particularly related to muscle and fat development and the later consequences in terms of body composition at slaughter. Studies have been carried out in a variety of species, including rats, guinea pigs, pigs, sheep and, more recently, cattle. This chapter will concentrate on the evidence for effects of prenatal environment on development of muscle and adipose cells in ruminant species, the possible mechanisms for these effects and the long-term consequences relating to postnatal growth and body composition.

4.1 Prenatal Development of Carcass Tissues

All tissues within the body develop from the single cell formed when the ovum is fertilised by a sperm. That single cell goes through thousands of cell cycles in order to replicate (proliferate) and form the thousands of cells within each tissue in the developing fetus and resulting offspring. The rates of cell proliferation are dependent upon the balance between factors that stimulate and those that inhibit cell proliferation. Often those factors are proteins and include hormones (e.g. insulin) and growth factors (e.g. epidermal growth factor (EGF) and platelet derived growth factor (PDGF)). Hence cell proliferation is needed to produce the numbers of cells required to make up a whole organism, but the specialisation of those cells into specific, functional cell types involves the process of cell differentiation. In order for cells to terminally differentiate they must exit the cell cycle and therefore, in general, factors that stimulate proliferation will inhibit differentiation and vice versa. The

---

J.M. Brameld (✉)
Division of Nutritional Sciences, University of Nottingham, School of Biosciences, LE12 5RD, Loughborough, UK
e-mail: john.brameld@nottingham.ac.uk

P.L. Greenwood et al. (eds.), *Managing the Prenatal Environment to Enhance Livestock Productivity*, DOI 10.1007/978-90-481-3135-8_4,

majority of differentiated cell types (e.g. hepatocyte, adipocyte and muscle fibre) are therefore unable to proliferate unless they are able to de-differentiate into a precursor cell type. The process of differentiation always involves the switching on of cell- or tissue-specific genes via activation of transcription factors that induce the molecular and morphological changes that result in that cell becoming a specific cell-type. As for proliferation, a variety of factors regulate differentiation, both positively and negatively. Again various hormones (e.g. insulin, thyroid hormones) and growth factors (e.g. Transforming Growth Factor β, insulin-like growth factors I and II) are involved, but also some nutrients (e.g. vitamin A), act as ligands for nuclear hormone receptors and thereby regulate gene transcription in a similar manner to transcription factors [111].

4.1.1 Skeletal Muscle Development (Myogenesis)

Probably the best studied cell type in terms of regulation of proliferation and differentiation is muscle cell lineage. Commitment or determination of cells to the muscle cell lineage is mainly via a group of muscle-specific transcription factors, collectively known as the muscle or myogenic regulatory factors (MRFs). These include myf-5, myoD, myogenin and MRF4, with the first two shown to be important for the commitment to muscle cell type and early stages of differentiation and the latter two more important for inducing formation of muscle fibres and later stages of differentiation (for reviews see [6, 17, 21, 26, 28, 88, 96]). Skeletal muscle starts to develop at a very early stage of embryonic development, with all muscle cells initially located in the myotome and dermomyotome regions of the somite. Commitment to the muscle cell lineage appears to be initiated by wnt and shh signalling peptides [121], which result in the switching on of gene expression for one of the two MRFs, MyoD and myf-5. The expression of one or other of these is enough to induce the cell to commit to muscle-lineage, resulting in the cell becoming a myoblast. At this stage the cells are still mononuclear and are able to migrate to other sites within the embryo and proliferate in response to various growth factors [17, 26]. Initiation of gene expression for a third MRF, myogenin, results in the alignment and fusion of the myoblasts and their differentiation into myotubes and, later, into muscle fibres. The latter stages of differentiation also involve the fourth MRF (MRF4), require some degree of innervation, and result in the formation of large multinuclear cells. Hence, myogenesis involves combinations of four muscle specific transcription factors regulating commitment to become a myoblast and subsequently differentiation into muscle fibres. Recent in vitro studies [67] indicate that various unsaturated fatty acids are able to dose-dependently stimulate or inhibit muscle cell differentiation, by an as yet unidentified mechanism. The monounsaturated fatty acid, oleic acid, and the polyunsaturated fatty acids, linoleic acid and the cis9, trans11 isomer of conjugated linoleic acid (CLA), all stimulated differentiation; whereas the trans10, cis12 isomer of CLA inhibited differentiation. This

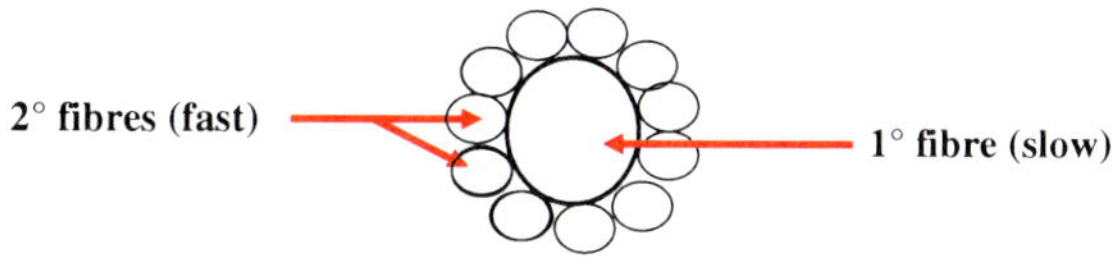

Fig. 4.1 Schematic representation of the formation of secondary (2°) muscle fibres on the surface of primary (1°) myofibres (myotubes) during fetal life

indicates one possible mechanism for the direct effects of diet on muscle development and fibre formation.

The numbers of muscle fibres within a specific muscle are thought to be set at around the time of birth in most mammals. Rodents are the exception in this with muscle fibre formation continuing through the neonatal period, with the numbers of fibres apparently fixed from the time of weaning at around 4 weeks of age [122]. Postnatal growth of muscle therefore involves increases in fibre size (hypertrophy) rather than numbers of fibres. Hence the processes of myogenesis predominantly take place in utero. The formation of muscle fibres takes place in two main waves, with primary fibres being formed first and secondary fibres developing around the primaries (Fig. 4.1) [88, 96, 121]. There may be specific myoblast precursors for these different populations of fibres; with embryonic myoblasts appearing to form primary muscle fibres in early to mid gestation and fetal myoblasts forming secondary muscle fibres in mid to late gestation (Table 4.1). In general, the primary fibres tend to become slow oxidative (type I) muscle fibres, while the secondary fibres tend to become faster fibre types (types IIA, IIB, etc.). However, some degree of plasticity is seen prenatally as maturation of muscle occurs in preparation for postnatal life [55, 86, 105], and postnatally, such that primary fibres can become fast fibres in "fast" muscles and secondary fibres can become slow fibres in "slow" muscles [88, 96]. In some species, tertiary muscle fibre formation has been described either during mid to late gestation or the early postnatal period [88, 123]. Initially, these fibres are closely associated with secondary fibres and appear to form both fast and slow fibre types, possibly being derived from different populations of satellite cells [74].

Although postnatal muscle growth mainly relates to an increase in fibre size and protein content, this is dependent upon an increase in the number of myonuclei and DNA content, which results from another population of muscle precursor cells

Table 4.1 Stages of gestation when different generations of muscle fibres appear in various mammalian species [adapted from 21, 96]

Species	Primary	Secondary	Tertiary	Length of gestation	References
Rat	14–16 df	17–19 df	–	22 days	[123]
Guinea pig	30 df	30–35 df	–	68 days	[37]
Pig	35 df	55 df	0–15 dpn	114 days	[81]
Sheep	32 df	38 df	62–76 df	145 days	[124]
Cattle	<47 df	90 df	110 df	278–283 days	[50, 104]

df: days of fetal life; dpn: days of postnatal life.

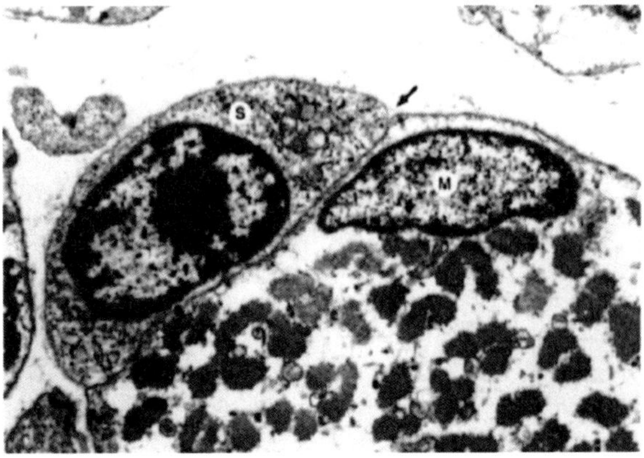

Fig. 4.2 Electron micrograph of a transverse section of 85 day fetal ovine *M. soleus*, showing a myofibre and myonucleus (M), and an associated cell with the morphology of a satellite cell (S). The basal lamina surrounding the myofibre and presumptive satellite cell is indicated by an arrow. Magnification approximately 10,000 ×. [56]

called satellite cells (for reviews see [6, 21, 26, 28, 88, 96]). Satellite cells can be considered as myoblasts that did not fuse to form fibres and survive into adulthood. They reside between the basal or external lamina and the sarcolemma (Fig. 4.2) and are able to proliferate and fuse with existing fibres during late prenatal and postnatal growth (fibre hypertrophy) in livestock or in response to muscle damage. Activity of muscle satellite cells and rate of increase in myonuclei and DNA in muscle is regulated by nutrient supply and growth of the fetus during late gestation [56] and during postnatal life [57]. These satellite cells might also relate to the myoblast population that forms the tertiary fibres in some species. The numbers present under the basement membrane of adult muscle fibres gradually decrease with age, so that the capacity for muscle growth or regeneration declines with age.

4.1.2 Adipose Cell Development (Adipogenesis)

Interestingly there are strong links between muscle and fat cell development. Striking evidence for such a developmental association comes from gene knockout studies to identify the roles of the myogenic regulatory factors (MRF), MyoD and myf-5 [102]. Mice carrying null mutations for both these genes were born alive, but were immobile and died soon after birth. Immunohistochemical analysis showed a complete lack of muscle (both myoblast precursor cells and muscle fibres), but "the spaces normally occupied by skeletal muscle contained either amorphous loose connective tissues or expanded areas of adipose tissue" [102]. This suggests that if cells are not committed to become muscle cells then they default to adipose cells. Fat cells (adipocytes) therefore share a common mesenchymal origin with skeletal muscle cells [23, 61, 64, 106]. Unlike muscle cells, it is a lack of exposure to signalling factors (e.g. wnts [101]) that appears to result in that cell becoming a fat cell. Hence it might be postulated that the adipocyte lineage is the default, if the cell is not committed to some other cell type. Adipogenesis involves determination of the stem-like precursor cell to become an adipoblast, followed by sequential differentiation steps to become a preadipocyte and then an adipocyte. The adipoblast and preadipocyte stages are associated with the capacity to proliferate and therefore increase in number. However, terminal differentiation into an adipocyte is associated with exit from the cell cycle and therefore loss of the capacity to increase in number. Once formed, adipocytes appear not to be lost (although this is a contentious issue) and are able to increase (or decrease) in size according to energy intake and expenditure, but with limits to the degree of hypertrophy possible. The molecular regulation of adipogenesis involves a cascade of transcription factors, switching on gene expression of each other and the various genes involved in nutrient uptake, and lipogenesis and lipolysis. The most important amongst these, in terms of differentiation of preadipocyte into adipocyte, are believed to be two CCAAT enhancer binding proteins (CEBP alpha and beta) and PPAR gamma [23, 61, 65, 106]. Unlike muscle, where the number of muscle fibres is set at birth, the capacity for preadipocytes to proliferate and form adipocytes appears to persist throughout life (although this is very difficult to prove). Hence there appears to be a limitless capacity to store fat. The effects of hormones, growth factors and nutrients in regulating adipogenesis have been reviewed previously [23, 61, 64–65, 106].

4.2 Factors Affecting Fetal Tissue Development

A variety of genetic and environmental factors have been found to affect development of tissues in the fetus and the interplay between all these factors will be important in determining the phenotype of the resulting offspring.

4.2.1 Maternal and Fetal Genotype

In a general sense, fetal genotype is most important in determining fetal growth during early and mid pregnancy, whereas maternal genotype is more important in determining fetal growth during late pregnancy when most fetal growth normally occurs and is increasingly subject to external influences mediated via the dam. The effect of fetal and maternal genotype on fetal growth has been most convincingly demonstrated in cattle by Ferrell [43] who implanted Charolais (heavier birth weight) or Brahman (lighter birth weight) embryos into Charolais (large frame size) and Brahman (small frame size) cows. At 232 days of pregnancy each fetal genotype was similar in size, irrespective of dam breed. However, by 274 days of gestation Charolais fetuses in Brahman cows were 7 kg lighter than those in Charolais cows. In contrast, Brahman fetuses in Charolais cows were only 2 kg heavier than those in Brahman cows. In relation to muscle development, weight of *semitendinosus* muscle was affected to a similar extent as was fetal weight by fetal and maternal genotype.

One of the most striking indications of the genetic regulation of muscle fibre number is the phenomenon of double muscling in cattle [6, 17, 21, 26]. The Belgian

Blue breed is a classic example and studies have demonstrated that the increase in numbers of muscle fibres is related to increased levels of mitogenic growth factors in the fetal circulation, resulting in increased rates of myoblast proliferation. The time period during gestation when myoblast proliferation occurs is also extended, along with a delay in the timing of differentiation. The delay in differentiation is associated with a delay in local expression of IGF-II [51], an important local regulator of myoblast differentiation via its positive effects on myogenin gene expression. Disruption in the gene for myostatin, a member of the transforming growth factor beta (TGF β) family, has been demonstrated in these cattle [6]. Myostatin (GDF-8) was originally identified in mice [92] and like other TGF β family members has been shown to decrease both proliferation and differentiation of myoblasts in vitro. Hence, the gene mutation results in loss of functional myostatin protein, eliminating a local inhibitor of myoblast proliferation and resulting in increased myoblast proliferation, and thereby increased numbers of muscle fibres. Interestingly, a number of gene mutations have been described in cattle and more recently pigs, with varying effects on muscling, including no effect on muscle fibre number in some cattle [107] and all the porcine studies to date [72–73]. This suggests that the lack of myostatin is not the only factor responsible for double muscling, as recently evidenced by, for example, studies on the over-expression of the myostatin binding protein, follistatin [80].

The *Callipyge* locus on chromosome 18 [30, 48] is associated with extreme hypertrophy of hind-quarter muscles and reduced fatness in sheep [49, 70–71, 76], improved feed efficiency [69], but markedly increased toughness in affected muscles [36, 49, 70, 76]. The callipyge phenotype is expressed during postnatal growth [36, 69] in all heterozygote offspring when the allele responsible for muscle hypertrophy is inherited from the sire, a mode of inheritance known as polar over-dominance [31, 52]. In callipyge lambs, muscle hypertrophy is associated with an increase in the average size of myofibres, but not in the number of myofibres in affected muscles [76]. In these sheep, increased myofibre size is due to a greater proportion and size of type 2B/2X (fast glycolytic) muscle fibres, and there is a reduction in the percentage of type 1 (slow oxidative) and 2A (fast oxidative-glycolytic) myofibres and an increase in size of type 2A myofibres present [27, 76, 82]. According to Koohmaraie et al. [76], hypertrophic muscles in these sheep also have an increased concentration of RNA, greater mass of DNA, RNA and protein, increased RNA to DNA, and reduced protein to RNA in affected muscles at equivalent carcass weights compared to non-callipyge lambs, suggesting increased muscle satellite cell number and/or activity and greater transcriptional and translational capacity and/or efficiency in callipyge lambs.

Molecular regulation of the callipyge phenotype has been postulated to involve an important role for *Dlk1*, one of a cluster of paternally imprinted genes located in the region of the causative single-nucleotide polymorphism (SNP) on ovine chromosome 18 [120]. In support of their postulate, these authors reported spatial and temporal concordance of overexpression of *DLK1* with emergence of hypertrophy in callipyge-affected muscles. They also observed that the *DLK1* protein was associated with putative satellite cells during prenatal and early postnatal development preceding appearance of the callipyge phenotype at 12 weeks post partum. Microarray-based transcriptional profiling has identified numerous genes that are differentially expressed in *longissimus dorsi* muscle from callipyge lambs at birth and/or 12 weeks of age, of which eight, including *Dlk1*, were over- or underexpressed at both stages of development [116]. These findings were integrated into a proposed network of genes and histone epigenetic modifications thought to underpin the changes in muscle fibre type and hypertrophy characteristic of the callipyge phenotype, with *Dlk1* ascribed a primary effector role [116].

4.2.2 Ontogeny of Endocrine Regulation

Major differences in endocrine status are found when comparing the circulating hormone levels in the fetus with those in neonatal and adult animals. One of the main hormonal systems that undergo dramatic changes is the growth hormone-insulin like growth factor (GH-IGF) axis (for review see [22]). Although the circulating levels of GH in the fetus are high, prenatal or fetal growth has long been considered to be largely independent of the influence of GH. This is because the expression of the GH-receptor (GHR) is very low, resulting in a loss of GH sensitivity, particularly in relation to its stimulation of IGF-I expression in the liver. Hence it has been suggested that the main growth factor for fetal growth is IGF-II, which is GH-independent. Humoral IGF-I therefore appears to have little effect on fetal growth, being more important in postnatal, GH-dependent growth [22]. However, this view is now being questioned. Loss of either IGF-I or IGF-II via gene knockout technologies results in impaired fetal growth and growth retarded offspring at birth [25], while only the loss of IGF-I has any effect on postnatal growth. The growth-regulatory effects of both IGFs is via the IGF type 1 receptor (IGF1R) and knockout of the IGF1R gene results in extreme growth retardation and death. The role of IGF-I in postnatal growth has been investigated further, and loss of liver-specific IGF-I expression (via cre-lox technology) results in reduced circulating IGF-I levels, but has no effect on postnatal growth rates [25]. Interestingly, the loss of hepatic and, thereby, circulating IGF-I, results in increased plasma GH levels, presumably due to the loss of negative feedback of IGF-I on GH production by the pituitary, and also results in mice developing insulin resistance [125]. Hence, the role of hepatic-derived IGF-I in the fetus is being questioned and brings in the possibility that other tissue sources (e.g. adipose tissue or skeletal muscle) may either compensate for the lack of circulating IGF-I or may be more important contributors.

The apparent GH-independence of fetal growth might also be questioned, since this has normally been assumed because of the lack of GHR expression by the fetal liver, but we do not know what the levels of GHR expression are in many other tissues [22]. For example, hypophysectomy of fetal sheep retarded bone growth [93] and substantially increased fat deposition in late pregnancy

[109]. The latter response was readily abolished with physiological replacement of GH, implying that the late-gestation sheep fetus possesses functional GH receptors in adipose tissue. There is certainly plenty of GH and IGF-II in fetal blood, as well as some IGF-I. We have studied changes in IGF-I and IGF-II mRNA expression in developing skeletal muscle, both in vivo [40] and in vitro [19]. Both genes are expressed at low levels when myoblasts are proliferating. Expression of IGF-II increases and peaks at the same time as early markers of myoblast differentiation (myogenin or creatine kinase), suggesting that it might be involved in inducing early differentiation [19, 40]. In contrast, IGF-I expression increases slightly later than IGF-II and appears to be a consequence of muscle fibre formation rather than an inducer of it. Once myofibres are formed, the level of IGF-I expression remains relatively constant thereafter, including during postnatal life [15]. This skeletal muscle IGF-I expression may then have an impact on postnatal growth rates, since the numbers of muscle fibres correlates strongly with postnatal growth rates in a variety of breeds of pig: we have found reduced skeletal muscle IGF-I expression in the slower growing breeds of pig [16].

There is increasing evidence that the GH-IGF axis is affected by nutritional manipulation in a similar manner in both the fetus and the pregnant mother [7]. Hence, reduced levels of nutrition are associated with increased GH and decreased IGF-I in the blood of both, at least during mid-to-late pregnancy in sheep [21]. We have shown reduced expression of both GHR and IGF-I mRNA in mid-gestation fetal sheep liver after a period of reduced nutrition to the mother [20]. Hence, there appears to be a functional GH-IGF axis in fetal life; however major changes do take place during late fetal and early neonatal life that result in the switching from relative GH-independence to GH-dependence, which may be somewhat delayed in severely growth-retarded fetuses and newborns [58, 99–100]. The cortisol surge has been shown to result in a coordinated induction of GHR and IGF-I and a reduction in IGF-II expression in the fetal liver [22]. This may also involve thyroid hormones, since the cortisol surge also induces hepatic expression of the iodothyronine deiodinase activating enzymes. This also agrees with our in vitro studies that demonstrated stimulatory effects of T3 and dexamethasone on GHR mRNA, which resulted in an increased response of IGF-I to GH [14]. At the time of birth there are relatively low levels of GHR and IGF-I expression in the liver, while blood GH concentrations are high. There is then a rapid decrease in GH production by the pituitary and a gradual increase in both GHR and IGF-I expression by the liver. These gradual changes in hepatic GHR are matched by increased responsiveness in GH-stimulated IGF-I [22] associated with increased hepatic expression of the acid-labile subunit (ALS) [99] which, with IGFBP-3, is necessary for formation of the ternary binding complex that accounts for most of circulating IGF-1 in postnatal life [13]. Just prior to weaning in pigs, hepatic IGF-I expression starts to plateau [15], whereas GHR expression and plasma IGF-I levels continue to rise, the latter probably as a result of increased IGFBP-3 production by the liver, which increases the half-life of IGF-I in the blood.

4.2.3 Maternal Under- or Over-Nutrition at Different Stages of Pregnancy

The fetal origins hypothesis [5] initially was based on human epidemiological evidence that exposure to different levels of nutrition during fetal development might alter the risk of developing various chronic diseases in later life. This has resulted in numerous animal studies, mostly on rodents, investigating the effects of under- or over-feeding of pregnant animals on the resulting offspring [77]. Most studies have tended to measure gene expression in a variety of tissues, but mainly in young offspring. The magnitude and significance of any effect appears to depend upon the level of nutritional insult to the mother, the timing of that insult in terms of stage of development of the fetus and the age or stage of development at which the offspring are studied. It is only recently that studies have been carried out aimed at testing hypotheses for mechanisms relating to development of specific tissues. It appears that the closer to the timing of the insult the offspring are studied, the greater the likelihood of observing an effect, which is why most studies are in relatively young offspring. However, recent studies on livestock species have investigated longer-term effects of maternal diet on the offspring and resulted in fewer, much more subtle differences than previously expected, suggesting that real (i.e. long-term) fetal programming might be less of an issue. Examples of effects of maternal nutrition on development of muscle fibres and other cell types will be given in the next section.

4.2.4 Placental Insufficiency

Placental weight and associated capacity for maternal-fetal nutrient transfer are powerful determinants of fetal growth during late gestation in all species studied. This has been most persuasively demonstrated by controlled manipulation of placental size and/or functional capacity using pre-mating carunclectomy [1], heat-induced placental stunting [2], or uteroplacental vascular embolisation [32] in sheep. Natural variations in fetal weight due to varying litter size in prolific ewes are strongly correlated with placental mass per fetus [58, 98]. In addition, the quite profound growth retardation of fetuses in overfed, primiparous ewes has been attributed to a primary reduction in placental growth [117]. Placental weight and birth weight are also highly correlated in cattle [4, 38, 125].

The probably common aetiology of intrauterine growth retardation (IUGR) in experimentally-induced and natural cases of placental insufficiency is illustrated by the similar patterns of association between fetal and placental weights in pregnant ewes with varying conceptus weights due to carunclectomy, heat stress, litter size, and overfeeding of primiparous dams [9]. In each case, severe growth retardation was associated with chronic fetal hypoxaemia and hypoglycaemia during late gestation [8, 32, 63, 118]. A detailed assessment of influences on placental transport of nutrients is beyond the scope of this review, but is provided by Bell et al. [9].

Nutritional manipulations during early-mid pregnancy have produced variable changes in placental growth and structure (for reviewed see [75]). For example, severe undernutrition of ewes from mating to day 90 of gestation caused placental and fetal stunting at day 90 [39] while feeding 50–60% of ME requirements between days 28 and 80 resulted in restricted placentome growth and a smaller placenta, but had no effect on fetal weight at either mid- or late-gestation [29]. On the other hand, moderate underfeeding during early-mid pregnancy increased [42] or had little effect [41] on placental growth. In addition to effects of severity of undernutrition, these discrepancies in response may be reconciled by variations in the dam's body condition at mating and capacity to later draw on body reserves as a compensatory mechanism to protect or enhance placental growth [91].

4.3 Prenatal Effects on Postnatal Tissue Growth and Development

4.3.1 Effects on Muscle Fibre Development

Nutrition of the pregnant mother has been shown to affect the numbers of secondary fibres in the muscles of resulting offspring in a variety of species, including rats, guinea pigs, pigs and sheep [21]. The variability in response is thought to be dependent upon the timing and magnitude of the nutritional insult, as well as the age at which the offspring are studied. A 70% maternal nutrient restriction (relative to controls) in sheep from 30 days prior to breeding until day 100 of gestation tended to reduce numbers of muscle fibres (at 140 days of gestation), but this putative effect did not reach statistical significance [95]. A study comparing muscles of single or twin lambs and lambs born in the spring or autumn [90] indicated possible effects of nutrition. Single lambs born in the autumn had reduced numbers and cross-sectional area of fibres in *semitendinosus* muscle compared with spring born lambs, but there was no effect in *plantaris* muscle [90]. The comparison of singles and twins indicated no effects on numbers of fibres, but reduced cross-sectional area of fibres in *semitendinosus*, *plantaris* and *gastrocnemius* muscles from twins [90]. This may relate to a greater degree of undernutrition in the twins during late gestation, after muscle fibre formation, but when fibre hypertrophy is taking place.

More severe, chronic fetal growth retardation commencing during mid to late gestation as a result of placental insufficiency results in profound differences in muscle weights of late fetal and newborn lambs and down-regulates muscle satellite cell activity and accumulation of myonuclei and muscle DNA (Figs. 4.3 and 4.4) [56–57]. Similarly, severe acute maternal nutritional restriction also reduces fetal muscle satellite cell activity [56]. By contrast, significant effects on number of muscle fibres in various anatomical muscles were not observed in these lambs, consistent with the notion that severe nutritional restriction during early to mid gestation is required to reduce myofibre number (see below). During postnatal growth,

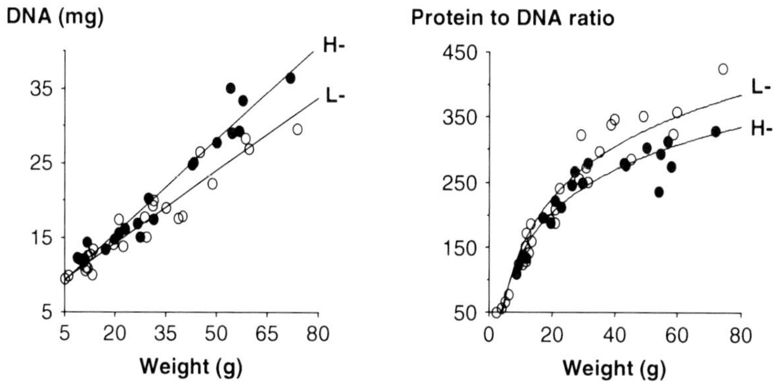

Fig. 4.3 Accretion of DNA and ratio of protein to DNA in normal (H-, *solid circles*, 4.8 kg) and low (L-, *open circles*, 2.2 kg) birth weight sheep *semitendinosus* muscle during postnatal growth to 20 kg live weight, demonstrating reduced muscle cellularity in the severely growth-retarded newborns [from 57]. Results are presented relative to weight of *M. semitendinosus*

the amount of DNA was reduced in the IUGR compared to normal lambs (Fig. 4.3), and the ratio of protein to DNA, indicative of cell size in syncytial tissue such as muscle, was greater in the IUGR lambs [57] which were also fatter at equivalent postnatal live weights [55].

Hence, it appears that a prolonged delay in the commencement of true muscle hypertrophy, as indicated by a rapid increase in the ratio of protein to DNA which commences around day 115 of gestation in normal fetal sheep (Fig. 4.4) [56], may have adverse postnatal consequences for muscle growth potential.

On the other hand, in IUGR cattle grown to 30 months of age within a pasture-based system, at equivalent weights during postnatal life differences in muscle mass were not evident [54, 60]. This finding suggests that activity of muscle satellite cells

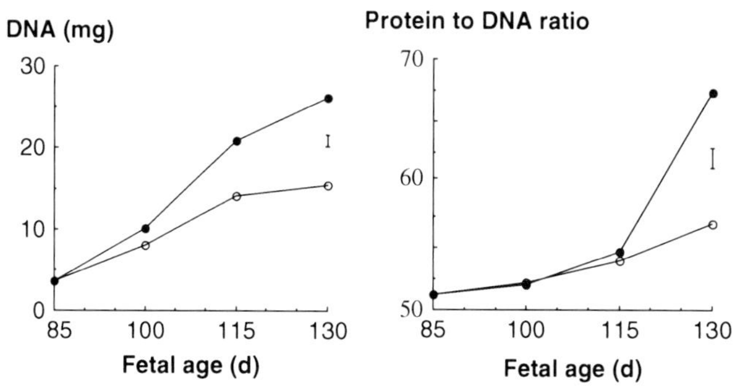

Fig. 4.4 DNA accretion and ratio of protein to DNA in normal (*solid circles*) and growth-retarded (*open circles*) fetal sheep *semimembranosus* muscle during growth from 85 to 130 days of pregnancy, demonstrating the commencement of muscle hypertrophy at ~115 days of age [from 56]

may help animals attain normal, or near normal, muscle mass and body composition, albeit at older ages.

A few studies have investigated the effects of manipulations during the periconception period and very early pregnancy on muscle development in lambs. A dramatic increase in numbers and sizes of muscle fibres and an increase in the secondary to primary myofibre ratio has been demonstrated in late gestation fetal lambs that underwent short-term embryo culture prior to implantation [88]. Embryo culture in serum-supplemented media appears to significantly stimulate growth and this growth advantage is retained throughout gestation. Singleton sheep fetuses (day 75 of gestation) resulting from superovulated donor ewes which had been fed at high or low intake (150% or 50% of maintenance) from 18 days before until 6 days after ovulation showed no differences in the numbers of primary muscle fibres formed, but the numbers of secondaries and the secondary to primary fibre ratio were increased in the fetuses of high intake ewes [97].

Studies at Nottingham set out to determine the timing for major muscle fibre formation in the fetal sheep and then test the hypothesis that maternal undernutrition during the myoblast proliferation stage, immediately before differentiation and the period of major fibre formation, would reduce the numbers of fibres formed. The first study [40] demonstrated a peak of myogenin and IGF-II mRNA expression at day 85 of gestation (Fig. 4.5) which, together with the histochemical analyses, indicated that the majority of fibres in the leg muscles were being formed at this time. This would agree with previous sheep studies (Table 4.1), suggesting that secondary/tertiary fibres are the major fibres being formed at this time. We then tested the hypothesis by comparing offspring from adequately-fed, control ewes with those from ewes that were nutrient restricted (50% of intake of controls) before (days 30–70), during (days 55–95) or after (days 85–115) the peak for fibre formation.

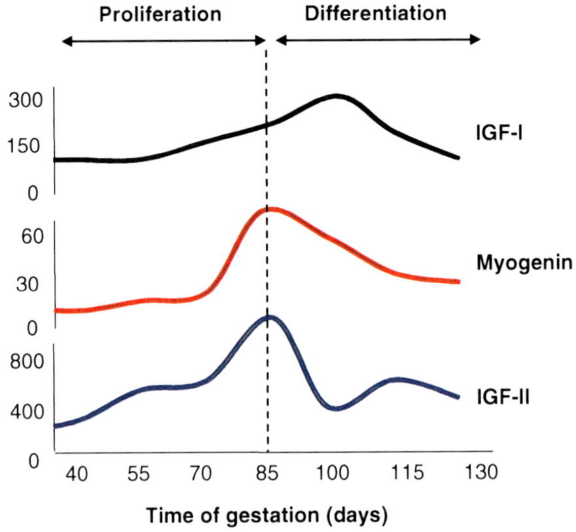

Fig. 4.5 Changes in gene expression during fetal sheep muscle development [data from 40]. The peaks in myogenin and IGF-II mRNA expression indicate when the majority of muscle fibres are being formed and were confirmed by histochemical analyses

In agreement with our hypothesis, reduced proportions of fast fibres and increased proportions of slow fibres, indicating a reduction in the number of secondary fibres, were observed only in neonatal lambs (14 days post partum) whose mothers were restricted immediately prior to major fibre formation (days 30–75) [41]. The interpretation of a reduction in the number of fast fibres is due to the combination of a reduction in proportion and increase in size of fast fibres, together with an increase in proportion but no change in size of slow fibres. A similar 50% maternal nutrient restriction [126] over a similar time period (days 28–78) also has been shown to reduce the numbers of secondary fibres in fetal sheep muscle at day 78 of gestation. We have subsequently [87] tested our hypothesis further in young rats in which the timing for fibre formation is very different to that for other mammals (Table 4.1). In agreement with our hypothesis, the timing for the nutritional insult (maternal low protein intake in this study) that mainly resulted in reduced secondary fibre formation at weaning was days 8–14 (mid-gestation) of the 22 day gestation [87], immediately before secondary fibre formation, which starts around days 17–19 in the rat (Table 4.1). Interestingly, there were no effects of feeding a maternal low protein diet throughout pregnancy [87]. However, there were differences between muscles in the fibre types affected, since the total number and density of fast and glycolytic fibres were reduced in *soleus* muscle; whereas it was the density of oxidative fibres that was reduced in the *gastrocnemius* muscle. We hypothesized that the muscle adapts such that any changes relate to the fibre types that are least important for the function of the muscle being studied. In both the sheep and rat studies at Nottingham, the magnitude of the effect differed between different muscles, indicating that different muscles develop at slightly different times and therefore the timing of when they are susceptible to nutritional manipulation is also slightly different. All these observed effects of nutrition are dependent upon the nutritional manipulation taking place prior to major fibre formation, which differs between species (Table 4.1). These effects are not normally associated with changes in muscle weights, which tend to be affected by nutritional manipulations at later stages of gestation, when fetal growth rate is greatest.

Recent studies have investigated the long-term consequences of maternal nutritional manipulations on muscle development in sheep. Despite the fact that previous studies targeting the period immediately before major fibre formation have shown effects on muscle fibre development in fetal [126] or neonatal lambs [41], when sheep are grown to conventional slaughter weights (40–45 kg or 17 weeks of age [34]) or beyond (24 wks [34] or 280 days [127]) those differences are lost and even a slight increase in proportions of fast glycolytic (IIB) fibres are observed, especially in older animals (24 weeks [34]; and 280 days [127]). Although the plasticity of muscle fibre type composition appears to be able to overcome the early effects of targeted maternal nutrient restriction on muscle fibre development, there may be long term effects on postnatal growth and body composition, that may or may not be related to the muscle development (see later).

Further to the above findings in sheep, *longissimus* myofibre characteristics of low birth weight cattle (28.6 vs. 38.8 kg) from cows severely undernourished from early gestation to parturition, differed at birth (reduced percentage of type 1, increased percentage of type 2A, and smaller cross-sectional area of type 2X myofibres) compared to high birth weight cattle (38.8 kg) from well-nourished dams. However, there was no apparent effect of prenatal growth on myofibre number, and effects on other myofibre characteristics due to prenatal growth were not evident at weaning at 7 months of age or at 30 months of age [54]. These findings were consistent with a general lack of effect of prenatal growth of cattle on carcass and yield characteristics at equivalent carcass weights (380 kg), or on meat quality characteristics [54, 60], although the low birth weight cattle were significantly smaller (647 vs. 703 kg live weight) at 30 months of age.

4.3.2 Effects on Nephron Number

As for numbers of muscle fibres, the numbers of nephrons within the kidneys are thought to be fixed around the time of birth. Interestingly, sheep studies also indicated that the numbers of nephrons in the kidneys of 4 week old sheep were reduced when the pregnant mothers were undernourished (50% of control intake) from days 30 to 75 of gestation, but not at other times [79]. Studies in rats to which a maternal low protein diet was fed at different stages of pregnancy have shown the critical timing for effects on nephron numbers is mid-late gestation [78], suggesting that the critical timing for nephrogenesis is very similar to that for myogenesis in both sheep and rats.

4.3.3 Effects on Adipose Tissue Development

Unlike muscle fibres and nephrons, there is no evidence that the number of adipocytes is fixed at any stage of life. This is consistent with the relatively unlimited capacity of adipose tissue to store fat during periods of energy excess and the need to prevent toxic accumulation of non-esterified fatty acids (NEFA) in the circulation. However, a number of studies have shown effects of maternal nutrition on amounts of adipose tissue in the resulting offspring, mainly in younger animals. Although the mechanisms for the effects are far from clear, studies in rats have shown that either undernutrition throughout pregnancy or overnutrition during late pregnancy result in increased adiposity of resulting offspring compared to ad libitum fed controls [44]. Malnutrition (30% of controls) throughout pregnancy in rats, followed by adequate nourishment before weaning, also has been shown to result in increased adiposity of offspring, particularly when they were exposed to a hypercaloric diet after weaning [112]. This is at least partly due to the malnourished offspring being hyperphagic, which might indicate an effect on development of the hypothalamus and its major role in the regulation of appetite and energy expenditure (see later).

In sheep severely growth restricted before birth due to placental insufficiency, relative feed intake during the first 2 weeks after birth exceeds that of normal birth weight sheep, as measured during artificial rearing under ideal conditions [55].

These lambs also exhibited relative immaturity of the GH-IGF system during the periparturient period [59, 99–100], and had lower maintenance energy requirements but used energy less efficiently for tissue deposition during the initial weeks following birth compared to normal newborns [55]. Despite their slower absolute rates of growth during the first 2 weeks post-partum, the low birth weight lambs had greater rates of fat deposition compared to normal lambs, and remained fatter at any given liveweight up to 20 kg [55]. Similar increases in fatness also have resulted at heavier weights due to placental insufficiency and low birth weight in lambs [83, 114].

The protein content of the maternal diet (or possibly the balance of nutrients) appears to be important, since prenatal exposure to a high (40%) protein diet (compared to an adequate, 20% protein diet) throughout pregnancy results in increased adiposity of young rat offspring [33]. This was associated with a reduction in energy expenditure and not food intake, since the rats were pair-fed. Rats subjected to a mild protein restriction (9% protein vs. 18% protein in controls) throughout gestation exhibit increased central fat deposition as adults [10–11] and this is associated with changes in food preference, with early gestation appearing to be a critical period.

Some studies in ruminants have investigated the effects of maternal nutrition on adiposity of offspring, with most being in fetal or neonatal lambs. A 70% nutrient restriction of ewes from 30 days before breeding until 100 days gestation had no effect on any of the measures of adiposity used at slaughter (58.5 kg) of the progeny [95]. A greater nutrient restriction (50% of control) throughout pregnancy increased perirenal adipose tissue weights at 110 days gestation [53], with no effect on fetal or placental weights. There appears to be a marked difference relating to the timing of the restriction, since undernutrition (50% of control) during early-to-mid gestation (28–80 days gestation) results in increased perirenal adipose tissue weights at 145 days gestation [12], whereas a 50% restriction during late gestation (from 115 days gestation) results in decreased perirenal adipose tissue weights at the same age [24]. However this reduction of fat with reduced nutrition in late pregnancy may simply be because all the energy is utilised for fetal growth rather than being deposited in adipose tissue.

A few studies have investigated effects of maternal undernutrition on adipose tissue development in older offspring. Studies at Nottingham showed that a 50% maternal nutrient restriction in early gestation (days 30–70) resulted in small but significant increases in intramuscular fat in *semitendinosus* and *longissimus dorsi* muscles at 24 weeks of age, particularly in male lambs [34]. The mean diameter of adipocytes in perirenal fat was also increased and the weight of perirenal fat (relative to body weight) tended to be increased at 24 weeks of age [34]. In general, small non-significant increases in many of the measures of adiposity were observed in the nutrient restricted offspring, particularly in the older (24 week old) lambs. Similar findings have been described by the Wyoming group, with the triglyceride content of the *longissimus dorsi* [127] and the weight of kidney and pelvic fat [46] being higher in maternal nutrient restricted wether lambs at 280 days. As in the Nottingham studies, most measures of adiposity tended to be higher in nutrient restricted lambs, although the effects were not always statistically significant. Hence, undernutrition during early pregnancy in sheep appears to be associated with long-term increases in adiposity of the offspring. The limited data available appears to suggest that the older the animals are, the greater the effect. This is the same critical period for muscle fibre development (and nephron development) and therefore the question of whether the effects are inter-related needs further investigation. One possible explanation is that a reduction in the numbers of muscle fibres formed might result in a small reduction in basal metabolic rate and thereby result in the excess energy needing to be stored and therefore more or larger adipocytes being formed. This would be predicted to accumulate with age and the magnitude of the effect to be greater in older animals.

4.3.4 Effects on Hypothalamic Cell (Neuron) Development

Work mainly in rodents has appeared to indicate effects of prenatal nutritional status on the regulation of appetite, possibly via effects on neuronal development in the hypothalamus. Severe malnutrition (30% of controls) throughout pregnancy in rats, followed by adequate nourishment before weaning, resulted in increased food intake (hyperphagia) of offspring, particularly when they were exposed to a hypercaloric diet after weaning [112]. Exposure to a maternal low protein diet (9% vs. 18% dietary protein) throughout pregnancy increased energy intake in rats at 12 weeks of age, due to an increased preference for high fat and reduced preference for high carbohydrate foods, but this effect was lost by 30 weeks of age [10]. Induction of gestational diabetes by the administration of a single dose of streptozotocin to rats on day 1 of pregnancy results in maternal and fetal hyperglycaemia, until countered by the transplantation of pancreatic islets into the mothers on day 15 of pregnancy [47]. Offspring from sham-transplanted diabetic mothers (i.e. hyperglycaemic throughout pregnancy) had reduced body weights and body lengths at 21 days of age, despite no differences in birth weight, and developed what the authors described as "hypothalamic malorganisation" [47]. This included increased immunostaining for the orexigenic neuropeptides, neuropeptide Y (NPY) and agouti-related peptide (AgRP), decreased immunostaining for the anorexigenic neuropeptide, alpha-melanocyte stimulating hormone (MSH) and no change in the MSH precursor protein, proopiomelanocortin (POMC). Whether these changes really represent changes in development of neuropeptide-specific neurons or simply reflect a response to reduced body weight and/or food intake is unclear.

A study in sheep [94] investigated effects of a high level of nutrition (160% vs. 100% of maintenance) during late gestation (day 115 to parturition) on adiposity and hypothalamic gene expression in 30 day old offspring. The-well fed offspring had similar birth weights and body weights at 30 days, but increased weights of subcutaneous and perirenal fat, particularly in females. This was associated with increased relative milk intake during the first 3 of the 4 weeks before slaughter, although there was no difference in the last week before slaughter. Analyses of hypothalamic expression of neuropeptides at 30 days showed an increased expression of the anorexigenic precursor, POMC, but no effect on the orexigenic neuropeptides, cocaine and amphetamine regulated transcript (CART), NPY and AgRP. Once again, whether these changes really represent changes in development of neuropeptide-specific neurons or simply reflect a response to the increased fat mass or previously increased food (milk) intake, needs further study.

4.3.5 Effects on Pituitary Cell and Liver Development

Alterations in the GH-IGF axis have been suggested as a mechanism for the long-term effects of IUGR or prenatal environment on reduced growth rates [84] and therefore a few studies have investigated effects on cells expressing components of the GH-IGF axis. Somatotropes are the cells within the pituitary that produce GH and therefore are important in regulating growth. At 135 days of gestation, fetal lambs exposed to a 50% maternal nutrient restriction from days 28 to 78 of gestation were found to have reduced densities and proportions of somatotropes relative to controls [84], but there was no effect on the density of all pituitary cells. This suggests a decrease in the proportion of pituitary cells that differentiated into somatotropes (or an increase in differentiation into another cell type), rather than an effect on pituitary cell proliferation per se. The same group [85] then demonstrated that this effect was not due to differences in the numbers of hypothalamic neurons expressing the GH-stimulatory factor, growth hormone releasing hormone (GHRH), nor on the density and percentage of somatotropes expressing the GHRH-receptor. The GH produced by the somatotropes regulates circulating IGF-I levels by binding to GH-receptors present on hepatocytes in the liver and stimulating IGF-I expression and secretion [14]. Liver weights and hepatic expression of GHR, but not IGF-I and IGF-II mRNA, were reduced in 3 year old offspring from ewes nutrient restricted (50% of maintenance) from conception to day 95 of gestation [68]. However, a similar nutrient restriction (50% from days 30 to 70 or days 30 to 85) had no effect on liver weights in younger lambs (24 and 17 weeks respectively [34]).

4.3.6 Effects on Feed Intake and Growth Patterns in Sheep

It has long been recognised that total nutrient availability during the pre-weaning and post-weaning periods are highly correlated with nutrient intake, digestibility and efficiency of nutrient utilisation. However, the extent to which the prenatal environment and variation in fetal growth affects these parameters has received relatively little attention, with most studies focussing on consequences of birth weight until relatively recently, as detailed below.

Positive weight balance and a net gain in weight takes longer to achieve in low birth weight lambs (2.3 kg) compared to those of normal birth weight (4.8 kg) immediately postpartum, even when reared in an optimal environment [55]. During the early postpartum period, digestibility of dry matter, organic matter, energy, crude protein, calcium and inorganic phosphorus is positively correlated with birth weight [66]. Birth weight is also correlated with digestibility of C16:0 and C18:0 fatty acids, suggesting lipids comprising medium-chain fatty acids may be more suitable for low birth weight lambs than those with long-chain fatty acids [66]. However, amounts and activity of digestive enzymes were not strongly associated with birth weight in lambs or calves, although there were some associations [62].

Relative feed intake was greater by about 20% (45 g/kg live weight per day) for low birth weight compared to normal birth weight lambs ad libitum fed during the early postpartum period, and remained higher until about 25 days of age [55]. However, at equivalent live weights during ad libitum or restricted feeding to 20 kg live weight, there were no differences due to birth weight in relative feed intake [55]. In the studies of Villette and Theriez [113], relative intake was not affected significantly by birth weight during the first five weeks of postnatal life, although was positively correlated with birth weight during week six of postnatal life. Relative intake did not differ due to birth weight in lambs reared on their dams during the first week of postnatal life [115].

Low birth weight lambs had lower feed to gain ratios (higher efficiency), overall, during milk-feeding to 20 kg live weight than normal birth weight lambs. However, this was due, at least in part, to differences in efficiency during the early postpartum period when the ratio of gain to maintenance was higher in the small compared to normal size lambs. Feed to gain ratios differed little from 10 to 20 kg live weight [55]. Retention of ingested dry matter, nitrogen, lipid, ash and energy from birth to 20 kg live weight did not differ between low and normal birth weight lambs, although slowly reared low birth weight lambs tended to deposit more fat per unit of fat ingested than the normal lambs [55]. Feed efficiency did not differ with birth weight during the first week of life during rearing on the dam [115] or from birth to weaning during artificial rearing [113].

Estimated maintenance energy requirements are lower from birth to 10 kg live weight in very low compared to normal birth weight lambs, however, no difference in maintenance energy requirements due to birth weight were apparent by 20 kg [55]. In contrast, low birth weight lambs utilised energy less efficiently for protein and fat deposition to 10 kg live weight although, again, differences due to birth weight were no longer evident by 20 kg [55].

During the post-weaning period, birth weight was not related to feed intake and feed efficiency to 35 kg live weight [113]. Furthermore, voluntary feed intake from weaning to 2 years of age was not affected by moderate maternal nutrient restriction at 0.7–0.8 of estimated energy requirements during the final 6 weeks of pregnancy that resulted in reduced birth weights (3.2 vs.3.5 kg) compared to ewes fed to requirements [104].

Recent studies in sheep indicate that the maternal nutrient restriction during early gestation, aimed at altering muscle fibre development, had variable effects on rates of growth, food intake and/or feed efficiency. In a study by the Wyoming group [46] male lambs (wethers) from nutrient restricted ewes (days 28–78 of gestation) were heavier than controls from around 16 weeks of age through to slaughter at 280 days, with body weights increasing with age in parallel. This is despite no differences being observed in birth weights or weight at 8 weeks of age [46]. No data for food intake or feed efficiency are included [46], so the mechanism for this very early difference in growth rates, which results in increased body weight in nutrient restricted wethers at 35 weeks of age, is unclear, although may be due to some imbalance in litter size between experimental groups. Similar studies at Nottingham in which only twin lambs were studied are in contrast to the Wyoming study. At 24 weeks of age, both male and female lambs exposed to maternal nutrient restriction (50%) during early pregnancy (days 30–70) weighed less at slaughter, despite no difference from controls in birth weights. The nutrient restricted lambs therefore had reduced growth rates [34], but there were no significant differences in feed intake or feed efficiency between 12 and 24 weeks of age [34], suggesting the effects must be much earlier. In contrast, a similar study in lambs grown to 17 weeks showed no differences in birth or slaughter weights, growth rates, feed intake or feed efficiency [34]. Thus the effects on growth rates are variable, but seem to suggest that the neonatal period before weaning is an important time if any effects are to be observed. Whether this relates to nutrition of the ewe during lactation and/or milk composition is unclear.

In cattle, differences in feed efficiency and net feed intake were not evident in feedlot at 26–30 months of age, following divergent growth during gestation resulting in low and high birth weights nor were there interactions between prenatal growth and pre-weaning growth or sire-genotype [54]. The low birth weight cattle which were smaller at feedlot entry than their high birth weight counterparts consumed less feed, although when the data were adjusted for differences in feedlot entry weight, no difference due to birth weight was evident. This latter finding is consistent with studies on twin cattle which tended to consume less feed in feedlot than singletons, due primarily to their lower live weight, [35]. Similarly, provision of supplement to cows for three months prepartum had no significant post-weaning effects on ADG, feed intake and feed efficiency in steers [108] or heifers [89] that were individually fed following weaning, although the heifers of supplemented cows tended to have greater absolute and residual feed intakes during individual feeding for 84 days post-weaning.

4.4 Physiological Mechanisms

Throughout the review we have tried to identify mechanisms that might be involved in the responses described. Studies at Nottingham [34, 40–41] have obviously related to the hypothesis that prenatal effects on muscle fibre development are via the regulation of muscle cell proliferation and/or differentiation. Hence factors that increase myoblast proliferation and/or inhibit or delay differentiation are predicted to result in increased formation of muscle fibres. However, this will be limited by the genetic potential of the animal, so that there will be a maximum number of fibres that can be formed related to the genotype of the animal. In contrast, factors that decrease myoblast proliferation and/or stimulate or induce early differentiation are predicted to result in decreased formation of muscle fibres.

There are a number of hormones and growth factors known to regulate myoblast proliferation and differentiation (for reviews see [6, 17, 21, 26, 88, 96]). Importantly, myostatin is a known inhibitor of muscle cell proliferation and differentiation; whereas the IGFs (IGF-I and -II) stimulate both proliferation and differentiation [45]. Recent studies indicate that the adipocyte produced hormone, leptin, stimulates muscle cell proliferation at physiological concentrations (Brameld et al., unpublished data) and insulin is known to have similar effects to the IGFs, albeit at much higher concentrations [17]. Nutrients have also been shown to directly regulate muscle cell differentiation, with vitamin A (retinoic acid) stimulating differentiation [17, 26] and unsaturated fatty acids stimulating (oleic, linoleic and c9,t11 CLA) or inhibiting (t10, c12 CLA) differentiation [3, 67]. Another mechanism whereby nutrition might regulate muscle cell proliferation and differentiation, particularly the carbohydrate and protein components of the diet, is via their effects on the GH-IGF axis. It has long been recognised that high levels of dietary energy and protein increase growth. Molecular mechanisms for these effects have been identified, including stimulatory effects of glucose and specific amino acids on expression of GH-receptor and IGF-I mRNA respectively in cultured pig hepatocytes [14, 18]. This work was carried out in the pig, in which dietary manipulation is easier, but similar studies using cultured sheep hepatocytes have also shown stimulatory effects of amino acids on IGF-I expression [111, 119], suggesting similar mechanisms.

The mechanisms responsible for the prenatal regulation of development of other tissues by maternal nutritional status are not as well understood, although the critical times during gestation appear to be very similar to those for myogenesis. This suggests that the factors that regulate proliferation and differentiation of those cell types might be similar to those that regulate muscle cells. In relation to long-term effects on adipose tissue development, the effects of prenatal environment are likely to be indirect via effects on development of tissues that play a key role in regulating appetite or energy expenditure. The latter could even relate to the observed effects on muscle.

4.5 Conclusions

We have summarised compelling evidence for prenatal effects on early postnatal development of tissues in ruminants, particularly sheep, with emphasis on muscle fibres and amounts of body fat in young offspring. Nevertheless, the research to date suggests that, given enough time, the animal is able to overcome or compensate for most of these early differences, resulting in only small (if any) residual effects on body composition at later stages of growth. Postnatal plane of nutrition may be an important influencing factor in later muting of prenatal and early postnatal carryover effects on carcass tissue growth and development. Most studies thus far of prenatal effects on older offspring have been on animals that were well fed during postnatal development. However, we have demonstrated that adverse effects of prenatal growth restriction of lambs on postnatal muscle mass at the same weaning weight are exacerbated by restricted nutrition from birth to weaning [55], and have observed additive effects of prenatal and pre-weaning nutrition and growth on subsequent live and carcass weights and meat yield characteristics in cattle [54, 60]. Therefore, there is a need for further investigation of the extent to which an animal can compensate later in life following variable maternal nutrition and/or growth during prenatal life if subjected to additional nutritional insults or other stresses during postnatal life. However, since most ruminants are slaughtered for meat at relatively young ages, there appear to be few serious consequences of concern to the meat producer or processor. Hence, such studies may be of more importance for breeding livestock that remain in the flock or herd for longer periods and as models for animals with longer lives, particularly humans.

References

1. Alexander, G. 1964. Studies of the placenta of the sheep (*Ovis aries* L.): effect of surgical reduction in the number of caruncles. *J. Reprod. Fertil.* **7**:307–322.
2. Alexander, G. and D. Williams. 1971. Heat stress and development of the conceptus in domestic sheep. *J. Agric. Sci., Camb.* **76**:53–72.
3. Allen, R.E., L.S. Luiten, and M.V. Dodson. 1985. Effect of insulin and linoleic acid on satellite cell differentiation. *J. Anim. Sci.* **60**:1571–1579.
4. Anthony, R.V., R.A. Bellows, R.E. Short, R.B Staigmiller, C.C. Kaltenbach, and T.G. Dunn. 1986. Fetal growth of beef calves. II. Effects of sire on prenatal development of the calf and related placental characteristics. *J. Anim. Sci.* **62**:1375–1387.
5. Barker, D.J.P. (2004) The developmental origins of chronic adult disease. *Acta Paediatrica Suppl.* **446**:26–33.
6. Bass, J.J., M. Sharma, J. Oldham, and R. Kambadur. 2000. Muscle growth and genetic regulation, p. 227–236. *In* P.B. Cronje (ed.), Ruminant Physiology: Digestion, Metabolism, Growth and Reproduction, CABI Publishing, Wallingford.
7. Bauer, M.K., B.H. Breier, J.E. Harding, J.D. Veldhuis, and P.D. Gluckman. 1995. The fetal somatotropic axis during long term maternal undernutrition in sheep: evidence for nutritional regulation in utero. *Endocrinology* **136**:1250–1257.
8. Bell, A.W., R.B. Wilkening, and G. Meschia. 1987. Some aspects of placental function in chronically heat-stressed ewes. *J. Dev. Physiol.* **9**:17–29.
9. Bell, A.W., P.L. Greenwood, and R.A. Ehrhardt. 2005. Regulation of metabolism and growth during prenatal life, p. 3–34. *In* D.G. Burrin, and H.J. Mersmann (eds.), Biology of Metabolism in Growing Animals, Elsevier, Amsterdam.
10. Bellinger, L., C. Lilley, and S.C. Langley-Evans. 2004. Prenatal exposure to a maternal low-protein diet programmes a preference for high-fat foods in the young adult rat. *Br. J. Nutr.* **92**:513–520.
11. Bellinger, L., D.V. Sculley, and S.C. Langley-Evans. 2006. Exposure to undernutrition in fetal life determines fat distribution, locomotor activity and food intake in ageing rats. *Int. J. Obesity* **30**:729–738.
12. Bispham, J., J. Dandrea, A. Mostyn, J.M. Brameld, P.J. Buttery, T. Stephenson, and M.E. Symonds. 2002. Impact of maternal nutrient restriction in early to mid gestation on leptin, insulin-like growth factor-I (IGF-I) and growth hormone receptor (GHR) mRNA abundance in adipose tissue of the fetal lamb. *Early Human Dev.* **68**: 135–136.

13. Boisclair, Y.R., R.P. Rhoads, I. Ueki, J. Wang, and G.T. Ooi. 2001. The acid-labile subunit (ALS) of the 150 kDa IGF-binidng protein complex: an important but forgotten component of the circulating IGF system. *J. Endocrinol.* **170:**63–70.
14. Brameld, J.M. 1997. Molecular mechanisms involved in the nutritional and hormonal regulation of growth in pigs. *Proc. Nutr. Soc.* **56:**607–619.
15. Brameld, J.M., P.A. Weller, J.M. Pell, P.J. Buttery, and R.S. Gilmour. 1995. Ontogenic study of insulin-like growth factor-I and growth hormone receptor mRNA expression in porcine liver and skeletal muscle. *Anim. Sci.* **61**:333–339.
16. Brameld, J.M., J.L. Atkinson, T.J. Budd, J.C. Saunders, J.M. Pell, A.M. Salter, R.S. Gilmour, and P.J. Buttery. 1996. Expression of insulin-like growth factor-I (IGF-I) and growth hormone receptor (GHR) mRNA in liver, skeletal muscle and adipose tissue of different breeds of pig. *Anim. Sci.* **62:**555–559.
17. Brameld, J.M., P.J. Buttery, J.M. Dawson, and J.M.M. Harper. 1998. Nutritional and hormonal control of skeletal muscle cell growth and differentiation. *Proc. Nutr. Soc.* **57:** 207–217.
18. Brameld, J.M., R.S. Gilmour, and P.J. Buttery. 1999. Glucose and amino acids interact with hormones to control expression of insulin-like growth factor-I (IGF-I) and growth hormone receptor (GHR) mRNA by cultured pig hepatocytes. *J. Nutr.* **129:**1298–1306.
19. Brameld, J.M., H. Smail, N. Imram, N. Millard, and P.J. Buttery. 1999. Changes in expression of IGF-I, IGF-II and GH-receptor (GHR) mRNA during differentiation of cultured muscle cells derived from adult and fetal sheep. *S. Afr. J. Anim. Sci.* (ISRP) **29:**307–310.
20. Brameld, J.M., A. Mostyn, J. Dandrea, T.J. Stephenson, J.M. Dawson, P.J. Buttery, and M.E. Symonds. 2000. Maternal nutrition alters expression of insulin-like growth factors in fetal sheep liver and skeletal muscle. *J. Endocrinology* **167**: 429–437.
21. Brameld, J.M., A.J. Fahey, S.C. Langley-Evans, and P.J. Buttery. 2003 Nutritional and hormonal control of muscle growth and fat deposition. *Arch. Anim. Breed.* (special issue) **46:**143–156.
22. Breier, B.H., M.H. Oliver, and B.W. Gallaher. 2000. Regulation of growth and metabolism during postnatal development, p. 187–204. *In* P.B. Cronje (ed.), Ruminant Physiology: Digestion, Metabolism, Growth and Reproduction, CABI Publishing, Wallingford.
23. Brun, R.P., J.B. Kim, E. Hu, S. Altiok, and B.M. Spiegelman. 1996. Adipocyte differentiation: a transcriptional regulatory cascade. *Curr. Opin. Cell Biol.* **8:**826–832.
24. Budge, H., A. Bryce, J.A. Owens, T. Stephenson, M.E. Symonds, and I.C. McMillen. 2002. Differential effects of nutrient restriction in late gestation and placental restriction throughout gestation on uncoupling protein 1 expression in fetal perirenal adipose tissue. *Early Human Dev.* **66**:43.
25. Butler, A.A. and D. LeRoith. 2001. Minireview: Tissue-specific versus generalized gene targeting of the igf1 and igf1r genes and their roles in insulin-like growth factor physiology. *Endocrinology* **142:**1685–1688.
26. Buttery, P.J., J.M. Brameld, and J.M. Dawson. 2000. Control and manipulation of hyperplasia and hypertrophy in muscle tissue, p. 237–254. *In* P.B. Cronje (ed.), Ruminant Physiology: Digestion, Metabolism, Growth and Reproduction, CABI Publishing, Wallingford.
27. Carpenter, C.E., D.R. Owen, N.E. Cockett, and G.D. Snowder. 1996. Histology and composition of muscles from normal and Callipyge lambs. *J. Anim. Sci.* **74:**388–393.
28. Chang, KC. 2007. Key signalling factors and pathways in the molecular determination of skeletal muscle phenotype. *Animal* **1:**681–698.
29. Clarke, L., L. Heasman, D.T. Juniper, and M.E. Symonds. 1998. Maternal nutrition in early-mid gestation and placental size in sheep. *Br. J. Nutr.* **79:**359–364.
30. Cockett, N.E., S.P. Jackson, T.L. Shay, D. Nielson, S.S. Moore, M.R. Steele, W. Barendse, R.D. Green, and M. Georges. 1994. Chromosomal localization of the callipyge gene in sheep (*Ovis aries*) using bovine DNA markers. *Proc. Natl. Acad. Sci. USA* **91:**3019–3023.
31. Cockett, N.E., S.P. Jackson, T.L. Shay, F. Famir, S. Bergmans, G.D. Snowder, D.M. Nielson, and M. Georges. 1996. Polar overdominance at the ovine Callipyge locus. *Science* **273:** 236–238.

32. Creasy, R.K., C.T. Barrett, M. De Swiet, K.V. Kahanpää, and A.M. Rudolph. 1972. Experimental intrauterine growth retardation in the sheep. *Am. J. Obstet. Gynecol.* **112:**566–573.
33. Daenzer, M., S. Ortmann, S. Klaus, and C.C. Metges. 2002. Prenatal high protein exposure decreases energy expenditure and increases adiposity in young rats. *J. Nutr.* **132:**142–144.
34. Daniel, Z.C.T.R., J.M. Brameld, J. Craigon, N. Scollan, and P.J. Buttery. 2007. Effect of maternal dietary restriction on lamb carcass characteristics and muscle fiber composition. *J. Anim. Sci.* **85:**1565–1576.
35. de Rose, E.P. and J.W. Wilton. 1991. Productivity and profitability of twin births in beef cattle. *J. Anim. Sci.* **69:**3085–3093.
36. Duckett, S.K., G.D. Snowder, and N.E. Cockett. 2000. Effect of the callipyge gene on muscle growth, calpastatin activity, and the tenderness of three muscles across the growth curve. *J. Anim. Sci.* **78:**2836–2841.
37. Dwyer, C.M., A.J.A. Madgwick, S.S. Ward, and N.C. Stickland. 1995. Effect of maternal undernutrition in early gestation on the development of fetal myofibres in the guinea-pig. *Reprod. Fertil. Dev.* **7:**1285–1292.
38. Echternkamp, S.E. 1993. Relationship between placental development and calf birth weight in beef cattle. *Anim. Reprod. Sci.* **3:**1–13.
39. Everitt, G.C. 1964. Maternal undernutrition and retarded fetal development in Merino sheep. *Nature* **201:**1341–1342.
40. Fahey, A.J., J.M. Brameld, T. Parr, and P.J. Buttery. 2005. Ontogeny of factors associated with proliferation and differentiation of muscle in the ovine fetus. *J. Anim. Sci.* **83:** 2330–2338.
41. Fahey, A.J., J.M. Brameld, T. Parr, and P.J. Buttery. 2005. The effect of maternal undernutrition before muscle differentiation on the muscle fiber development of the newborn lamb. *J. Anim. Sci.* **83:**2564–2571.
42. Faichney, G.J. and G.A. White. 1987. Effects of maternal nutritional status on fetal and placental growth and on fetal urea synthesis in sheep. *Aust. J. Biol. Sci.* **40:**365–377.
43. Ferrell, C.L. 1991. Maternal and fetal influences on uterine and conceptus development in the cow: I. Growth of the tissues of the gravid uterus. *J. Anim. Sci.* **69:**1945–1953.
44. Fiorotto, M.L., T.A. Davis, P. Schoknecht, H.J. Mersmann, and W.G. Pond. 1995. Both maternal over- and under-nutrition during gestation increase the adiposity of young progeny in rats. *Obesity Res.* **3:**131–141.
45. Florini, J.R., D.Z. Ewton, and S.A. Coolican. 1996. Growth hormone and the insulin-like growth factor system in myogenesis. *Endocrine Rev.* **17:**481–517.
46. Ford, S.P., B.W. Hess, M.M. Schwope, M.J. Nijland, J.S Gilbert, K.A. Vonnahme, W.J. Means, H. Han, and P.W. Nathanielsz. 2007. Maternal undernutrition during early to mid-gestation in the ewe results in altered growth, adiposity, and glucose tolerance in male offspring. *J. Anim. Sci.* **85:**1285–1294.
47. Francke, K., T. Harder, L. Aerts, K. Melchior, S. Fahrenkrog, E. Rodekamp, T. Ziska, F.A. Van Assche, J.W. Dudenhausen, and A. Plagemann. 2005. Programming of orexigenic and anorexigenic hypothalamic neurons in offspring of treated and untreated diabetic mother rats. *Brain Res.* **1031:**276–283.
48. Freking, B.A., J.W. Keele, C.W. Beattie, S.M. Kappes, T.P.L. Smith, T.S., Sonstegard, M.K. Nielson, and K.A. Leymaster. 1998. Evaluation of the ovine *Callipyge* locus: I. Relative chromosomal position and gene action. *J. Anim. Sci.* **76:**2062–2071.
49. Freking, B.A., J.W. Keele, M.K. Nielson, and K.A. Leymaster. 1998. Evaluation of the ovine *Callipyge* locus: II. Genotypic effects on growth, slaughter and carcass traits. *J. Anim. Sci.* **76:**2549–2559.
50. Gagniere, H., B. Picard, and Y. Geay. 1999. Contractile differentiation of fetal cattle muscles: intermuscular variability. *Reprod. Nutr. Dev.* **39:**637–655.
51. Gerrard, D.E. and A.L. Grant. 1994. Insulin-like growth factor-II expression in developing skeletal muscle of double muscled and normal cattle. *Domest. Anim. Endocrinol.* **11:** 339–347.

52. Georges, M., C. Charlier, and N. Cockett. 2003. Polar overdominance at the ovine Callipyge locus supports trans interaction between the products of reciprocally imprinted genes. *Proc. Assoc. Advmt. Anim. Breed. Genet.* **15:**178–181.
53. Gopalakrishnan, G., S.M. Rhind, T. Stephenson, C.E. Kyle, A.N. Brooks, M.T. Rae, and M.E. Symonds. 2001. Effect of maternal nutrient restriction at defined periods in early to mid gestation on placento-fetal, kidney and adipose tissue weights at 110 days gestation in sheep. *Early Human Dev.* **63:**58–59.
54. Greenwood, P.L. and L.M. Cafe. 2007. Prenatal and pre-weaning growth and nutrition of cattle: long-term consequences for beef production. *Animal* **1:**1283–1296.
55. Greenwood, P.L., A.S. Hunt, J.W. Hermanson, and A.W. Bell. 1998. Effects of birth weight and postnatal nutriton on neonatal sheep: I. Body growth and composition, and some aspects of energetic efficiency. *J. Anim. Sci.* **76:**2354–2367.
56. Greenwood, P.L., R.M. Slepetis, J.W. Hermanson, and A.W. Bell. 1999. Intrauterine growth retardation is associated with reduced cell cycle activity, but not myofibre number, in ovine fetal muscle. *Reprod. Fertil. Dev.* **11:**281–291.
57. Greenwood, P.L., A.S. Hunt, J.W. Hermanson, and A.W. Bell. 2000. Effects of birth weight and postnatal nutrition on neonatal sheep: II. Skeletal muscle growth and development. *J. Anim. Sci.* **78:**50–61.
58. Greenwood, P.L., R.M. Slepetis, and A.W. Bell. 2000. Influences on fetal and placental weights during mid and late gestation in prolific ewes well nourished throughout pregnancy. *Reprod. Fertil.Dev.* **12:**149–156.
59. Greenwood, P.L., A.S. Hunt, R.M. Slepetis, K.D. Finnerty, C. Alston, D.H. Beermann, and A.W. Bell. 2002. Effects of birth weight and postnatal nutriton on neonatal sheep: III. Regulation of energy metabolism. *J. Anim. Sci.* **80:**2850–2861.
60. Greenwood, P.L., L.M. Cafe, H. Hearnshaw, D.W. Hennessy, J.M. Thompson, and S.G. Morris. 2006. Long-term consequences of birth weight and growth to weaning for carcass, yield and beef quality characteristics of Piedmontese- and Wagyu-sired cattle. *Aust. J. Exp. Agric.* **46:**257–269
61. Gregoire, F.M., S.M. Smas, and H.S. Sul. 1998. Understanding adipocyte differentiation. *Physiol. Rev.* **78:**783–809.
62. Guilloteau, P., T. Corring, R. Toullec, Y. Villette, and J. Robelin. 1985. Abomasal and pancreas enzymes in the newborn ruminant: Effects of species, breed, sex and weight. *Nutr. Rep. Int.* **31:**1231–1236.
63. Harding, J.E., C.T. Jones, and J.S. Robinson. 1985. Studies on experimental growth retardation in sheep. The effects of a small placenta in restricting transport to and growth of the fetus. *J. Dev. Physiol.* **7:**427–442.
64. Hausman, D.B., M. Digirolamo, T.J. Bartness, G.J. Hausman, and R.J. Martin. 2001. The biology of white adipocyte proliferation. *Obes. Rev.* **2:**239–254.
65. Hausman, G.J., J.T. Wright, R. Dean, and R.L. Richardson. 1993. Cellular and molecular aspects of the regulation of adipogenesis. *J. Anim. Sci.* **71**(supplement 2):33–55.
66. Houssin, Y. and M.J. Davico. 1979. Influence of birthweight on the digestibility of a milk-replacer in newborn lambs. *Ann. Rech. Vet.* **10:**419–421.
67. Hurley, M.S., C. Flux, A.M. Salter, and J.M. Brameld. 2006. Effects of fatty acids on skeletal muscle cell differentiation *in vitro*. *Br. J. Nutr.* **95:**623–630.
68. Hyatt, M.A., G.S. Gopalkrishnan, J. Bispham, S. Gentili, I.C. McMillen, S.M. Rhind, M.T Rae, C.E. Kyle, A.N. Brooks, C. Jones, H. Budge, D. Walker, T. Stephenson, and M.E. Symonds. 2007. Maternal nutrient restriction in early pregnancy programs hepatic mRNA expression of growth-related genes and liver size in adult male sheep. *J. Endocrinol.* **192:** 87–97.
69. Jackson, S.P., R.D. Green, and M.F. Miller. 1997. Phenotypic characterization of Rambouillet sheep expressing the *Callipyge* gene: I. Inheritance of the condition and production characteristics. *J. Anim. Sci.* **75:**14–18.
70. Jackson, S.P., M.F. Miller, and R.D. Green. 1997. Phenotypic characterization of Rambouillet sheep expressing the *Callipyge* gene: II. Carcass characteristics and retail yield. *J. Anim. Sci.* **75:**125–132.

71. Jackson, S.P., M.F. Miller, and R.D. Green. 1997. Phenotypic characterization of Rambouillet sheep expressing the *Callipyge* gene: III. Muscle weights and muscle weight distribution. *J. Anim. Sci.* **75:**133–138.
72. Jiang, Y.L., N. Li, X.Z. Fan, L.R. Xiao, R.L. Xiang, X.X. Hu, L.X. Du, and C.X. Wu. 2002. Associations of T→A mutation in the promoter region of myostatin gene with birth weight in Yorkshire pigs. *Asian-Australasian J. Anim. Sci.* **15:**1543–1545.
73. Jiang, Y.L., N. Li, G. Plastow, Z.L. Liu, X.X. Hu, and C.X. Wu. 2002. Identification of three SNPs in the porcine myostatin gene (MSTN). *Anim. Biotech.* **13:**173–178.
74. Kalhovde, J.M., R. Jerkovic, I. Sefland, C. Cordonnier, E. Calabria, S. Schiaffino, and T. Lomo. 2005. Fast and slow muscle fibres in hindlimb muscles of adult rats regenerate from intrinsically different satellite cells. *J. Physiol.* **562:**847–857.
75. Kelly, R.W. 1992. Nutrition and placental development. *Proc. Nutr. Soc. Aust.* **17:**203–211.
76. Koohmaraie, M., S.D. Shackelford, T.L. Wheeler, S.M. Lonergan, and M.E. Doumit. 1995. A muscle hypertrophy condition in lamb (Callipyge): Characterization of effects on muscle growth and meat quality traits. *J. Anim. Sci.* **73**:3596–3607.
77. Langley-Evans, S.C. 2006. Developmental programming of health and disease. *Proc. Nutr. Soc.* **65:**97–105.
78. Langley-Evans, S.C., S.J.M. Welham, and A.A. Jackson. 1999. Fetal exposure to a maternal low protein diet impairs nephrogenesis and promotes hypertension in the rat. *Life Sci.* **64:**965–974.
79. Langley-Evans, S.C., A. Fahey, and P.J. Buttery. 2003. Early gestation is a critical period in the nutritional programming of nephron number in the sheep. *Pediatric Res.* **53:**30A.
80. Lee, S.-J. 2007. Quadrupling muscle mass in mice by targeting TGF-β signaling pathways. *PLoS One* **2:**e789. doi:10.1371/journal.pone.0000789.
81. Lefaucheur, L., F. Edom, P. Ecolan, and G.S. Butler-Browne. 1995. Pattern of muscle fiber type formation in the pig. *Dev. Dynamics* **203:**27–41.
82. Lorenzen C.L., M. Koohmaraie, S.D. Shackelford, F. Jahoor, H.C. Freetly, T.L. Wheeler, J.W. Savell, and M.L. Fiorotto. 2000. Protein kinetics in Callipyge lambs. *J. Anim. Sci.* **78:**78–87.
83. Louey, S., M.L. Cock and R. Harding. 2005. Long-term consequences of low birthweight on postnatal growth, adiposity and brain weight at maturity in sheep. *J. Reprod. Devel.* **51:** 59–68.
84. Lutz, L., L. Dufourny, and D.C. Skinner. 2006. Effect of nutrient restriction on the somatotropes and substance P-immunoreactive cells in the pituitary of the female ovine fetus. *Growth Hormone IGF Res.* **16:**108–118.
85. Lutz, L., N. Schoefield, C. Crowe, L. Dufourny, and D.C. Skinner. 2007. No effect of nutrient restriction from gestational days 28 to 78 on immunocytochemically detectable growth hormone-releasing hormone (GHRH) neurons and GHRH receptor colocalization in somatotropes of the ovine female fetus. *J. Chem. Neuroanatomy* **33:**34–41.
86. Maier, A., J.C. McEwan, K.G. Dodds, D.A. Fischman, R.B. Fitzsimons, and A.J. Harris. 1992. Myosin heavy chain composition of single fibres and their origins and distribution in developing fascicles of sheep tibialis cranialis muscles. *J. Muscle Res. Cell Motil.* **13:** 551–572.
87. Mallinson, J.E., D.V. Sculley, J. Craigon, R. Plant, S.C. Langley-Evans, and J.M. Brameld. 2007. Fetal exposure to a maternal low-protein diet during mid-gestation results in muscle-specific effects on fibre type composition in young rats. *Br. J. Nutr.* **98:** 292–299.
88. Maltin, C.A., M.I. Delday, K.D. Sinclair, J. Steven, and A.A. Sneddon. 2001. Impact of manipulations of myogenesis in utero on the performance of adult skeletal muscle. *Reproduction* **122:**359–374.
89. Martin, J.L., K.A. Vonnahme, D.C. Adams, G.P. Lardy, and R.N. Funston. 2007. Effects of dam nutrition on growth and reproductive performance of heifer calves. *J. Anim. Sci.* **85:**841–847.

90. McCoard, S.A., S.W. Peterson, W.C. McNabb, P.M. Harris, and S.N. McCutcheon. 1997. Maternal constraint influences muscle fibre development in fetal lambs. *Reprod. Fertil. Dev.* **9:**675–681.
91. McCrabb, G.J., A.R. Egan, and B.J. Hosking. 1992. Maternal undernutrition during mid-pregnancy in sheep: variable effects on placental growth. *J. Agric. Sci., Camb.* **118:**127–132.
92. McPherron, A.C., A.M. Lawlerand, and S.J. Lee. 1997. Regulation of skeletal muscle mass in mice by a new TGF-β superfamily member. *Nature* **387:**83–90.
93. Mesiano, S., I.R. Young, R.C. Baxter, R.L. Hintz, C.A. Browne, and G.D. Thorburn. 1987. Effect of hypophysectomy with and without thyroxine replacement on growth and circulating concentrations of insulin-like growth factors I and II in the fetal lamb. *Endocrinology* **120:**1821–1830.
94. Muhlhausler, B.S., C.L. Adam, P.A. Findlay, J.A. Duffield, and I.C. McMillen. 2006. Increased maternal nutrition alters development of the appetite-regulating network in the brain. *FASEB J.* **20:**E556–E565.
95. Nordby, D.J., R.A. Field, M.L. Riley, and C.J. Kercher. 1987. Effects of maternal undernutrition during early pregnancy on growth, muscle cellularity, fiber type and carcass composition in lambs. *J. Anim. Sci.* **64:**1419–1427.
96. Picard, B., L. Lefaucheur, C. Berri, and M.J. Duclos. 2002. Muscle fibre ontogenesis in farm animal species. *Reprod. Nutr. Dev.* **42:**415–431.
97. Quigley, S.P., D.O. Kleemann, M.A. Kakar, J.A. Owens, G.S. Nattrass, S. Maddocks, and S.K. Walker. 2005. Myogenesis in sheep is altered by maternal feed intake during the peri-conception period. *Anim. Reprod. Sci.* **87:**241–251.
98. Rhind, S.M., J.J. Robinson, and I. McDonald. 1980. Relationships among uterine and placental factors in prolific ewes and their relevance to variations in fetal weight. *Anim. Prod.* **30:**115–124.
99. Rhoads, R.P., P.L. Greenwood, A.W. Bell, and Y.R. Boisclair. 2000. Organization and regulation of the gene encoding the sheep acid-labile subunit of the 150 kDa-binding protein complex. *Endocrinology* **141:**1425–1433.
100. Rhoads, R.P., P.L. Greenwood, A.W. Bell, and Y.R. Boisclair. 2000. Nutritional regulation of the genes encoding the acid-labile subunit and other components of the circulating insulin-like growth factor system in the sheep. *J. Anim. Sci.* **78:**2681–2689.
101. Ross, S.E., N. Hemati, K.A. Longo, C.N. Bennett, P.C. Lucas, R.L. Erickson, and O.A. MacDougald. 2000. Inhibition of adipogenesis by Wnt signalling. *Science* **289:**950–953.
102. Rudnicki, M.A., P.N.J. Schnegelsberg, R.H. Stead, T. Braun, H.H. Arnold, and R. Jaenisch. 1993. MyoD or Myf-5 is required for the formation of skeletal muscle. *Cell* **75:** 1351–1359.
103. Russell, R.G. and F.T. Oteruelo. 1981 An ultrastructural study of the differentiation of skeletal muscle in the bovine fetus. *Anat. Embryol.* **162:**403–417.
104. Sibbald, A.M. and G.C. Davidson. 1998. The effect of nutrition during early life on voluntary food intake by lambs between weaning and 2 years of age. *Anim. Sci.* **66:**697–703.
105. Sivachelvan, M.N. and A.S. Davies. 1981. Antenatal anticipation of postnatal muscle growth. *J. Anat.* **132:**545–555.
106. Smas, C.M. and H.S. Sul. 1995. Control of adipocyte differentiation. *Biochem. J.* **309:** 697–710.
107. Smith, J.A, A.M. Lewis, P. Wiener, and J.L. Williams. 2000. Genetic variation in the bovine myostatin gene in UK beef cattle: Allele frequencies and haplotype analysis in the South Devon. *Anim. Genetics* **31:**306–309.
108. Stalker, L.A., D.C. Adams, T.J. Klopfenstein, D.M. Feuz, and R.N. Funston. 2006. Effects of pre- and postpartum nutrition on reproduction in spring calving cows and calf feedlot performance. *J. Anim. Sci.* **84**:2582–2589.
109. Stevens, D. and Alexander, G. 1986. Lipid deposition after hypophysectomy and growth hormone treatment in the sheep fetus. *J. Dev. Physiol.* **8:**139–145.

110. Stubbs, A.K., N.M. Wheelhouse, M.A. Lomax, and D.G. Hazlerigg. 2002. Nutrient-hormone interaction in the ovine liver: methionine supply selectively modulates growth hormone-induced IGF-1 gene expression. *J. Endocrinol.* **174:**335–341.
111. Taylor, P.M. and J.M. Brameld. 1999. Mechanisms and regulation of transcription and translation, p. 25–50. *In* G.E. Lobley, A. White, and J.C. MacRae (eds.), Protein Metabolism and Nutrition: Proceedings of the VIIIth International Symposium on Protein Metabolism and Nutrition, EAAP publication No. 96, Wageningen Pers.
112. Vickers, M.H., B.H. Breier, W.S. Cutfield, P.L. Hofman, and P.D. Gluckman. 2000. Fetal origins of hyperphagia, obesity and hypertension and its postnatal amplifications by hypercaloric nutrition. *Am. J. Physiol.* **279:**E83–E87.
113. Villette, Y. and M. Theriez. 1981. Influence of birth weight on lamb performances. I. Level of feed intake and growth. *Ann. Zootech.* **30:**151–168.
114. Villette, Y. and M. Theriez. 1981. Influence of birth weight on lamb performances. II. Carcass and chemical composition of lambs slaughtered at the same weight. *Ann. Zootech.* **30:**169–182.
115. Villette, Y. and M. Theriez. 1983. Milk intake in lambs suckled by their dams during the first week of life. *Ann. Zootech.* **32:**427–440.
116. Vuocolo, T., K. Byrne, J. White J, S. McWilliam, A. Reverter, N.E. Cockett, and R.L. Tellam. 2007. Identification of a gene network contributing to hypertrophy in callipyge skeletal muscle. *Physiol. Genomics* **28:**253–272.
117. Wallace, J.M., D.A. Bourke, R.P. Aitken, R.M. Palmer RM, P. Da Silva, and M.A. Cruickshank. 2000. Relationship between nutritionally-mediated placental growth restriction and fetal growth, body composition and endocrine status in adolescent sheep. *Placenta* **21:** 100–108.
118. Wallace, J.M., D.A. Bourke, R.P. Aitken, N. Leitch, and W.W. Hay Jr. 2002. Blood flows and nutrient uptakes in growth-restricted pregnancies induced by overnourishing adolescent sheep. *Am. J. Physiol.* **282:**R1027–R1036.
119. Wheelhouse, N.M., A.K. Stubbs, M.A. Lomax, J.C. MacRae, and D.G. Hazlerigg. 1999. Growth hormone and amino acid supply interact synergistically to control insulin-like growth factor-I production and gene expression in cultured ovine hepatocytes. *J. Endocrinol.* **163:**353–361.
120. White, J., T. Vuocolo, M. McDonough, M.D. Grounds, G.S. Harper, N.E. Cockett, and R. Tellam. 2007. Analysis of the callipyge phenotype through skeletal muscle development; association of Dlk1 with muscle precursor cells. *Differentiation* **DOI:**10.1111/j.1432-0436.2007.00208.x
121. Wigmore, P.M. and D.J.R. Evans. 2002. Molecular and cellular mechanisms involved in the generation of fiber diversity during myogenesis. *Int. Rev. Cytology* **216:**175–232.
122. Wilson, S.J., J.J. Ross, and A.J. Harris. 1988. A critical period for formation of secondary myotubes defined by prenatal undernourishment in rats. *Development* **102:**815–821.
123. Wilson, S.J., J.C. McEwan, P.W. Sheard, and A.J. Harris. 1992. Early stages of myogenesis in a large mammal: formation of successive generations of myotubes in sheep tibialis cranialis muscle. *J. Muscle Res. Cell Motil.* **13:**534–550.
124. Yakar, S., J.L. Liu, A.M. Fernandez, Y. Wu, A.V. Schally, J. Frystyk, S.D. Chernausek, W. Mejia, and D. LeRoith. 2001. Liver-specific igf-1 gene deletion leads to muscle insulin insensitivity. *Diabetes* **50:**1110–1118.
125. Zhang, W.C., T. Nakao, M. Moriyoshi, K. Nakada, T. Ohtaki, A.Y. Ribadu, and Y. Tanaka. 1999. The relationship between plasma oestrone sulphate concentrations in pregnant dairy cattle and calf birth weight, calf viability, placental weight and placental expulsion. *Anim. Reprod. Sci.* **54:**169–178.
126. Zhu, M.J., S.P. Ford, P.W. Nathanielsz, and M. Du. 2004. Effect of maternal nutrient restriction in sheep on the development of fetal skeletal muscle. *Biol. Reprod.* **71:**1968–1973.
127. Zhu, M.J., S.P. Ford, W.J. Means, B.W. Hess, P.W. Nathanielsz, and M. Du. 2006. Maternal nutrient restriction affects properties of skeletal muscle in offspring. *J. Physiol.* **575:** 241–250.

Chapter 5
Mechanistic Aspects of Fetal Development Relating to Postnatal Fibre Production and Follicle Development in Ruminants

C. Simon Bawden, David O. Kleemann, Clive J. McLaughlan, Gregory S. Nattrass, and Stephanie M. Dunn

Introduction

More than 50 years ago, it was noted that wool follicle development in sheep fetuses suffering growth restriction in utero (e.g. twin lambs or single lambs born to maiden ewes) was significantly impaired [105]. A little later, observations of the postnatal wool production performance in lambs from ewes undernourished during gestation indicated that restriction of fetal development had permanent effects upon wool follicle development and long-lasting effects upon postnatal wool fibre production [107]. Though not described in this way at the time, these were early indicators that programming of the fetus for development takes cues from the "maternal environment" during gestation. The cues can be in operation as early as the first few days in embryonic life and possibly earlier [69, 120]. Fetal programming, and influences on gene expression and ultimately development imposed upon the fetus in utero, has become a topic of increasing interest over the last decade. This is especially so since lifetime consequences of such programming have been demonstrated in humans [6, 73]. Perturbed fetal programming can occur through different mechanisms and the effects of one such mechanism, intra-uterine growth retardation (IUGR), on development of the progeny have been reported in many publications. Further, experiments in animals have defined many of the consequences of IUGR related to subsequent growth and development of the progeny (for review see [45]) and the endocrine system has emerged as a major determinant of intra-uterine programming [37]. Reported effects on the progeny include restriction of the growth of fetal organs and tissues including impaired development of tissues at the cellular level. As the physiological processes leading to lifetime consequences for the organism have been discussed in detail elsewhere (and in other sections of this same volume), they will only be given cursory attention here.

C.S. Bawden (✉)
SARDI Livestock and Farming Systems, 5371, Roseworthy, SA, Australia
e-mail: simon.bawden@sa.gov.au

P.L. Greenwood et al. (eds.), *Managing the Prenatal Environment to Enhance Livestock Productivity*, DOI 10.1007/978-90-481-3135-8_5,

While features such as postnatal cardiovascular and respiratory fitness [7, 24] and general growth are known to be casualties of perturbed fetal programming in many species, production animals suffer defects in the development of tissues that determine economically important traits. For example, skeletal muscle and wool follicle development in sheep are each affected by prenatal growth restrictions imposed by the uterine environment during gestation [46, 47, 107] with significant impacts upon postnatal productivity. Clearly, altered development of other fetal organs, tissues and systems in the prenatal period will have a bearing upon development of the skin and follicles and on subsequent productivity of the follicles. In this case, it is currently believed that the prenatal and postnatal effects reported are mediated through either alteration of blood and nutrient supply or endocrine inputs, or by developmental manipulation of other metabolic events that result in altered physiology in the individual as a juvenile and adult (e.g. modified partitioning of protein synthesis between the muscle and skin compartments [75]). However, research of the mechanisms of fetal development relating to follicle formation and postnatal fibre production in ruminants lags behind that of other production species. Despite a paucity of data from intra-uterine growth restriction experiments that document gestational morphometric and physiological effects on hair and wool growth, research that has made definitive findings of prenatal and postnatal effects upon follicle development and subsequent fibre production in the progeny is discussed in detail in this chapter. In addition, other developmental impositions provided to the fetus by the "maternal environment" during gestation and known to have an impact on fibre production are summarised. Finally, current studies on gene expression associated with fetal skin development are discussed and the potential to define the prenatal mechanisms that have most impact on postnatal wool production is canvassed.

5.1 Fibre Production in Ruminants

In ruminants, the hair or wool follicle population is created in a number of stages. Some of the follicles begin to develop (initiate) early in gestation and some late. In addition, while all of the follicles are formed prenatally, some do not begin to produce a fibre (mature) until after birth, in the early postnatal period. Thus between wool follicle initiation and maturation, there are points in both prenatal and early postnatal development where these processes are subject to environmental influences and can be perturbed. Given the data gathered over many decades in relation to fibre production in the ovine compared to other fibre-producing ruminant species, the text below will focus on wool production in sheep. It will first detail the cellular and molecular events that result in formation of wool follicles producing wool fibres, then will give examples of ways in which these events have been shown to be susceptible to environmental influences, specifically via the fetus' and lambs' "maternal environment". Lastly, examples of attempts to define the lifetime production issues resulting from perturbation of the normal prenatal and postnatal processes involved in follicle development and maturation will be discussed.

5.1.1 Wool Growth in Sheep

Three parameters have a major impact upon the returns from wool growth in adult sheep. First is the size of the wool follicle population in wool-bearing areas of the sheep's skin, which affects overall yield and qualities of the fibres, including fibre diameter. Second is the efficiency of maturation of the follicles in the population, which determines the number of follicles producing emergent wool fibres for harvesting. Third is the number of developing follicle cells that become committed to a fibre fate that is to be used to produce either wool fibre cortex or cuticle cells. Interestingly, all of these parameters are determined prior to adulthood. While the entire wool follicle population is established prior to birth in a process that takes a substantial part of gestation, follicle maturation spans a significant portion of late gestation and the early postnatal period. Hence there are opportunities for extensive prenatal and some postnatal management of follicle development to alter production of wool. This fact was realised in quite early studies of wool follicle development in which lifetime performance of lamb progeny was found to be affected by poor nutrition of the pregnant ewe during gestation [107]. The findings from these studies and more recent research will be presented here.

Discussion throughout this text is predominantly of sheep that produce large quantities of fine wool fibres, which are destined for use at the top end of the market. Indeed, of greatest interest are the Merino and Merino crossbreeds, which have larger follicle populations, including many more of the so-called secondary fibres (S) than primary fibres (P) accompanied by minimal difference between the diameter of S and P fibres [22]. The difference between formation of the primary and secondary wool follicles and their maturation to produce wool fibres will be described in detail below. In general, sheep other than Merinos (sheep with coarser wool) have far fewer secondary wool fibres. As a result of developmental differences and subsequent interfollicular competition for keratin-forming substrates, primary fibres of significantly higher diameter than the secondary fibres are formed. In such cases, the combination of higher mean fibre diameter and the large diameter difference between primary and secondary fibres reduces the value of the fleeces, that are more often used in processes at the lower end of the production market (e.g. for carpet vs. fine wool suit manufacture). With this in mind, one could expect that depending upon the particular constitution of primary and secondary follicles in the skin, the results discussed in the examples of Merino and other breed manipulations below would be applicable to most wool-producing sheep breeds. While literature describing gestational manipulations of follicle development in other major fibre-producing ruminants, namely Angora and Cashmere goats, is virtually non-existent, it would be of interest to determine whether the principles of treatments known to affect the ovine also apply to these caprine breeds.

To begin to discuss the topic, a full understanding of the developmental progression of establishment of the Merino wool follicle population during gestation and follicle maturation for fibre production in gestation and the early postnatal period, is required. This follows below.

5.1.2 Establishment of the Wool Follicle Population During Gestation

We have recently undertaken a comprehensive analysis of the morphological changes occurring in skin and follicles of Merino sheep throughout gestation, specifically to define the temporal boundaries of establishment of the wool follicle population. This will be published in detail elsewhere. However, our observations are in accord with earlier published work [41, 50, 106, 111]. As is the case in other mammals including mouse and humans, establishment of the follicle population in sheep proceeds in multiple stages, involving the production of distinctly different follicle types, in a wave of growth that begins anteriorly and ends posteriorly. Apart from formation of the large follicles that become the vibrissae and eyelashes in skin of the head region, formation of the follicle population in the true wool-bearing regions of the sheep occurs in three stages. Between days 50–60 of gestation, development of the first group of follicles, termed the primary follicles is initiated, and proceeds for some 15 days. During this period, successive addition of the so-called primary central X, primary central Y and primary lateral X (2 per central X follicle) and lateral Y follicles (2 per central Y follicle), form follicle trios that denote the primary margins of follicle groups. Beyond this point, all primary follicles develop a sweat gland first, then a bilobed sebaceous gland and later a contractile muscle band, the arrector pili muscle. Following maturation of the primary follicles to produce cell layers including the hardened cortex and cuticle, the first wool fibres emerge from the skin surface at around day 100 of gestation.

The wool follicle population is expanded in two further inter-related developmental stages. From day 80–85 of gestation, development of wool follicles termed the secondary original follicles is initiated, whilst from day 100 onwards, expansion of the follicle population is continued through branching of the secondary original follicles to form secondary-derived follicles. Follicle branching is a phenomenon more prevalent in highly prolific wool producers such as the Merino, while in some sheep breeds it is virtually non-existent. Secondary follicles are readily distinguished from primary follicles in that while they feature a sebaceous gland, neither a sweat gland nor an arrector pili muscle are formed. Similar to primary follicles, maturation of secondary original follicles proceeds over approximately 40 days. While most of the secondary-derived follicles are initiated by day 135 of gestation [19] the remainder are initiated before birth. Interestingly, maturation of the secondary follicles occurs in two waves; through late gestation and the early postnatal period, reflecting the staged formation of secondary original and secondary-derived wool follicles [41]. Indeed, a small number of wool follicles do not produce emergent fibres until 6 months after birth.

The timing of formation of the primary (1^0), secondary (2^0) and secondary-derived (2^d) follicles and their appendages is illustrated diagrammatically in Fig. 5.1. Considering that before birth, all follicles have been initiated, the ratio of secondary follicles to primary follicles (S/P ratio) is maximal at birth. Moreover, in the Merino, the total number of primary fibres is fairly constant at different ages and between individuals but the total number of secondary follicles is quite variable between strains and individuals. Hence the S/P fibre ratio is commonly used to ascertain differences in fibre number and degree of maturation of the secondary follicle population. This method of measurement will be referred to frequently in the discussion of data to follow.

5.1.3 Cellular and Molecular Activity During Follicle Formation

A detailed description of the nature and timing of cellular and molecular events that occur in sheep skin during wool follicle development follows, with reference to the findings made in other species in which epithelial-mesenchymal interactions dictate formation of similar epidermal appendages. This discussion illustrates the complexity of the events involved in production of wool follicles and wool fibres and identifies the genes/gene pathways that may be involved in developmental fetal programming events that alter follicle formation and fibre production during gestation and the early postnatal period.

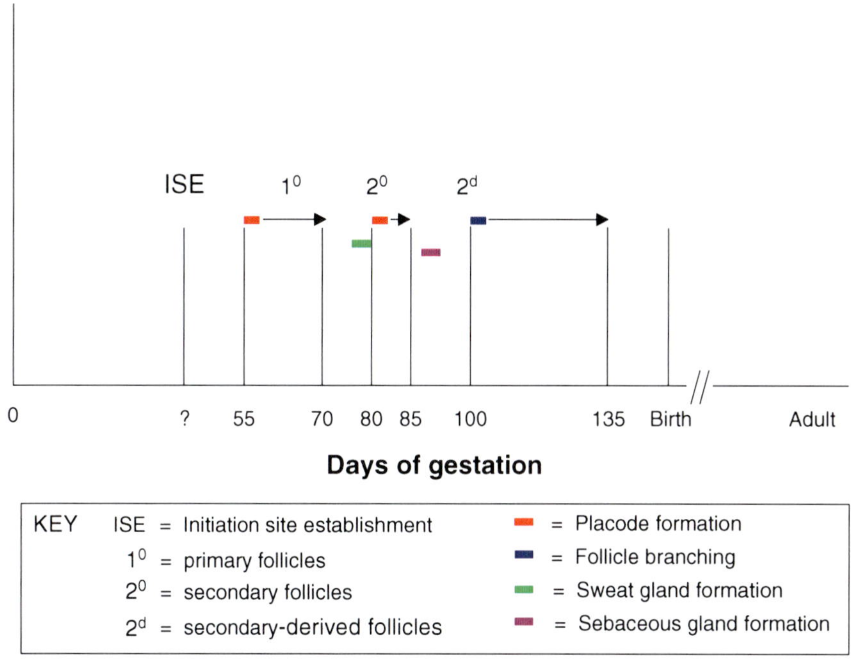

Fig. 5.1 Timing of the development of Merino wool follicles

5.1.3.1 Formation of the Primary Follicles

From day 35 to 50 of gestation, the skin is comprised of quite thin epidermal and dermal layers. The epidermis layer is only one to two cells thick and nuclei of fibroblasts in the dermis are sparsely arranged, with no sign of collagen formation. In this period, the dermis thickens somewhat, with the overall density of fibroblast nuclei also visibly increasing. At day 50, three epithelial cell layers are visible in some regions and the basal layer becomes quite distinct. The first sign of establishment of the sites of initiation of primary follicles is visible at around day 55 of gestation, as very infrequent thickenings in the basal layer of the epidermis (Fig. 5.2, panel a). In zones of remodelling at the initiation sites, the close proximity of nuclei in neighbouring cells indicates some compaction of the epithelium and the commencement of formation of structures termed epidermal placodes. In other systems in which epidermal placode formation has been well characterised, notably in the developing hair follicle (for reviews see [12, 83, 108]) the activity of cells in the epithelium has been found to be dictated by molecular signals from the underlying dermis (mesenchyme). Signalling reportedly involves both Wnt

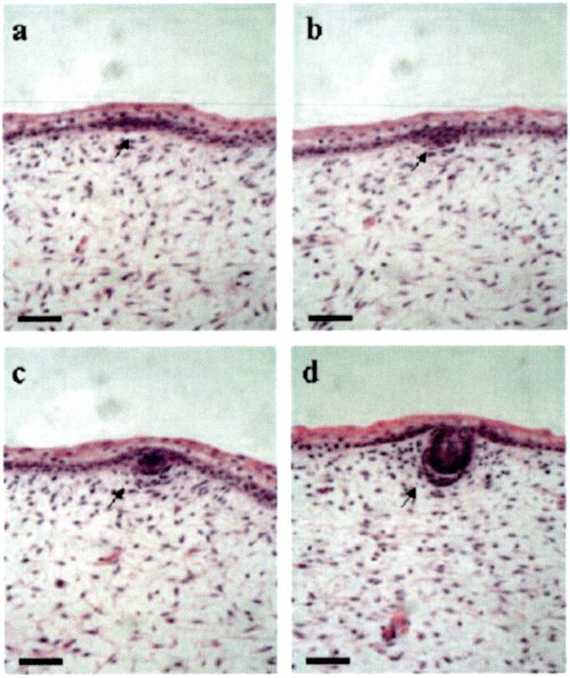

Fig. 5.2 Initiation and down-growth of a primary wool follicle in Merino fetal skin. Shown are photomicrographs of Merino fetal skin sections prepared in the longitudinal orientation and stained with Haematoxylin and Eosin, with the epidermal layer uppermost (pink-stained cells, *top*) and dermal layer beneath (diffuse purple-stained nuclei, *bottom*). *Arrows* indicate the position of an initial epithelial thickening (panel a; day 55 of gestation), the commencement of aggregation of dermal cells beneath the developing epithelial placode (panel b; day 57 of gestation), further aggregation of dermal cells to produce the dermal condensate (panel c; day 60 of gestation) and down-growth of the epidermal bud into the dermis, pushing the dermal condensate ahead of it (panel d; day 65 of gestation). The *scale bar* indicates a length of 50 μm

[Wingless/Int] and FGF [Fibroblast Growth Factor] molecules produced in the dermis, acting either sequentially or in concert. In the chicken, development of the dorsal feather-inducing dermis has been shown to be reliant upon a signal from the dorsal neural tube and the signal can be substituted by Wnt-1 [93]. Further, initiation of the epidermal placodes that precede chick feather formation has recently been demonstrated to require FGF-10 [77, 116] and FGF-10 is known to be required for whisker development in mouse [90]. The result of Wnt and FGF action is induction of other molecules in both the dermis and epidermis that promote epidermal placode formation. These include Wnt6 and the Tumour Necrosis Factor family signalling molecules (Ectodysplasin [EDA] and the Ectodysplasin Receptor [EDAR]; [74]), additional Wnt molecules (Wnt 10a, 10b; [103]), members of the TGFß superfamily of factors (i.e. the Bone Morphogenetic Proteins BMP-2, -4, -7; [14, 57]) and their regulators (e.g. Noggin) and members of the cell fate-determining Notch signalling pathway (Notch-1, -2, Delta-like, Jagged-1, -2, Lunatic Fringe; [32, 100, 119]). Without exception, we have demonstrated the presence of transcripts encoding molecules in each of these molecular signalling pathways in developing Merino midside fetal skin via quantitative RT-PCR.

Within the next day, distinct aggregations of dermal cells begin to form in the dermis underlying the epithelial thickenings, evidenced by readily visible clusters of dermal cell nuclei (Fig. 5.2, panel b). Dermal cell clustering occurs in response to new molecular signals from the epidermis, including further Wnt and growth factor signals (Platelet-derived growth factor alpha; PDGF-α; [64]). Where epithelial placodes are forming, a basement membrane layer produced predominantly by the epithelium, provides a distinct separation between epithelial and dermal cells. Despite the vigorous communication that occurs between the epidermis and dermis throughout further development of the follicle, separation of these cell types by interposition of the basement membrane is maintained. By day 60, condensation of epithelial cells to form distinct placodes is accompanied by even closer aggregation of the dermal cells immediately beneath the placode structures (Fig. 5.2, panel c). These form dermal condensates, precursors to the dermal papillae found in the mature follicles.

Initiation of new primary follicles, indicated by the formation of new epithelial thickenings/placodes continues until about day 70 of gestation. All initiated primary follicles proceed from the placode stage to the "germ" or "epithelial bud" stage, in which the rounded mass of epithelial cells pushes down into the dermis, bounded on the dermal surface by a dermal condensate. Down-growth of the epithelial portion of the bud into the dermis, pushing the dermal condensate ahead of it, is first visible from day 65 for primary follicles (Fig. 5.2, panel d) and by day 75, all follicles have proceeded to this stage of bud down-growth. This process is promoted by a member of the Hedgehog signalling factor family (Sonic Hedgehog; Shh; [114]) that supports both condensation of the dermal cells and formation of the dermal papilla of the follicle by induction of another of the Wnt factors (Wnt5a; [103]). The process of follicle down-growth is also directed by additional signals from the dermis, provided by another TGFß family molecule induced by Shh (ActivinβA; [89]) and the cytokine Hepatocyte Growth Factor/Scatter Factor (HGF/SF; [76]). Elongation of the primary follicles by additional proliferation in the epithelium continues until late in follicle development, when invagination of the epithelium at the growing follicle base encloses cells of the dermal condensate to complete formation of the dermal papilla (Fig. 5.4).

During the early period of primary follicle formation, another notable feature is an increase in circulatory capacity of the dermis. The frequency of capillaries in the dermis increases up until day 60 and soon after, the first signs of arteriole formation near the base of the dermal layer are visible. Subsequently, the metabolic needs of the epidermis and dermis layer are served by an ever-expanding network of interfollicular capillaries (upper dermis) and arterioles (lower dermis). This is of some consequence considering that the synthetic capacity of the expanding follicle population increases substantially as development proceeds. In our own study of wool follicle development, the expression of genes with known effects in circulatory development (vascular endothelial growth factor A (VEGFA) and hypoxia inducible factor-1 (HIF-1); [101]) has recently been detected in sheep skin samples spanning the entire gestational period and RNA in situ hybridisation will soon be used to spatially map activity of the genes.

As well as a complex circulatory system, the developing hair follicle is supplied with significant input from the peripheral nervous system. In mouse, it has been shown that neurotrophin receptors in the epidermal placode (TrkC; [16]) and dermal condensate (p75 neurotrophin receptor; [15]) and the ligand neurotrophin-3 [13] affect the rate of hair follicle morphogenesis in the earliest stages. Further, by the time follicles have matured, three dermal nerve plexi (the subepidermal, deep cutaneous and subcutaneous plexi) feed into two different perifollicular nerve bundles that are in close contact with the follicle [97].

5.1.3.2 Sweat and Sebaceous Glands

A defining feature of primary follicle development is the formation of both sweat glands and sebaceous glands associated with the follicle structure. The timing of formation of sweat glands is described here in consideration of the problems posed by fleece rot and fly strike and the potential for perturbation of sweat gland development by gestational intervention treatments. After day 75, the initiation of sweat gland formation is visible as very small protrusions of cells from the necks of developing primary follicles. These protrusions lead to the expanded sweat gland buds seen by day 80 (Fig. 5.3a). In samples from subsequent gestational time points, elongation of the sweat gland bud is evident, with the most advanced approaching half the length of their originating primary follicles at day 85 (Fig. 5.3c). After this point, no new sweat glands are formed. A few days later, the first evidence of sebaceous gland formation is visible in some primary follicles, as bilateral bulges in the follicle neck region (Fig. 5.3d). Expansion of the sebaceous glands as bilobed structures is obvious by day 95 and in the same period, many sweat glands have undergone rapid elongation to be nearly the length of their respective follicles. Within two days, extended ducts of the more developed sweat glands end in visible lumen structures (Fig. 5.3e).

5.1.3.3 Formation of Secondary Follicles

The first sign of secondary original follicle initiation is the formation of new epidermal thickenings/placodes at sites between existing primary follicles in fetal skin, around day 80 of gestation (Fig. 5.3b). Further placode formation is evident up until day 85 and follicle development then proceeds through the bud down-growth stage in a fashion similar to the events described for primary follicle formation. The secondary follicles do not form sweat glands, only sebaceous glands, but the similarities between the early stages of formation of the primary and secondary original follicles suggest strongly that many of the molecules responsible for the development of these two wool follicle populations in sheep are in common. While this has generally been shown to be the case in multi-stage hair follicle development in other species, with some specific molecular differences between primary and secondary follicle formation identified in mouse [14, 108], it is yet to be determined for sheep. After day 90, evidence of secondary original follicle formation is only visible as early peg stage follicles.

Formation of secondary-derived follicles, visible as budding of the epithelium very high up on the neck of developing secondary original follicles, occurs at about day 100 of gestation (Fig. 5.3f). At this stage, the secondary original follicles themselves are only two thirds of the length of adjacent primary follicles and quite narrow

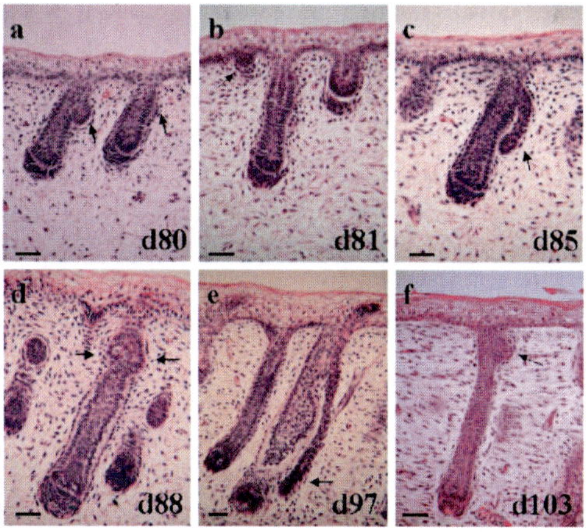

Fig. 5.3 Formation of the sweat and sebaceous glands and secondary original and secondary-derived follicles in Merino fetal skin. Shown are photomicrographs of Merino fetal skin sections prepared in the longitudinal orientation and stained with Haematoxylin and Eosin, with the epidermal layer uppermost (skin surface) and dermal layer beneath. *Arrows* indicate the positions of sweat gland buds developing in two primary follicles (panel a; day 80 of gestation), an initiating secondary original follicle (panel b; day 81 of gestation), an elongating sweat gland with associated primary follicle (panel c; day 85 of gestation), bilobed sebaceous gland buds developing in a primary follicle (panel d; day 88 of gestation), a well developed primary follicle sweat gland duct with lumen (panel e; day 97 of gestation) and a secondary-derived follicle initiating on the neck of a secondary original follicle (panel f; day 103 of gestation). The scale bar indicates a length of 50 μm

by comparison. Given that the process of formation of secondary-derived wool follicles proceeds via a morphological branching mechanism, the molecular events involved are likely to be somewhat different to those involved in formation of the primary and secondary original follicles. Considering the importance of formation of secondary-derived follicles to the overall follicle density achieved (see below), the molecular mechanisms required for wool follicle branching are currently attracting extensive investigation.

5.1.4 Issues Related to Establishment of the Wool Follicle Population

A number of issues of importance to the production of wool in sheep arise from a consideration of the developmental progression of establishment of the wool follicle population itself. First, the size of the wool follicle population determines follicle density, which in turn influences both clean fleece weight and fibre diameter [17, 18, 19]. In the earliest stages of development in skin, commitment of cells to epidermal and dermal fates, then subsequently to follicular or interfollicular fates has a bearing upon the total wool follicle population size. So it would appear that the cell fate decisions made very early in gestation, prior to the formation of follicles themselves, will have a bearing upon the total size of the wool follicle population. In this regard, one might expect that for any given sheep breed, there may be a propensity to form a certain number of follicle precursor cells, both of epidermal and dermal origin, that will then dictate the ultimate follicle forming capacity of the breed.

This exact notion has been expressed in a theory termed the "founder cell theory" [84, 85] that attempts to explain the different follicle densities and fibre diameters evident in different sheep breeds. The theory focuses on the dermal follicle precursor cell component and takes into account the fact that the number of cells in the dermal papilla (dermal papilla volume) determines hair volume [59]. It proposes that even for a fixed population of follicle forming dermal stem cells in the skin of one breed, the skin could incorporate either a large number of the precursor cells into a small number of formed follicles (in which case follicle density would be low and fibre diameter high) or a small number of precursor cells into a large number of follicles (in which case follicle density would be high and fibre diameter low). Extending this, in breeds able to make greater numbers of finer fibres, once formation of the primary follicle population has concluded, there are larger numbers of follicle precursor cells remaining in the dermis able to be recruited for formation of secondary original and secondary-derived follicles. Hence this theory proposes interdependence between follicle density and fibre diameter and also between the formation of primary follicles and secondary follicles. Further, it indicates a potential role for the self-aggregation ability of dermal precursor cells in determination of follicle size and fibre diameter, where breeds whose precursor cells can form larger aggregates will ultimately have fewer, coarser wool fibres.

Interestingly, contrary to the notion of papilla cell numbers being fixed at the time of follicle initiation, an increase in dermal papilla cell number after initiation has been observed [1]. An approximate doubling of cells, by cell division, occurs between the first and last stages of primary follicle development. Following this finding, close analysis of follicle attributes in skin from individual sheep across prenatal and postnatal development found that the bulk of the inverse correlation between follicle density and fibre diameter in an adult sheep was actually established in the gestational period after follicle initiation [2]. Overall, whether determined at follicle initiation or later, follicle size could clearly be influenced by any constraints upon development.

Another important consideration is the timing of formation of different components of the wool follicle population in gestation, in relation to consolidation of fetal support systems and to the formation of other fetal tissues and organ systems. As will become apparent later, from the limited data available, it appears that it may be more difficult to perturb the early stages of establishment of the wool follicle population, specifically formation of the primary follicles, though this clearly requires further investigation. Results to date suggest the multiplicity of differentiation events combined with the sheer magnitude of synthetic activity later in gestation allow more ready manipulation of later stage formation and maturation of the follicle population, namely the secondary follicles.

5.2 Maturation of the Follicle Population for Fibre Production

Once formed in the skin, the primary, secondary original and secondary-derived follicles that have grown down into the dermis each progress through a maturation phase that culminates in fibre production. The efficiency of maturation in these different follicle types is governed by both genetics and environment. In this section we describe the cellular events and molecular inputs critical for wool follicle maturation and fibre production and the timing of follicle maturation in all follicle types.

5.2.1 Cellular Correlates of Follicle Maturation

Mature wool follicles in Merino sheep are highly complex structures formed from many unique cell layers, that exhibit cell-specific and tightly regulated gene expression patterns (for review see [99]; see Fig. 5.4). Arising from an epithelial downgrowth into the dermis as previously described, the mature fibre-producing wool follicle includes both a dermal and an epidermal component. The dermal component consists of two parts; (i) the dermal papilla, a body of specialised dermal fibroblasts derived from the initial dermal condensate, that resides within the base of the follicle and acts as a control centre for follicle activity, and (ii) the dermal sheath, a connective tissue layer that is contiguous with the dermal papilla, lies immediately beneath the basal epidermal layer and encases the follicle. The epidermal component can include up to nine different cell types. From the outermost to the innermost layer, these are cells of the outer root sheath and companion layer, inner root sheath

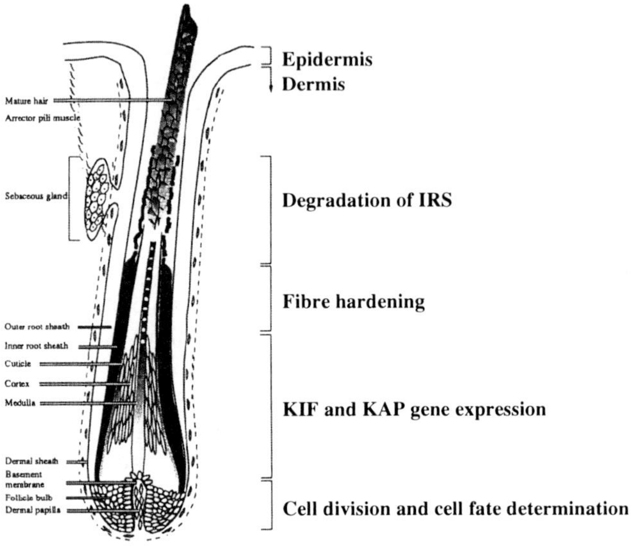

Fig. 5.4 Cell compartments of the mature hair follicle. Diagrammatic representation of a hair follicle, including cell compartments comprising the follicle and ancillary structures associated with the follicle. Regions delineating the epidermis, dermis and zones in which specific cellular activities occur are indicated on the right (adapted from [99])

(Henle, Huxley and cuticle layers), fibre cuticle, fibre cortex (orthocortex, mesocortex, paracortex) and medulla (not present in fine wool Merino fibres).

As follicle maturation proceeds, cells of the epidermal layers, notably the cortex, cuticle and inner root sheath (IRS), undergo a process of terminal differentiation, in which the fibre is produced. Continuous cell division in the base of the follicle (follicle bulb region) supports movement of the concentrically arranged inner root sheath, cuticle and cortical cell layers upwards toward the skin surface. During this upward movement, cells of the inner root sheath become filled with protein and begin to harden, via covalent cross-linking of the deposited proteins. A similar process occurs in the cuticle and cortex but cross-linking of the proteins and cell hardening does not occur until somewhat later, when cells have progressed higher up in the follicle. While not part of the final fibre, the result of early hardening of the IRS is that it acts as a rigid mould in which the fibre is formed. Cells of the cuticle and cortex harden, move upwards while encased within the IRS and are finally extruded from the skin surface, as the mature wool fibre. Cells of the IRS become detached from the fibre as it nears the skin surface and are ultimately degraded.

5.2.2 Stem Cell Compartments in the Mature Follicle

Adding to the complexity described, it has recently been shown that both the dermal and epidermal components of mammalian hair follicles contain stem cell populations able to contribute to establishment of entirely new follicles (and other tissues) under the right conditions [10, 61, 88]. While it is known that follicle neogenesis is an extremely involved process, both at the cellular and molecular level, a question remains as to the ability of Merino skin to form even larger numbers of wool follicles, via use of available stem cells resident in the epidermis, dermis and follicle structures described. Hair follicle densities achieved in other species suggest it may be possible though this would depend on the ability of sheep skin to support additional wool follicles. In addition, similar to the proposed programmed use of the follicle epidermal and dermal stem cells for follicle initiation [85], it is possible that stem cell populations required to supply the follicle for other aspects of development or normal wool growth may be diverted to other activities by inappropriate environmental cues. Along these lines, given the fate of the IRS described above, and that in Merinos as much as 75–90% of the cells of the follicle bulb are committed to production of this layer [9], there is potential for a great improvement in wool fibre protein output from the follicle if more cells in the developing follicle could be diverted to a fibre cell (cortex or cuticle cell) fate.

5.2.3 Molecular Determinants of Follicle Maturation

Similar to the molecular processes leading to follicle initiation, many regulatory molecules are known to be involved in maturation (terminal differentiation) of hair follicles. In the early stages, the Notch signalling pathway is involved in determining the fate paths of cells that leave the follicle bulb region and differentiate into specific cell types of the follicle and yet another of the Wnts [Wnt3a] directs terminal differentiation of cells destined to become the fibre cortex and cuticle. Further, transcription factors GATA3 and TCF-3 play an early role in formation of the IRS, while BMP-2 and -4 [72, 123] regulate the expression of other transcription factors including those that control expression of the keratin intermediate filament (KIF) genes (e.g. FoxN1; winged helix-forkhead) and keratin-associated proteins (KAPs) genes (e.g. HoxC13; Homeobox factor C13). While previous RNA in situ experiments have defined the expression zones of many of these molecules in adult mouse and sheep skin [11, 44, 48, 62, 78], we have recently confirmed the presence of transcripts encoding some of these molecules in fetal sheep skin during gestation by quantitative RT-PCR [Bawden et al., unpublished].

5.2.4 Timing of Maturation of Primary, Secondary Original and Secondary-Derived Follicles

In the most developed primary follicles just prior to day 100 of gestation, the first signs of terminal differentiation within the follicles appear, namely formation of the IRS layer and development of some hardening fibre cells. As developing layers of the IRS make their way upwards in the follicle, within the bounding outer root sheath (ORS), an advancing and widening conical structure termed the "hair cone" is seen. The hair cone is comprised of developing IRS and contains the precursors of cuticle and cortical cells. As cell movement proceeds upwards, the apex of the hair cone eventually "opens" within the upper region of the follicle neck to form a hardened cylindrical structure within which the cuticle and cortical cells continue to develop. Terminal differentiation of cells in the cuticle and cortex, involving synthesis and cross-linking of the KIF proteins and KAPs leads to formation of a hardened fibre. Within two to three days of formation of the first visible hair cones, wool fibre tips are seen emerging from the skin surface for a small proportion of the primary follicles while by day 110 of gestation, fibre production is evident in the majority of primary follicles.

In later gestational samples, maturation of the secondary original follicles proceeds as for the primary follicles and earlier reports suggest that most are fully mature and producing fibre prior to birth [50]. In contrast, the majority of secondary-derived follicles, initiated from approximately day 100 until day 135 of gestation, are reported to complete the maturation process in the first four to five weeks after birth [41], with a small number only fully maturing after that, even as late as 6 months of age.

5.3 Follicle Behaviour Late in Gestation and the Early Postnatal Period

Production of the hardened wool fibre requires constant and rapid cell division in the follicle bulb region, terminal differentiation including synthesis and cross-linking of KIF and KAP proteins of the cortex and cuticle and production of the IRS. In the process, there is a large expenditure of energy and raw materials. Given that the hair follicle is one of the most metabolically active sites in the body, it is one that is logically most sensitive to environmentally imposed limitations in these commodities. This will become obvious as the effects of experimentally determined influences on follicle development during gestation and the postnatal period are described. However, a description of some of the expected behaviours of sheep wool follicles under normal conditions is given first.

5.3.1 Competition Between Follicle Types During Development

In Romney Marsh sheep, that are of the "long wool" variety, where lengths of the primary and secondary original fibres are comparable, the annual fleece features fibres of some 17.5 cm in length with a regular staple crimp and only rare medullation. Two separate mutations, designated N and nr, were identified by Dry in the Romney breed [29] that alter development of the primary follicles. The mutations result in primary follicles that develop faster and that attain a greater terminal size than the primary follicles of normal Romney sheep. Interestingly, the mutations do not affect the general timing of development of the follicle population, the frequencies of the different follicle types, the ratios of secondary to primary follicles measured in newborn lambs (S/P ratio; 2:1 for N, nr and normal sheep) nor the density of the follicle population [39]. However, concomitant with the increase in growth rate (38%) and size of the primary follicles, decreases of similar magnitude in growth rate (28%) and size of the secondary-derived follicles are found in the mutant sheep [39]. Clearly the N and nr mutations act by increasing the competitive efficiency of the primary follicles. At the same time, the size and activity of the secondary original follicles is unaffected. Indeed, the situation in these mutants is reminiscent of the case for wool follicles and fibres in the "carpet" variety sheep.

These findings illustrate two important points in relation to establishment and maturation of the wool follicle population. The first is that genetic mechanisms exist that are able to define the fibre length and diameter distribution of the fleece. This is significant given the different uses to which fibre and fleeces of the long wool and carpet sheep varieties are put, and that in the Merino, minimal differences in fibre diameter across the fleece are desirable. Second, it illustrates clearly the competition that exists between developing wool follicles during gestation. It is apparent that the nature (size and activity) of primary follicles laid down early in gestation can have a measurable influence on the development and subsequent productivity of follicles initiated later in gestation, specifically the secondary-derived follicles. The effect of enlarged and more highly active primary follicles is thought to be restricted largely to the secondary-derived follicles as these are initiated in skin between the primary and secondary original follicles, where the competition for growth substrates is maximal. One final finding of interest is that in this study, no consistent difference in fleece weight (mean greasy fleece weight) was measured for the differing N, nr and wild-type genotypes.

An additional study of the diameter and spatial relationships of follicles in the skin of sheep breeds with contrasting fibre and fleece characteristics (Lincoln, Ryeland and medium strong Merino; [43]) supports the notion that competition exists between neighbouring wool follicles. Indeed, there is a strong relationship between the diameter of any single fibre and those fibres in the skin surrounding it. In examination of skin cross-sections from all three breeds, it was shown that for any single fibre (denoted the "central" fibre), there is a negative correlation between the diameter of the fibre and the number, diameter and displacement distance of the fibres adjacent to it. Further, this correlation was shown to hold for a maximum distance defined by the diameter of the central fibre. Given the average diameter of central fibres measured for the different sheep (Lincoln 43.8 μm, Ryeland 36.3 μm, Merino 26.2 μm), the maximum distances over which the competitive effect operated were determined to be 283, 173 and 147 μm respectively. That is, the distance over which a follicle effectively competes with surrounding follicles decreases as the size/diameter of that follicle/fibre decreases.

These results again highlight the potential for developmental advantage of follicles that are formed first and which may compete more successfully for fibre-forming substrates. Follicles that compete more effectively for substrates early in development may indeed go on to be more productive, more efficient follicles in later life, to the preclusion of activity in neighbouring follicles. Evidence that such competition actually operates during development has come from examination of follicle development in varieties with extremely coarse wool [42]. In comparison to many other breeds with carpet-type fleeces, it was noted that a significantly lower proportion of the secondary follicles develop completely in Scottish Mountain Blackface sheep, a breed that features very large primary follicles. A similar effect was determined for two other carpet breeds, the Welsh Mountain and Swedish Landrace sheep, where a negative correlation between primary fibre diameter and the frequency of secondary follicles was evident. In the latter case, the result of the comparison of mean primary fibre diameter to S/P follicle ratio suggests that as primary follicle size is increased, fewer secondary follicles are initiated during development. In contrast to this situation, a most desirable feature of the Merino breed is its ability, through a precise developmental regime, to minimise differences between the size and growth efficiencies of the primary, secondary and secondary-derived wool follicles that produce the fleece. Indeed it is likely that developmental competition between primary and secondary follicles in Merino strains producing very fine wool is minimal or virtually non-existent.

5.3.2 Development of Follicles to Produce Fibres – Before and After Birth

As described previously, all primary follicles in the wool follicle population are fully mature and produce a wool fibre before birth of the lamb. In stark contrast, while some secondary follicles reach this level of maturity during gestation, the majority produce a wool fibre only after birth. This was shown to vary between breeds, with higher levels of maturation of secondary follicles after birth occurring in Merino than in coarse wool sheep [41, 42]. Comparison of S/P fibre ratios in the Merino with some British breeds at birth (Merino 3.05; British 2.2) and as adults (Merino 12–40; British 4.5–8) showed that while S/P fibre ratios at birth are not very different between breeds, there is a large difference in adulthood, with 75–95% of Merino secondary follicles estimated to mature to produce fibre after birth, compared to 60–80% for the British breeds.

In an effort to define the critical features of secondary follicle maturation (i.e. follicle maturation to produce a wool fibre), sampling and histological examination of midside skin from Merino lambs was carried out over a 22 week period from birth [41]. Changing S/P fibre ratios revealed that maturation of the secondary follicles occurs rapidly in the first 5 weeks then slows dramatically. Indeed it is evident that the majority of the adult secondary follicle population reaches maturity in the period from birth to 28–35 days postnatally. In earlier research, it was reported that maturation of the secondary follicle population could be influenced by adverse nutrition (imposed naturally by twinning; [105]). On the basis of this observation and characterisation of secondary follicle maturation, the author rightly suggested "it is feasible that manipulation of the level of nutrition for a short period could have a marked effect on the adult follicle population" [41]. Further, a suggestion this manipulation could either be prenatal, during the period of initiation of the secondary follicles, or postnatal, during the period of maturation of the secondary follicles, was made. This was based on earlier observations of potential S/P ratios in late gestation. Counts of total mature and immature primary and secondary wool follicles in late gestation Merino fetal skin [19] determined potential S/P ratios of 7.0 (day 125), 10.5 (day 135) and 19.0 (day145) that indicated the majority of secondary follicles had been formed before birth. Combined with the fact that by far the largest proportion of wool follicles in the follicle population of Merinos are secondary follicles, the suggestion that an opportunity exists for manipulation of development of the wool follicle population was insightful.

5.4 Investigation of Influences on the Mechanisms of Prenatal and Postnatal Development of the Wool Follicle Population and Postnatal Wool Fibre Production

Considerable information exists about development of the wool follicle population, including detailed characterisation of the morphological events involved in initiation and maturation of the primary, secondary original and secondary-derived wool follicles [19, 39–41, 43, 50, 105]. Armed with this information, researchers over the preceding five decades have investigated the effects of environmental influences, notably those imposed on the developing fetus and lamb by the ewe during pregnancy and lactation, on establishment and maturation of the wool follicle population. Further, the effects of prenatal and postnatal environmental influences upon the lifetime production of wool have also been investigated. A number of studies have defined the short-term and long-term outcomes for wool production in experimental lamb groups and these are described below. The duration and nature of treatments applied in these investigations are depicted diagrammatically in Fig. 5.5. It should be noted that depending upon the time of lambing and environmental conditions including feed availability, many of the treatments applied to the test animal groups bear close resemblance to field conditions in which they are normally found. Hence these treatments provide valuable predictions of the likely developmental and production outcomes.

5.4.1 Manipulation of Prenatal Development of the Wool Follicles

Knowledge of the nature and timing of the events of maturation of the wool follicle population and realisation of the potential for manipulation of these events [41, 105] led to new research ultimately testing the plasticity of wool follicle development. In Merino and Merino crosses, the early experiments were focussed upon

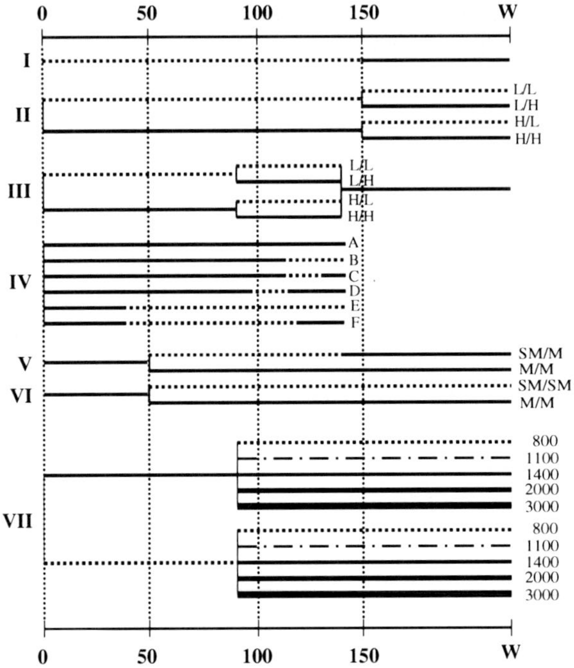

Fig. 5.5 Maternal feed restriction experiment timelines. The timelines at the top and bottom of the diagram represent the gestational period from the day of conception (0), through to birth (day 150) and weaning (W; between approximately 12 and 16 weeks postnatally) with completion of the early, mid and late gestational periods marked (*vertical dotted lines*). The horizontal experimental timelines in I to VII show the timing and duration of treatment periods and depict different levels of maternal nutrition (maintenance nutrition: *solid line*; sub-maintenance [restricted] nutrition: *dotted* and *dashed lines*) for the trial regimes described in the experiments of Short ([112]; I), Schinckel and Short ([107]; II), Taplin and Everitt ([117]; III), Hutchison and Mellor ([58]; IV), Kelly et al. ([65, 66]; V and VI) and Thompson and Oldham ([118]; VII). Where two levels of feed were offered consecutively, lettering indicates low (L) and high (H) or sub-maintenance (SM) and maintenance (M) feeding. In IV, lettering denotes experimental groups (A to F) while numbering in VII denotes levels of feed-on-offer in kg of dry matter per hectare for the different treatment groups

secondary wool follicle maturation, given that in even the least prolific wool production strains, secondary follicles formed by far the greatest proportion of wool follicles in the skin. In line with earlier findings that all secondary follicles that produce a fibre had been formed before birth, examination of secondary follicle development in the first six months postnatally showed that S/P follicle ratio at birth was virtually equivalent to S/P fibre ratio at 6 months of age (correlation value, $r = 0.97$; [111]). In the same study, 65% of secondary follicles were found to mature by postnatal day 28, with the maturation rate being greatest in the period from day 7 to day 21, decreasing thereafter. Similar findings were reported elsewhere [106] and all the observations suggested a clear postnatal time window in which secondary follicle maturation would be most susceptible to environmental influences. However, from the data gathered, it was unclear whether secondary follicle development (i.e. secondary follicle initiation itself) could be influenced via alteration of the prenatal environment.

Consequently, the possibility that prenatal formation and postnatal maturation of secondary wool follicles could each be influenced by environment, in this case adverse maternal nutrition achieved by severe restriction of feed intake of ewes throughout pregnancy (ad-lib feeding after the gestational period) was investigated ([112]; see Fig. 5.5, I). Comparison of ewes on an "intermediate" plane of nutrition during gestation, that gained more than 3.0 kg in bodyweight, with ewes on the low plane of nutrition throughout gestation that lost 3.5 kg in body weight and produced less fleece (0.7 kg less) over the 10 month experimental period from pre-treatment to lamb weaning, indicated the success of the nutritional restriction. At birth, lambs born to the feed-restricted ewes had significantly lower mean body weights, S/P fibre ratios and fibre densities (means $\pm$ S.D.; 2.8 $\pm$ 0.2 kg; 2.2 $\pm$ 0.2; 76.9 $\pm$ 6.1 fibres/mm^2) than lambs born to the intermediate-fed ewes (3.8 $\pm$ 0.2 kg; 3.1 $\pm$ 0.2; 90.0 $\pm$ 4.9 fibres/mm^2). However, S/P follicle ratios at birth were only slightly lower for lambs from feed restricted ewes and not significantly different from the S/P follicle ratio of lambs from intermediate-fed ewes (19.1 $\pm$ 0.1 vs. 21.2 $\pm$ 0.9). While the lamb data cited above were derived from a combination of male and female data, trends for the individual sexes were the same for all parameters. In the early postnatal period and up to 56 days, weight gain in the "low" group lambs was significantly lower than lambs in the "intermediate" group, most likely a carryover effect of poor gestational nutrition of the ewes, leading to lower milk production [3]. Subsequently the difference in weight gain between the lamb groups was not significant. The S/P fibre ratio and follicle density of low group lambs (9.8 $\pm$ 1.0; 47.7 $\pm$ 3.0 fibres/mm^2) remained significantly lower than those in intermediate group lambs (13.9 $\pm$ 0.9; 57.3 $\pm$ 2.2 fibres/mm^2) throughout the measurement period (up to postnatal day 168).

Overall these results indicate that adverse maternal nutrition imposed during the entire gestational period has a significant effect upon postnatal maturation of the secondary follicle population, which represents the majority of wool-producing follicles in sheep. Further, while of lesser magnitude, the data suggest there is an effect upon prenatal development of the follicle population, for despite similar S/P follicle ratios at birth (slightly fewer secondary follicles formed in the low treatment lambs), the resultant density of productive follicles in lambs as they matured was significantly lower in the low treatment lamb group, for both males and females. While the number of primary and secondary follicles in the skin of sheep is determined by developmental events that occur well before birth, one could argue that any interference to these events during gestation could well reduce the subsequent productive competency of wool follicles, by way of affecting development of any one of the multiplicity of cellular structures and functions essential for wool synthesis after birth. In this case, though the numbers of secondary follicles initiated in the low treatment group lambs was counted as only slightly lower than in intermediate treatment group lambs, it is likely that the development of far more of the secondary follicles to full productive competence was compromised by the adverse prenatal environment.

In relation to the effects upon wool production of the lambs from the two different treatment groups, it was determined that while the amount of wool produced by the lambs was not significantly different over the first 200 days after birth (clean fleece weights of 850 ± 78 g vs. 875 ± 75 g for low vs. intermediate lambs), fleeces of low treatment group lambs featured both coarser and longer fibres (mean FD 22.9 ± 0.6 μm; staple length 51 ± 2 mm) than the intermediate treatment group lambs (mean FD 20.7 ± 0.4 μm; staple length 47 ± 1 mm).

Interestingly, the combined male and female data showed the mean diameter of both primary and secondary fibres was higher in the low vs. intermediate treatment group lambs. So in addition to measured effects upon development and maturation of the secondary wool follicle population, these data indicate that economically significant qualitative features of the wool produced, namely fibre diameter and staple length, are also able to be influenced by the environment imposed upon the developing fetus, in this case through adverse maternal nutrition. Indeed, fibre diameter measurements suggest that underfeeding the dam throughout the entire gestation clearly affects development of the primary follicles as well as the secondary follicles. Moreover, the data support the earlier contention that follicles in skin compete for the available keratin-forming substrates, in that the changes in the follicle number and spatial relationships (i.e. fibre density) seen here affect the fibre diameter and length but not the overall quantity of wool produced.

In a separate study, it was found that high birth weights were associated with increased total numbers of primary follicles, that this was independent of skin area at birth [106] but was correlated with differences in size (but not follicle density) at day 90 of gestation. It was also noted that, while the S/P follicle ratio at birth provides a ceiling to development of follicle number in the postnatal period, prenatal growth conditions as judged from measured birth weight, can influence the number of secondary follicles initiated and the number that mature. Further, it was conceded that while the factors involved in determination of birth weight may also determine the number of follicles initiated, additional factors could be operating. Finally it was noted that lamb growth from birth to one month has a significant bearing upon the proportion of immature follicles that will ultimately mature to produce a wool fibre. This will be discussed further below.

5.4.2 Manipulation of Prenatal and Postnatal Development of the Wool Follicles

In order to extend earlier findings, a comprehensive investigation of the effects of prenatal and early postnatal nutrition on adult fleece and body characters was made in Merinos of the Medium Peppin strain [107]. Similar to the earlier work, the nutritional treatments included either high (H) or low (L) rations for ewes during the entire gestation period. This was followed by either high or low nutrition for singleton lambs from birth until 16 weeks of age, to produce four different treatment groups (HH, HL, LH, LL; Fig. 5.5, II). Postnatal nutritional treatments included hand rearing with a high or low allowance of cows' milk plus supplements, from three days after birth, combined with access to varying levels of dry feed supplements from 2 weeks onwards. All lambs then underwent a "recovery period" from 16 weeks until 48 weeks of age, during which they were given the same nutritional ration. Finally, production measurements were made between 48 weeks and the conclusion of the trial at 180 weeks, during which the nutritional ration, adjusted for individual body weight, was the same for all sheep.

In addition to significant differences in mean birth weights between lambs from ewes fed high and low nutrition rations (mean ± S.D.; $H = 3910 \pm 436$ g; $L = 2570 \pm 294$ g), mean body weights at 16 weeks also demonstrated large differences (HH 18.6 kg; HL 10.3 kg; LH 15.8 kg; LL 8.6 kg) though by 108 weeks, these differences represented a much lower proportion of total body weight (HH 53.2 kg; HL 48.1 kg; LH 48.5 kg; LL 44.3 kg). Comparison of midside skin growth between birth and 108 weeks showed expansion to be between 5-fold (H group) and 7-fold (L group), with greater relative postnatal skin growth occurring in the animals that were smaller at birth. Though not different statistically, the mean total number of primary follicles in H and L lambs at birth (4.2 ± 0.2 million and 3.8 ± 0.2 million, respectively) indicate that despite higher follicle densities, the smaller L lambs do have fewer primary follicles. In agreement with previous data, lambs from H ewes had significantly more mature fibre-producing secondary follicles (79.2/mm^2 vs. 53.5/mm^2) and a higher S/P fibre ratio at birth (3.86 vs. 2.05) than lambs from L ewes. Estimates of the total number of follicles at birth suggested lambs from high feed intake ewes had 15% more follicles than lambs from low feed intake ewes (89 ± 4.7 million follicles vs. 76 ± 4.7 million follicles).

With respect to the ability to produce wool subsequent to the treatment period, it is of great significance that up to 16 weeks, secondary follicle maturation in all lambs maintained on low nutrition in this postnatal period was markedly reduced (40% mature for LL lambs, 29% mature for HL lambs) in comparison to lambs fed a high nutrition ration postnatally (75% mature for HH lambs, 70% mature for LH lambs). Secondary follicle maturation increased substantially in all lamb groups during the recovery period (weeks 16–48) and there was little additional increase in S/P ratio in any group after 48 weeks. Notably, while S/P fibre ratios of HH, HL and LH lambs were quite similar at 48 weeks of age and even more so at 108 weeks of age (22.3 ± 1.8, 21.2 ± 0.9 and 23.0 ± 1.9, respectively), the S/P fibre ratio in skin of LL lambs was significantly lower (16.5 ± 1.0 at 108 weeks of age). Given a similarly high proportion of immature secondary follicles (Si) versus secondary and primary follicles producing fibres (Sf and Pf) were present in the skin of LH and LL lambs at birth (Si/Sf + Pf was 5.78 vs 5.23, respectively), it is evident that the disadvantage imposed on follicle development in the fetus by low ewe nutrition during gestation is exacerbated by poor nutrition in the early postnatal period. Fewer of the immature secondary wool follicles present at birth are able to fully mature to produce a fibre if early postnatal nutrition in the lamb is also poor.

The major effect of the prenatal low nutrition ewe treatment, a reduction in fetal follicle numbers presumably through interference with follicle initiation, is ultimately manifested as a lower number of wool fibres in the fleeces of the resulting

LH and LL lambs (approximately 58 and 52 million fibres, respectively; average 55 million fibres) compared to their high nutrition HH and HL lamb counterparts (approximately 64 and 69 million fibres, respectively; average approximately 67 million fibres). While daily wool growth measurements made at three separate intervals (weeks 70–110, 145–155 and 165–180) after the nutritional treatments had concluded consistently showed that HH sheep produce the most wool and LL sheep the least, a highly significant association between body weight and daily wool growth was evident. This lends weight to the previous assertion that follicle maturation is closely linked with postnatal lamb growth in the first month [106] and indicates the long-term nature of the effect on wool production.

It is of interest that in a similar study of the effects of pre and postnatal nutrient deprivation in ewes on wool production in single born Merinos, it was reported that despite a significant difference in greasy fleece weight produced in the first year, there were no long-term effects on any characteristic of wool production as a result of the treatments [26]. In this work, the pregnant and lactating ewe mothers were offered high and low planes of nutrition from day 100 of gestation until weaning (groups HH, HL, LH and LL; as described for Schinckel and Short above [107]), and lambs were run together under commercial conditions after weaning. First, prenatal nutrition deprivation had no significant influence on progeny birth weight even though significant differences in weight gain in the ewes occurred. Second, although low postnatal nutrition reduced S/P fibre ratio and total wool fibre number up to weaning, these differences were not evident at 1 year of age. Finally, greasy fleece weights at the year two and three shearings were not significantly different between the groups and there were no treatment effects on staple length, fibre diameter or yield. Given the higher severity and longer duration of nutritional treatments imposed by Schinckel and Short and that these caused larger differences in ewe live weights, significant differences in lamb birth weights and permanent effects on wool production, comparison of these studies highlights the need to bear in mind nutritional impositions that may reasonably occur in the field. An illustration of this point is given in a description of the effect of different pre-weaning farm environments on adult wool production in which it was concluded that poor pre-weaning environmental conditions are unlikely to permanently reduce adult wool production in the Merino [27].

Deleterious effects of prenatal and postnatal nutritional restriction on wool follicle development were also reported for Scottish Blackface sheep carried either as singles or twins, then reared either naturally or artificially as singles or twins in the early postnatal period [28]. This included a reduction in the number of active secondary follicles at birth and consequent reduction in the S/P fibre ratio, attributed to reduced prenatal nutrition of the ewe. However, the treatments used did not have a permanent effect upon wool production if postnatal nutrition was adequate. While this is different to findings in the Merino, it may be attributed to differences in the duration and severity of treatments employed and to breed differences, both in developmental timing of follicle formation and in response to seasonal influences.

In a variation of the experiment of Schinckel and Short [107], South Australian Merino ewes in a moderate condition were joined with Merino rams, then subjected to controlled grazing to provide either a high (H) or low (L) plane of nutrition for the first 90 days of gestation and either fed the same ration (HH and LL groups) or allocated to a group receiving the opposite nutritional ration (HL and LH groups) until day 140 of gestation [31, 117]. The maternal nutrition regime is summarised in Fig. 5.5, III. Following this, all ewes were fed a supplementary concentrate offered ad libitum and grazed together with their singleton progeny after lambing. Lambs were weaned onto pasture at 12 weeks of age, with midside skin sampled at birth, 12 weeks and 18 months of age and fleeces collected by shearing at 2, 16, 24 and 33 months after weaning. As data presented by the author for male and female progeny within treatment groups generally showed similar trends to data averaged over the groups, only the latter is discussed below.

As a result of the nutritional differences of ewes between conception and day 90, H ewes were significantly heavier than L ewes (8.3 kg; $p<0.001$). Mean post-partum body weights of HH, HL, LH and LL ewes (45.1, 38.8, 41.4 and 33.6 kg, respectively) specifically highlighted the weight loss suffered by LL ewes. Significantly reduced birth weights of lambs from underfed ewes indicated that both early and late gestational underfeeding could influence birth weight and the effects were cumulative. Interestingly, while a maternal nutritional deficit early in gestation affected lamb growth in early postnatal life, the effect lessened with age and was not significant 18 weeks after weaning. By contrast, maternal nutritional deficit in late pregnancy had a much greater impact on lamb growth that persisted throughout the experiment (up to approximately 3 years). Finally, as previously reported, the effects of maternal undernutrition on lamb birth weight and subsequent lamb growth were accompanied by commensurate changes in the wool follicle population and wool production. First, primary follicle density at birth was significantly greater in lambs born to ewes undernourished either early or late in gestation (and was accounted for by differential lamb size) while the density of mature secondary wool follicles was lower only in lambs born to ewes undernourished late in gestation. In addition, lambs from ewes on a low nutritional plane in late gestation had fewer mature primary and secondary follicles at birth and the difference persisted until the last midside sampling, when the estimated total follicle population in progeny from HH and LH ewes (averages of 62.6 and 61.8 million follicles, respectively) was significantly higher than progeny from HL and LL ewes (averages of 57.8 and 48.3 million follicles, respectively). Finally, it was noted that production of wool over the trial period was markedly reduced in lambs born to ewes undernourished for 140 days of gestation, with a smaller clean fleece weight measured for the LL lambs at each shearing and that the effect of undernutrition late in gestation played the most significant role in reduced wool production.

5.4.3 Manipulation of Follicle Development by Timed Prenatal Intervention

With evidence that prenatal nutritional restriction could influence secondary follicle initiation, it was then pertinent to determine distinct gestational periods in which such restrictions could influence follicle formation most dramatically. One such experiment conducted in the Scottish Blackface breed compared the effects of moderate and severe underfeeding of pregnant ewes on the initiation of secondary follicles in the fetuses [58]. In comparison to a well fed pregnant ewe control group (A), periods of nutritional stress imposed on other pregnant ewes varied and included short term severe underfeeding (B: days 112–142, C: days 112–131 with control feed levels from days 132–142; D : days 95–116 with control feed levels from days 117–142) and long term moderate underfeeding (E : days 35–142, F : days 35–119 with near control feed levels from days 120–142). Before the gestational underfeeding treatment periods began, ewes in groups B, C and D were fed the same nutritional ration as control ewes in group A. The maternal nutrition regime is summarised in Fig. 5.5, IV. Though no other detail is given, the authors state that each treatment group was homogenous with respect to the sex of the fetuses. According to our current knowledge of follicle initiation in Merinos, we would expect that in sheep, the earliest primary follicles initiate at around day 50–60 of gestation (completed by day 70), secondary original follicles initiate at around day 80 (completed by day 85) and secondary-derived follicles initiate at around day 100 (completed by day 135). Hence the periods of nutritional restriction in groups E and F encompass nearly the entire follicle initiation period (see Fig. 5.1), while group D encompasses the early-mid period of formation of secondary-derived follicles, and groups B and C the mid-late period of formation of secondary-derived follicles. It should be remembered that maturation of primary and secondary follicles also occurs during the timeframes covered by these nutritional impositions. As above, maturation of the follicles to produce fibre occurs between initiation and approximately day 100 of gestation for the earliest primary follicles and between initiation and until many weeks after birth for secondary and secondary-derived follicles.

In addition to significantly lower weights in fetuses from all nutritionally restricted ewes compared with fetuses from control ewes, sacrifice of lambs at day 142 and determination of skin S/P follicle ratios (total of mature and immature follicles) showed significantly lower S/P ratios in groups where moderate or severe underfeeding occurred in late gestation. Fetuses from groups B, C and E (days 112–131) had lower S/P follicle ratios (1.71, 1.42, 1.79, respectively) than control group A fetuses (2.42). Further, where feed levels were restored to those of control ewes, rescue of S/P ratios was only possible if the feed restriction was lifted between days 117–120 of gestation (groups D and F). Restoration of feed to ewes from day 132 of gestation (group C) was unable to avert the effect upon S/P ratio. This is not surprising, as the initiation of secondary follicles has been reported to be complete in many breeds, including the Romney, Drysdale, Wiltshire, Merino and Merino-Romney cross by around 130 days of gestation [52, 113]. It should be added that despite the pooling of data from singleton and twin fetuses, all S/P results go against a trend that might be expected if parity effects were operating. In light of this data and the information above, the following may be concluded. First, it is not possible to say whether primary follicle initiation or subsequent development was affected by any of the prenatal nutrition treatments imposed. Second, given the timing of the treatments, it is likely that formation of the secondary follicles was most affected in the experimental groups B, C and E, where lamb skin S/P follicle ratios were significantly lower than in control lambs. An interesting addition to this experiment would have been a severe feed restriction imposed on some pregnant ewes from day 50, with and without restoration of feed levels from day 75 onwards, to determine the influence of attempted interference to formation of the primary follicles upon subsequent S/P ratios in the fetuses.

Apart from measurement of fetal weights at day 142, some fetal morphometry was undertaken during the treatment periods, from day 90 until late gestation. Using crown-rump length growth rate as a measure of fetal growth, fetuses from ewes given short-term severe feed-restrictions (treatments B, C and D) exhibited slower growth rates (3.7 mm per day at day 115) compared to fetuses from well-fed control ewes (6.5 mm per day even at day 120; [80]). In treatment group B (ewes underfed from day 112 until day 142) fetal growth continued to decline to only 2.9 mm per day at day 132 (c.f. 3.7 mm per day for fetuses from control group A). The crown-rump length of fetuses in treatment group C (ewes underfed from day 112 until day 132) reduced to 3.6 mm per day but followed fetus controls from there onwards. Similarly, group D fetuses (ewes underfed from day 95 until day 116) showed immediate growth rate increases upon return of ewes to normal rations [81]. For fetuses from long-term underfed ewes (treatments E and F), measurement of fetal growth rates from day 90 to day 130 showed a reduction in crown-rump length growth from 5.9 mm per day (days 90–111) to only 3.6 mm per day (day 130). Lastly, as for fetal weight, by day 142, placental weight and fetal crown-rump length were each significantly lower in fetuses from feed-restricted ewes than in fetuses from maintenance fed ewes [82]. Though rudimentary in comparison to measurements that have been made previously, the data here indicate the nutritional restrictions imposed on the ewes in this study were of a magnitude to provide genuine intra-uterine growth restriction to the developing fetuses, with proportionate effects upon development of tissues, organs and systems in the developing fetuses being likely [4; 79].

5.4.4 *Observations of the Long-Term Effects of Prenatal and Postnatal Manipulation of Wool Follicle Development*

A more recent experiment in Merinos has defined the effects of strict maternal undernutrition for a prolonged period during gestation on the quantity and quality of wool production in the resulting lambs, initially up to 1.4 years of age [65]. Ewes were provided either maintenance (M; grain supplement in addition to available pasture) or sub-maintenance (SM; pasture only) nutrition from day 50 of gestation until 10 days before lambing (day 140), that created a mean weight loss of some 12.1 kg in the latter. Eleven days prior to lambing, all ewes were run together and given an equivalent grain supplement throughout the lambing period and until weaning at 12 weeks postnatal. Run in the same group, lambs were given no nutritional restriction, with access to their mothers and available grain until weaning. A summary of the maternal nutrition regime is given in Fig. 5.5, V. In order to minimise between-animal variation, lambs analysed in the study were clones, produced through transfer of bisected day 6 embryos. Ewes carrying paired clones were assigned to appropriate M and SM groups to allow comparison of effects on genetically identical siblings.

More sophisticated measurements than had been made previously gave a superior account of the effects of undernutrition on the nutritional status of both the ewes and fetuses during the treatment period. Measurement of placental development by determination of (i) average cotyledon diameters using ultrasound scanning, and (ii) blood flow resistance in maternal and umbilical arteries, indicated no significant differences between M and SM ewes at day 98. However, determinations of plasma metabolites in late pregnancy (day 133) were consistent with underfeeding of the pregnant ewes; significantly lower glucose and protein concentrations in SM ewes and their fetuses and higher ß-hydroxybutyrate concentrations in SM ewes. As seen in earlier such trials, mean SM lamb birth weights were significantly lower (5.0 vs. 5.5 kg [s.e.d. = 0.13 kg; $p<0.01$]) and weights remained significantly lower until lamb shearing (1.6–1.7 kg lower; $p<0.01$) and were still lower at hogget shearing (1.3–2.0 kg lower; $p<0.1$). In regard to wool production and again like earlier results, mean birth S/P fibre ratios for SM vs. M lambs were significantly lower (midside ratio 4.1 vs. 5.5 [s.e.d. = 0.24; $p<0.01$]) and this was also maintained as the animals aged (20.0 vs. 21.5 [s.e.d = 0.45; $p<0.01$] at day 150 [lamb shearing] and 21.2 vs. 19.1 [s.e.d. = 0.37; $p<0.01$] at hogget shearing). Further, lower S/P fibre ratios measured in SM lambs at the midside and other body sites (shoulder and hind leg) were borne out in lower clean fleece weights obtained at lamb shearing (0.1 kg less; $p<0.01$) and hogget shearing (0.14 kg less; $p<0.1$). Lastly, while there were no significant differences in staple length, staple strength, yield or percentage of fibres greater than 30 μm in diameter as a result of maternal undernutrition, measurements made at 5 sites determined the mean fibre diameter for SM hogget lambs to be 0.1 μm greater than for M hogget lambs. Interestingly, the higher mean fibre diameter for SM lambs included growth of slightly finer fibres (midside 18.3 vs. 18.5 μm [s.e.d. = 0.28; $p<0.1$]) when feed quality was low (December) and growth of much coarser fibres (midside 22.4 vs. 20.3 μm [s.e.d. = 0.5; $p<0.01$]) when feed quality was high (June). Whether the result of establishment of follicle populations with different metabolic capacities and support systems, the differential operation of epigenetic effects or a combination of both, the treatment has created a lasting difference in the way the mature follicles of SM lambs respond to environmental changes during later life, when compared to the mature follicles of their genetically identical M lamb counterparts. Given the lower clean fleece weights achieved and penalties applied to higher diameter fibres in general, while measurement of wool quality and quantity produced by lambs from underfed ewes was reported until the lambs reached the hogget stage in the study to this point, the early data made a strong case for management of ewe nutrition during mid to late pregnancy to prevent ewe weight loss and maximise progeny performance.

An even clearer picture of the long-term effects of maternal undernutrition during gestation has been presented in a most recent publication of the extension of this work [66]. In addition to presentation of the production data collected from the M and SM lambs described above over another 5 years (a total of 6.4 years; Experiment 1), long term production data (3.4–4.4 years) from additional M and SM lamb groups (Experiment 2) are presented. Lambs in the second experiment were produced essentially as described for the first experiment but with maternal undernutrition continuing until lamb weaning at 12 weeks postnatal. Comparison of the maternal nutritional regimes used in Experiments 1 and 2 can be made in Fig. 5.5, V and VI. In Experiment 2, most findings for ewes and their progeny up until lambs were 1.4 years of age mirrored those from Experiment 1, with the exception that the difference between mean weaning liveweights of M and SM lambs at 12 weeks was far greater (25.0 vs. 14.0 kg; $p<0.001$) where maternal undernutrition was continued until weaning. In Experiment 1, weaning liveweights for M and SM lambs were significantly different but quite similar by comparison (25.9 vs. 24.2 kg; $p<0.01$). As a result, throughout the rest of the trial period in Experiment 2, SM animals were on average 3.2 kg lighter than M animals. Further, as might be expected from the birth weight and all subsequent liveweight data, midside S/P fibre ratios were also lower for SM than M animals at birth (3.8 vs. 4.9), weaning (14.9 vs. 18.7) and at 3.4 years (18.4 vs. 20.9) with an overall lower average adult S/P ratio (16.9 vs. 19.5; s.e.d. = 0.41; $p<0.001$). As for lamb shearing data described in Experiment 1, SM lamb clean fleece weights were also significantly lower than for M lambs in Experiment 2 (0.4 kg less; $p<0.001$). In both experiments, mean annual adult clean fleece weights were significantly lower for SM than M animals (0.17 kg less [Experiment 1] and 0.24 kg less [Experiment 2]; $p<0.05$). Lastly, while most other wool quality traits measured in SM and M progeny as adults were not significantly different over the trial period, analysis of combined data for mean fibre diameters measured in the two experiments revealed that on average, wool fibres from SM animals were 0.3 μm broader than their M counterparts. From these data, the authors conclude that sub-maintenance feeding of ewes during gestation can affect fetal formation of the wool follicle population and have permanent effects on liveweight, wool production and wool quality of the progeny.

On the basis of ample evidence that ewe nutrition during pregnancy can have substantial impacts upon the production performance of the progeny, a long-term program to monitor effects upon wool quality and quantity, in addition to reproductive performance and body composition was established in 2001 by Thompson and Oldham [118]. In an effort to establish optimal performance in both medium wool Merino ewes and their progeny, the investigation sought to identify critical windows during the reproductive cycle of breeding ewes when the impacts of nutrition on fetal growth and development would maximise the production of meat and wool and per hectare returns for producers. In early pilot scale trials, adult Merino ewes (condition score 3) were split into two groups following AI and fed to either lose or maintain weight during early and mid pregnancy (achievement of condition scores of either 2 or 3 by day 90 of gestation). Each separate flock was then subdivided into groups that grazed different levels of pasture through late pregnancy and lactation. In this case, ewes were allocated one of five different levels of feed on offer (800, 1100, 1400, 2000 and 3000 kg green DM/ha) between day 90 and lamb weaning [33]. At lamb weaning, the nutritional treatment ceased and ewes were run as a single flock. The feeding regime for different ewe groups in this trial is summarised in Fig. 5.5, VII.

Preliminary data from this program indicates first that birth weight in singleton lambs was not affected by the plane of nutrition of ewes in the first 90 days of gestation [34], though progeny growth rates to 12 months benefited from higher ewe nutrition during this period [94]. In addition to higher clean fleece weights, staple strengths [95] and fecundity [91] for ewes fed at higher planes of nutrition between day 90 and lamb weaning, a significant increase in lamb birth weight was evident (0.2 kg increase per 1 unit increase in ewe condition score; $p<0.01$; [34]). Further, the level of feed on offer to ewes during lactation had a marked effect on progeny growth rates, with a large range in weaning liveweights achieved at two different trial sites (16–22 kg and 14–28 kg; [94]). At 12 months of age, differences in progeny liveweights still remained, though they were much lower (less than 3 and 4 kg at the sites, respectively). Earlier research has suggested that such lower lamb weights may well reduce subsequent wool production and quality [65]. Preliminary analysis of wool production in the progeny as hoggets has confirmed this, with progeny from ewes grazed at higher levels of food on offer producing more wool and finer wool [35]. It is expected that the impending publication of more detailed data from the Lifetime Wool program will reveal more on the ultimate long term effects of maternal undernutrition on wool production in the progeny in larger scale trials that are currently in progress [92].

5.5 Responses of the Developing Fetuses and Fetal Follicles to Other Influences

5.5.1 Lactation and Shearing of the Ewe

It is known that other short-term or long-term stresses imposed on the pregnant ewe either as a result of routine management practises (e.g. shearing) or of pregnancy itself (e.g. the birth process, lactation) have an effect upon development of fetal follicles. While the impact of lactation on lamb performance is more obvious and has been considered briefly in the context of ewe nutrition in the postnatal period until weaning [31, 65, 107, 118] studies of the effects of mid-pregnancy shearing on follicle characteristics in the lambs at birth have made contrasting findings in relation to wool follicle characteristics and wool production in the resultant lambs. In singleton and twin lambs of ewes shorn mid-pregnancy (day 70 of gestation) Revell et al. [104] reported that birth weights were on average 0.5 kg higher ($p<0.01$) than for lambs born to unshorn ewes. This finding is similar to those made by others [63, 86]. In addition, in near-term (day 140) fetuses from shorn ewes, there was an increase in the ratio of secondary to primary follicles in midside skin, commanding a 9% increase in secondary follicle density in each case. This is also consistent with the earlier finding of association between birth weight and S/P ratio in the skin [31, 107, 112]. Further, fleece data suggested a trend towards reduced mean fibre diameters in lambs from mid-pregnancy shorn ewes.

In other studies of the effects of mid-pregnancy shearing, Sherlock et al. [110] reported lower secondary follicle numbers and higher fibre diameters, while Kenyon et al. have separately reported higher numbers of secondary follicles with no change in fibre diameter [67] and also no effect on either lamb wool follicle characteristics or fibre diameter [68]. Interestingly, while timing, duration and magnitude of a maternal thyroid hormone response associated with mid-pregnancy shearing [87, 115] may be sufficient to alter the ratio of secondary to primary follicles, the conflicting findings described above suggest this is not directly responsible for changes in the follicle population observed in the lambs born [68].

5.5.2 *Fetal Endocrine Mechanisms – Thyroid Hormone and Cortisol*

Given the known importance of the fetal thyroid gland in maturation of the fetal wool follicle population [23, 56] and a possible role in initiation of the secondary follicles [122] it may be significant that in the experiment described by Revell et al. [104], weight of the thyroid gland in near-term fetuses extracted from mid-pregnancy shorn ewes was 67% greater than those of fetuses from unshorn ewes. Whether increased maternal thyroid hormones have an effect, the likely increased fetal thyroid gland function could itself account for the larger number of secondary follicles in fetuses from shorn ewes. This possibility is entirely consistent with the previous evidence that maternal nutritional restriction during gestation, which is known to cause retardation in development of the fetal thyroid gland, also causes a significant (and permanent) reduction in the size of the secondary wool follicle population of the resultant lamb.

An analysis of the effects of cortisol in late gestation sheep fetuses has shown that while fetal plasma cortisol levels usually increase approximately 9-fold in the period from 25 days prepartum to 5 days prepartum [38], administration of repeated doses of the cortisol analogue beta-methasone between days 104 and 124 of gestation, a period preceding the natural surge, causes a reduction in the ratio of secondary to primary follicles [53].

5.5.3 *Maternal Heat Stress*

It has been demonstrated in many studies that high ambient temperatures for ewes during gestation can result in low lamb birth weights [5, 109, 124]. Cartwright and Thwaites [20] described "fetal stunting" in progeny of ewes maintained under conditions of heat stress for the last two thirds of pregnancy and ascertained that despite appetite loss in the ewes, fetal undernutrition was not due to lower maternal feed intake per se, as pair-fed controls run at ambient temperatures produced lambs with normal birth weights. Analysis of the effects of heat stress on wool follicle development revealed a drastic reduction in the number of wool follicles in lambs from heat-stressed ewes, primarily the result of impaired development and maturation of the secondary wool follicles [21]. A more recent investigation of the effects of maternal heat stress imposed for the last two and a half months of gestation in sheep (day 64 until approximately day 141) has found a large reduction in placental weight in heat stressed ewes and in the weight of the developing fetuses (42 and 27% reductions, respectively; [8, 30]). In addition, the plasma concentration of cortisol was significantly reduced in heat-stressed ewes late in pregnancy, while triiodothyronine levels were markedly lower at all stages of pregnancy. It is clear from these studies that heat stress-induced fetal growth retardation, resulting from gross underdevelopment of the placenta and associated effects on maternal endocrine status, can impact heavily upon development of wool follicles during gestation and is likely to have significant effects on postnatal wool production.

5.5.4 Hypoxia

As an adjunct to what might be expected from other experiments that induce intrauterine growth retardation, the effects of hypoxia on development of the wool follicle population was studied, after long-term (day 30 – day 135 of gestation) and short-term (day 120 – day 140 of gestation) hypoxia treatments were applied to ewes carrying singleton fetuses [60]. While long-term hypoxia resulted in significantly reduced fetal weights at day 140, fetal weights were not affected by the short-term treatment. Interestingly, analysis of skin sections revealed that the secondary to primary follicle ratio (S/P) was lower in both groups of fetuses subjected to hypoxic conditions than in the respective control fetuses. While the effect on primary follicle formation is unclear, it may be concluded that the initiation of secondary wool follicles between days 120 and 140 of gestation is reduced by hypoxia. The reduced oxygen levels accompanied by reduced levels of fetal plasma thyroid hormones [49] imply a role for altered fetal thyroid function.

5.6 Summary

Follicle Development and Postnatal Fibre Production

As an example of follicle formation and fibre production in ruminants, we have described the essential features of wool follicle formation and maturation in the Merino sheep – that it is a three-stage process spanning all of gestation and the early postnatal period and produces distinct wool follicle populations, the primary and secondary follicles. Nomination of follicle initiation sites occurs at some time prior to day 50 while primary, secondary original and secondary-derived follicles begin to form at approximately day 55, day 80 and day 100 of gestation, respectively. We described that during development, competition exists between follicles, that there appear to be a finite number of follicle stem cell precursors that can be distributed in different proportions between the follicle types and that this has a bearing upon both the number of follicles formed, thus the follicle density and in turn, the diameter of resultant fibres. Further, we have highlighted the molecular complexity of follicle formation and maturation. Lastly, the critical genes facilitating epithelial-mesenchymal interactions required to establish the follicle population (Wnts) and cell fate determination (Notch pathway molecules) and the key regulatory morphogens (TGFß family members) and growth factors (FGFs) utilised in morphogenic processes required for follicle establishment and fibre synthesis functions, have been described.

Manipulation of Follicle Development and Postnatal Fibre Production

The outcomes of short-term and long-term sheep trials focussing upon effects of the "maternal environment" upon development of wool follicles and wool fibre production have been presented. While most have not illuminated the specific physiological mechanisms involved, these experiments undoubtedly provide early examples of the effects of intra-uterine growth retardation (IUGR). In the work discussed, it is more than likely that the physiological consequences of IUGR, as described for sheep in more recent publications [54, 55, others] have adversely impacted on development of the wool follicle population. The outcomes of manipulation of mechanisms governing both prenatal and postnatal development of the wool follicle population and postnatal wool fibre production have indicated the plasticity of these processes, defined some of the gestational and early postnatal periods within which these effects can act and quantified the ultimate effects they can have upon wool production.

Many of the experiments described have determined the effect of prenatal and postnatal maternal nutritional treatments upon the most economically significant traits of wool produced by the lamb progeny, namely clean fleece weight (decreased) and fibre diameter (increased). The effect of these treatments on another economically important trait, staple strength, has been found to be insignificant to date (but see below) and the same is true for staple length, yield and the proportion of fibres with diameters greater than 30 μm [65, 66]. Fibre crimp can change over time (i.e. reduces with age), can be influenced postnatally (e.g. by diet) and may ultimately depend more on genotype than environment, with recent proteomic evidence of an association of specific cysteine-rich keratin associated proteins with crimp frequency [98]. Staple strength can also be influenced postnatally and proteomic analysis of wool from different sheep strains and breeds [36] may shed light on the proteins that contribute to individual differences in this important trait. Nonetheless, considering the dependence of staple strength on low variability in fibre diameters (i.e. fleeces with a lower coefficient of variation in fibre diameter have higher staple strength) it would be surprising if follicle development, that includes the establishment of diameter, has no bearing upon staple strength.

Future Work

From the summary of research findings, it is obvious that a more systematic study of the influences of developmental aberrations on the processes of wool follicle development and wool fibre production is required. Although treatment periods have sometimes spanned the whole of gestation, experiments to date have been designed to assess the effects of maternal manipulations on later gestational and early post-natal events, predominantly the formation and maturation of the secondary wool follicle population. Experiments enabling determination of ultimate production outcomes have been especially valuable where the manipulations employed closely mimic field conditions facing the sheep. However, analysis of effects on fetal wool follicle development and subsequent productivity, via manipulations targeted to specific "windows" of the gestational period, has been neglected to this point. This applies especially to the early parts of gestation and to the gestational periods where initiation of the different wool follicle types is known to occur.

To this point, too few and insufficiently discrete gestational time windows have been assessed using manipulative treatments. Further, too few fetal samples have been extracted and analysed across gestation during the treatment periods reported to date to ascertain wool follicle-specific developmental consequences. By contrast, to date, ample organ and tissue samples have been collected from growth-restricted/growth-enhanced fetuses generated via multiple mechanisms and the morphometric and physiological consequences of the aberrant growth have been described in great detail. While there is no value in repeating such efforts per se, there is sufficient known of the molecular biology of follicle formation, from our own research and that of others in the hair research field, such that the skin (and other tissues) of fetal samples thus obtained could be scrutinised to determine the exact impacts of targeted gestational manipulations on wool follicle development at the molecular level. While we know of effects of some of the major endocrine pathways (for review see [121]) and growth factor molecules on wool growth in the adult, and a little about the influences of these on wool follicle development [23, 56, 122], a comprehensive gene expression profiling analysis similar to one we are currently undertaking in skin samples collected from Merino fetuses throughout gestation (day 35 until day 143) would identify key molecular targets of the particular gestational manipulations applied. An immediate priority in our own work is spatio-temporal characterisation of expression of the receptors for thyroid hormone and glucocorticoids in developing fetal sheep skin. If related to what is already known of the molecules responsible for initiation and maturation of the wool follicle, the findings would identify the likely windows of action of these key endocrine components and precisely define the developmental mechanisms with most impact upon wool follicle development and postnatal fibre production.

In an earlier consideration of the variation of wool growth in relation to physiological state, many of the observations described above, concerning prenatal and postnatal influences on lifetime wool production of the developing lamb are discussed, along with the effects of subsequent age, sex and reproduction on wool growth [25]. Interestingly, the author suggests that formation of the primary follicles early in gestation may not be prone to the effects of an adverse maternal environment. This is quite contentious for even though some of the periods of maternal undernutrition used in the studies described up until that time, and even more since, have included early gestation, the results in relation to primary follicle development are equivocal. This is because no fetal skin samples from consecutive early gestational time points have been collected or analysed in order to assess effects of manipulations on primary follicle development in real time. A study of fetal development and postnatal wool production in lambs from Merino ewes undernourished from mating until day 35 of gestation showed significantly lower fetal weights and sizes at day 35 but no significant differences in lamb fleece weights at the first and second shearings [96]. While the possibility of compensatory growth and development in late gestation may explain the apparent lack of effect of this early treatment on postnatal wool production, data with which to ascertain the exact effects upon wool follicle development are not available.

More recently, an investigation of the effects of maternal nutrient restriction early in gestation (days 20–77) on fetal and placental morphometry at term has been undertaken [51]. With nutrition-restricted and control ewes fed at maintenance from day 77 until term, final morphometry showed no differences in fetal weights or fetal organ weights between the groups, but larger placentae in feed-restricted ewes that featured a smaller fetal component in relation to placentae from controls. No data pertaining to wool follicle development was collected, though considering the correlation between fetal weight and total weight of the fetal component of the placentae reported for fetuses from feed-restricted ewes, it would be surprising if absolutely no wool follicle-related effects occurred. In another study, the effects of maternal nutrient restriction during early pregnancy (from mating to day 70 of gestation) on muscle development and meat quality in lambs were investigated [71]. This and many studies cited in other sections of this volume have described multiple effects of early maternal undernutrition on muscle development. As suggested above, considering the timing of the treatment in these cases and given the interrelationships between primary and secondary wool follicle formation, similar investigations including collection of consecutive fetal, as well as early postnatal, juvenile and adult skin samples, followed by analysis of postnatal wool fibre production, would be a most informative exercise. Other experiments that have perturbed the environment of embryos as early as the start of gestation (days 0–6) have also indicated downstream consequences for fetal development [69, 70, 102, 120]. While it is possible that such early intervention could ultimately impact upon wool follicle development in utero, these are the types of experiments required if all of the effects of gestational manipulation of fetal wool follicle development are to be unravelled.

References

1. Adelson, D.L., B.A. Kelley, and B.N. Nagorcka. 1992. Increase in dermal papilla cells by proliferation during development of the primary wool follicle. *Aust. J. Agric. Res.* **43:**843–856.
2. Adelson, D.L., D.E. Hollis, and G.E. Brown. 2002. Wool fibre diameter and follicle density are not specified simultaneously during wool follicle initiation. *Aust. J. Agric. Res.* **53:**1003–1009.
3. Alexander, G. 1956. Influence of nutrition upon duration of gestation in sheep. *Nature* **178:**1058–1059.
4. Alexander, G. 1974. Birth weight of lambs: influences and consequences, pp. 213–239. *In* K. Elliott, and J. Knight, (eds.), Size at Birth, Elsevier, Amsterdam.

5. Alexander, G. and D. Williams. 1971. Heat stress and growth of the conceptus in sheep. *Proc. Aust. Soc. Anim. Prod.* **6:**102–105.
6. Barker, D.J.P. 1998. Mothers, Babies and Health in Later Life. 2nd Ed. Churchill Livingstone, Edinburgh.
7. Barker, D. 2002. Fetal programming of coronary heart disease. *Trends Endocrinol. Metab.* **13:**364–368.
8. Bell, A.W., B.W. McBride, R. Slepetis, R.J. Early, and W.B. Currie. 1989. Chronic heat stress and prenatal development in sheep: I. Conceptus growth and maternal plasma hormones and metabolites. *J. Anim. Sci.* **67:**3289–3299.
9. Black, J.L. and P.J. Reis. 1979. Speculation on the control of nutrient partition between wool growth and other body functions, pp. 269–293. *In* J.L. Black, and P.J. Reis, (eds.), Physiological and environmental limitations to wool growth, University of New England Publishing Unit, Armidale, N.S.W., Australia.
10. Blanpain, C., W.E. Lowry, A. Geoghegan, L. Polak, and E. Fuchs. 2004. Self-renewal, multipotency and the existence of two cell populations within and epithelial stem cell niche. *Cell* **118:**635–648.
11. Blessing, M., L.B. Nanney, L.E.King, C.M. Jones, and B.L.M. Hogan. 1993. Transgenic mice as a model to study the role of TGF-ß related molecules in hair follicles. *Genes Dev.* **7:**204–215
12. Botchkarev, V.A. and R. Paus. 2003. Molecular biology of hair morphogenesis: Development and cycling. *J. Exp. Zoolog. B. Mol. Dev. Evol.* **298**:164–180.
13. Botchkarev, V.A., N.V. Botchkareva, K.M. Albers, C. van der Veen, G.R. Lewin, and R. Paus. 1998. Neurotrophin-3 involvement in the regulation of hair follicle morphogenesis. *J. Invest. Dermatol.* **111:**279–285.
14. Botchkarev, V.A., N.V. Botchkareva, O. Huber, K. Funa, and B.A. Gilchrest. 2002. Modulation of BMP signaling by noggin is required for induction of the secondary (non-tylotrich) hair follicles. *J. Invest. Dermatol.* **118:**3–10.
15. Botchkareva, N.V., V.A. Botchkarev, L.H. Chen, G. Lindner, and R. Paus. 1999. A role for p75 neurotrophin receptor in the control of hair follicle morphogenesis. *Dev. Biol.* **216:** 135–153.
16. Botchkareva, N.V., V.A. Botchkarev, M. Metz, I. Silos-Santiago, and R. Paus. 1999. Retardation of hair follicle development by the deletion of TrkC, high affinity neurotrophin-3 receptor. *J. Invest. Dermatol.* **113:**425–427.
17. Carter, H.B. and W.H. Clarke. 1957. The hair follicle group and skin follicle population of Australian Merino sheep. *Aust. J. Agric. Res.* **8:**91–108.
18. Carter, H.B. and W.H. Clarke. 1957. The hair follicle group and skin follicle population of some non-Merino breeds of sheep. *Aust. J. Agric. Res.* **8:**109–119.
19. Carter, H.B. and M.H. Hardy. 1947. Studies in the biology of the skin and fleece of sheep. 4. The hair follicle group and its topographical variations in the skin of the Merino fetus. *Coun. Sci. Industr. Res. Aust. Bull.* No. **215**.
20. Cartwright, G.A. and C.J. Thwaites. 1976. Fetal stunting in sheep. 1. The influence of maternal nutrition and high ambient temperature on the growth and proportions of Merino fetuses. *J. Agric. Sci.* **86:**573–580.
21. Cartwright, G.A. and C.J. Thwaites. 1976. Fetal stunting in sheep. 2. The effects of high ambient temperature during gestation on wool follicle development in the fetal lamb. *J. Agric. Sci.* **86:**581–585.
22. Champion, S.C. and G.E. Robards. 2000. Follicle characteristics, seasonal changes in fibre cross-sectional area and ellipticity in Australasian specialty carpet wool sheep, Romneys and Merinos. *Small Rumin. Res.* **38:**71–82.
23. Chapman, R.E., P.S. Hopkins, and G.D. Thorburn. 1974. The effects of fetal thyroidectomy and thyroxine administration on the development of the skin and wool follicles of sheep fetuses. *J. Anat.* **117:**419–432.

24. Cock, M.L., E.J. Camm, S. Louey, B.J. Joyce, and R. Harding. 2001. Postnatal outcomes in term and preterm lambs following fetal growth restriction. *Clin. Exp. Pharmacol. Physiol.* **28:**931–937.
25. Corbett, J.L. 1979. Variation in wool growth with physiological state, pp. 79–98. *In* J.L. Black, and P.J. Reis, (eds.), Physiological and environmental limitations to wool growth, University of New England Publishing Unit, Armidale, N.S.W., Australia.
26. Denney, G.D., K.J. Thornberry, and M.A. Sladek. 1988. The effect of pre and postnatal nutrient deprivation on live weight and wool production of single born Merino sheep. *Proc. Aust. Soc. Anim. Prod.* **17**:174–177.
27. Denney, G.D. 1990. Effect of pre-weaning farm environment on adult wool production of Merino sheep. *Aust. J. Exp. Agric.* **30**:17–25.
28. Doney, J.M. and W.F. Smith. 1964. Modification of fleece development in Blackface sheep by variation in pre- and post-natal nutrition. *Anim. Prod.* **6**:155–167.
29. Dry, F.W. 1940. Recent work upon the wool zoology of the N.Z. Romney. *N.Z. J. Sci. Tech.* **22:**209–220.
30. Early, R.J., B.W. McBride, I. Vatnick, and A.W. Bell. 1991. Chronic heat stress and prenatal development in sheep. II. Placental cellularity and metabolism. *J. Anim. Sci.* **69:**3610–3616.
31. Everitt, G.C. 1967. Residual effects of prenatal nutrition on the postnatal performance of Merino sheep. *Proc. N.Z. Soc. Anim. Prod.* **27:**52–68.
32. Favier, B., I. Fliniaux, J. Thelu, J.P. Viallet, M. Demarchez, C.A. Jahoda, and D. Dhouailly. 2000. Localisation of members of the notch system and the differentiation of vibrissa hair follicles: receptors, ligands, and fringe modulators. *Dev. Dyn.* **218:**426–437.
33. Ferguson, M., B. Paganoni, and G. Kearney. 2004. Lifetime Wool. 3. Ewe liveweight and condition score. *Anim. Prod. Aust.* **25:**242.
34. Ferguson, M., D. Gordon, B. Paganoni, T. Plaisted, and G. Kearney. 2004. Lifetime Wool. 6. Progeny birth weights and survival. *Anim. Prod. Aust.* **25:**243.
35. Ferguson, M., B. Paganoni, and G. Kearney. 2004. Lifetime Wool. 8. Progeny wool production and quality. *Anim. Prod. Aust.* **25:**244.
36. Flanagan, L.M., J.E. Plowman, and W.G. Bryson. 2002. The high sulphur proteins of wool: Towards an understanding of sheep breed diversity. *Proteomics* **2:**1240–1246.
37. Fowden, A.L. and A.J. Forhead. 2004. Endocrine mechanisms of intrauterine programming. *Reproduction* **127:**515–526.
38. Fowden, A.L., J. Szemere, P. Hughes, R.S. Gilmour, and A.J. Forhead. 1996. The effects of cortisol on the growth rate of the sheep fetus during late gestation. *J. Endocrinol.* **151:** 97–105.
39. Fraser, A.S. 1952. Growth of the N-type fleece. *Aust. J. Agric. Res.* **3:**435–444.
40. Fraser, A.S. 1952. Growth of wool fibres in sheep. *Aust. J. Agric. Res.* **3:**419–434.
41. Fraser, A.S. 1954. Development of the skin follicle population in the Merino sheep. *Aust. J. Agric. Res.* **5:**737–744.
42. Fraser, A.S. and M. Hamada. 1952. Observations on some British sheep. *Proc. Roy. Soc. Edinb.* B **64:**462–477.
43. Fraser, A.S. and B.F. Short. 1952. Competition between skin follicles in sheep. *Aust. J. Agric. Res.* **3:**445–452.
44. Godwin, A.R. and M.R. Capecchi. 1998. Hoxc13 mutant mice lack external hair. *Genes Dev.* **12:**11–20.
45. Greenwood, P.L. and A.W. Bell. 2003. Consequences of intra-uterine growth retardation for postnatal growth, metabolism and pathophysiology. *Reproduction Suppl.* **61:**195–206.
46. Greenwood, P.L., R.M. Slepetis, J.W. Hermanson and A.W. Bell. 1999. Intrauterine growth retardation is associated with reduced cell cycle activity, but not myofibre number, in ovine fetal muscle. *Reprod. Fertil. Dev.* **11:**281–291.
47. Greenwood, P.L., A.S. Hunt, J.W. Hermanson, and A.W. Bell. 2000. Effects of birth weight and postnatal nutrition on neonatal sheep. II. Skeletal muscle growth and development. *J. Anim. Sci.* **78:**50–61.

48. Haines, B.P. 1991. BMP-2 expression in follicle development. Hons. Thesis, University of Adelaide, South Australia.
49. Harding, J.E., C.T. Jones, and J.S. Robinson. 1985. Studies on experimental growth retardation in sheep. The effect of a small placenta in restricting transport to and growth of the fetus. *J. Dev. Physiol.* **7**:427–442.
50. Hardy, M.H. and A.G. Lyne. 1956. The pre-natal development of wool follicles in Merino sheep. *Aust. J. Biol. Sci.* **9**:423–441.
51. Heasman, L., L. Clarke, K. Firth, T. Stephenson, and M.E. Symonds. 1998. Influence of restricted maternal nutrition in early to mid gestation on placental and fetal development at term in sheep. *Pediatr. Res.* **44**:546–51.
52. Hocking Edwards, J.E., M.J. Birtles, P.M. Harris, A.L. Parry, E. Paterson, G.A. Wickham, and S.N. McCutcheon. 1996. Pre-and post-natal wool follicle development and density in sheep of five genotypes. *J. Agric. Sci.* **126**:363–370.
53. Hocking Edwards, J.E., J. Newnham, M. Ikegami, D. Polk, and A. Jobe. 1997. Betamethasone increases sweat gland formation and suppresses fibre formation in fetal sheep. *Aust. J. Dermatol.* **38**:A327.
54. Holst, P.J., I.D. Killeen, and B.R. Cullis. 1986. Nutrition of the pregnant ewe and its effects on gestation length, lamb birth weight and lamb survival. *Aust. J. Agric. Res.* **37**:647–655.
55. Holst, P.J., C.J. Allan, and A.R. Gilmour. 1992. Effects of a restricted diet during mid pregnancy of ewes on uterine and fetal growth and lamb birth weight. *Aust. J. Agric. Res.* **43**: 315–324.
56. Hopkins, P.S. and G.D. Thorburn. 1972. The effects of fetal thyroidectomy on the development of the ovine fetus. *J. Endocrinol.* **52**:55–66.
57. Huelsken, J., R. Vogel, B. Erdmann, G. Cotsarelis, and W. Birchmeier. 2001. ß-catenin controls hair follicle morphogenesis and stem cell differentiation in the skin. *Cell* **105**:533–545.
58. Hutchison, G. and D.J. Mellor. 1983. Effects of maternal nutrition on the initiation of secondary wool follicles in fetal sheep. *J. Comp. Path.* **93**:577–583.
59. Ibrahim, L. and E.A. Wright. 1982. A quantitative study of hair growth using mouse and rat vibrissal follicles. I. Dermal papilla volume determines hair volume. *J. Embryol. Exp. Morphol.* **72**:209–224.
60. Jacobs, R., J. Falconer, J.S. Robinson, and M.E.D. Webster. 1986. Effect of hypoxia on the initiation of secondary wool follicles in the fetus. *Aust. J. Biol. Sci.* **39**:79–83.
61. Jahoda, C.A.B., J. Whitehouse, A.J. Reynolds, and N. Hole. 2003. Hair follicle stem cells differentiate into adipocyte and osteogenic lines *Exp. Dermatol.* **12**:849–859.
62. Jave-Suarez, L.F., H. Winter, L. Langbein, M.A. Rogers, and J. Schweizer. 2001. HOXC13 is involved in the regulation of human hair keratin gene expression. *J. Biol. Chem.* **277**: 3718–3726.
63. Jopson, N.B., G.H. Davis, P.A. Farquhar, and W.E. Bain. 2002. Effects of mid-pregancy nutrition and shearing on ewe body reserves and fetal growth. *Proc. N.Z. Soc. Anim. Prod.* **62**:49–52.
64. Karlsson, L., C. Bondjers, and C. Betsholtz. 1999. Roles for PDGF-A and sonic hedgehog in development of mesenchymal components of the hair follicle. *Development* **126**:2611–2621.
65. Kelly, R.W., I. MacLeod, P. Hynd, and J. Greeff. 1996. Nutrition during fetal life alters annual wool production and quality in young Merino sheep. *Aust. J. Exp. Agric.* **36**:259–267.
66. Kelly, R.W., J.C. Greeff, and I. Macleod. 2006. Lifetime changes in wool production in Merino sheep following differential feeding in fetal and early life. *Aust. J. Agric. Res.* **57**:867–876.
67. Kenyon, P.R., R.G. Sherlock, and S.T. Morris. 2004. Are elevated maternal thyroid hormone concentrations post mid-pregnancy shearing responsible for changes in lamb fleece characteristics ? *Proc. N.Z. Soc. Anim. Prod.* **64**:272–276.
68. Kenyon, P.R., R.G. Sherlock, T.J. Parkinson, and S.T. Morris. 2005. The effect of maternal shearing and thyroid hormone treatments in mid pregnancy on the birth weight, follicle and wool characteristics of lambs. *N.Z. J. Agric. Res.* **48**:293–300.

69. Kleemann, D.O., S.K. Walker, and R.F. Seamark. 1994. Enhanced fetal growth in sheep administered progesterone during the first three days of pregnancy. *J. Reprod. Fertil.* **102:**411–417.
70. Kleemann, D.O., S.K. Walker, K.M. Hartwich, F. Lok, R.F. Seamark, J.S. Robinson, and J.A. Owens. 2001. Fetoplacental growth in sheep administered progesterone during the first three days of pregnancy. *Placenta* **22:**14 23.
71. Krausgrill, D.I., N.M. Tulloh, W.R. Shorthose, and K. Sharpe. 1999. Effects of weight loss in ewes in early pregnancy on muscles and meat quality of lambs. *J. Agric. Sci.* **132:**103–116.
72. Kulessa, H., G. Turk, and B.L. Hogan. 2000. Inhibition of Bmp signaling affects growth and differentiation in the anagen hair follicle. *EMBO J.* **19:**6664–6674.
73. Langley-Evans, S.C. 2001. Fetal programming of cardiovascular function through exposure to maternal undernutrition. *Proc. Nutr. Soc.* **60:**505–513.
74. Laurikkala, J., J. Pispa, H.-S. Jung, P. Nieminen, M. Mikkola, X. Wang, U. Saarialho-Kere, J. Galceran, R. Grosschedl, and I. Thesleff. 2002. Regulation of hair follicle development by the TNF signal ectodysplasin and its receptor Edar. *Development* **129:**2541–2553.
75. Liu, S.M., G. Mata, H. O'Donaghue, and D.G. Masters. 1998. The influence of live weight, live-weight change and diet on protein synthesis in the skin and skeletal muscle in young Merino sheep. *Br. J. Nutr.* **79:**267–274.
76. Lindner, G., A. Menrad, E. Gherardi, G. Merlino, P. Walker, B. Handjiski, B. Roloff, and R. Paus. 2000. Involvement of hepatocyte growth factor/scatter factor and met signaling in hair follicle morphogenesis and cycling. *FASEB J.* **14:**319–332.
77. Mandler, M. and A. Neubüser. 2004. FGF signaling is required for initiation of feather placode development. *Development* **131:**3333–3343.
78. Meier, N., T.N. Dear, and T. Boehm. 1999. Whn and mHa3 are components of the genetic hierarchy controlling hair follicle differentiation. *Mech. Dev.* **89:**215–221.
79. Mellor, D.J. 1983. Nutritional and placental determinants of fetal growth rate in sheep and consequences for the newborn lamb. *Br. Vet. J.* **139:**307–324.
80. Mellor, D.J. and L. Murray. 1981. Effects of placental weight and maternal nutrition on the growth rates of individual fetuses in single and twin bearing ewes during late pregnancy. *Res. Vet. Sci.* **30:**198–204.
81. Mellor. D.J. and L. Murray. 1982. Effects on the rate of increase in fetal girth of refeeding ewes after short periods of severe undernutrition during late pregnancy. *Res. Vet. Sci.* **32:** 377–382.
82. Mellor, D.J. and L. Murray. 1982. Effects of long term undernutrition of the ewe on growth rates of individual fetuses during late pregnancy. *Res. Vet. Sci.* **32:**177–180.
83. Millar, S. 2002. Molecular mechanisms regulating hair follicle development. *J. Invest. Dermatol.* **118**:216–225.
84. Moore, G.P.M., N. Jackson, and J. Lax. 1989. Evidence of a unique developmental mechanism specifying both wool follicle density and fibre size in sheep selected for single skin and fleece characters. *Genet. Res.* **53:**57–62.
85. Moore, G.P.M., N. Jacskon, K. Isaacs, and G. Brown. 1998. Pattern and Morphogenesis in Skin. *J. Theor. Biol.* **191:**87–94.
86. Morris, S.T. and S.N. McCutcheon. 1997. Selective enhancement of growth in twin fetuses by shearing ewes in early gestation. *Anim. Sci.* **65:**105–110.
87. Morris, S.T., S.N. McCutcheon, and D.K. Revell. 2000. Birth weight responses to shearing ewes in early to mid gestation. *Anim. Sci.* **70:**363–369.
88. Morris, R.J., Y. Liu, L. Marles, Z. Yang, C. Trempus, S. Li, J.S. Lin, J.A. Sawicki and G. Cotsarelis. 2004. Capturing and profiling adult hair follicle stem cells. *Nat. Biotechnol.* **22**, 411–417.

89. Nakamura, M., M.M. Matzuk, B. Gerstmayer, A. Bosio, R. Lauster, Y. Miyachi, S. Werner, and R. Paus. 2003. Control of pelage hair follicle development and cycling by complex interactions between follistatin and activin. *FASEB J.* **17:**497–499.
90. Ohuchi, H., H. Tao, K. Ohata, N. Itoh, S. Kato, S. Noji, and K. Ono. 2003. Fibroblast growth factor 10 is required for proper development of the mouse whiskers. *Biochem. Biophys. Res. Comm.* **302:**562–567.
91. Oldham, C.M. and A.N. Thompson. 2004. Lifetime Wool. 5. Carryover effects on ewe reproduction. *Anim. Prod. Aust.* **25:**291.
92. Oldham, C.M., P. Barber, M. Curnow, S. Giles, and J. Speijers. 2004. Lifetime Wool. 14. Putting it all together in the paddock. *Anim. Prod. Aust.* **25:**292.
93. Olivera-Martinez, I., J. Thelu, M.A. Teillet, and D. Dhouailly. 2001. Dorsal dermis development depends on a signal from the dorsal neural tube, which can be substituted by Wnt-1. *Mech. Dev.* **100:**233–244.
94. Paganoni, B.L., M. Ferguson, G. Kearney, and T. Plaisted. 2004. Lifetime Wool. 7. Progeny growth rates. *Anim. Prod. Aust.* **25:**295.
95. Paganoni, B.L., M. Ferguson, and G. Kearney. 2004. Lifetime Wool. 4. Ewe wool production and quality. *Anim. Prod. Aust.* **25:**294.
96. Parr, R.A., A.H. Williams, I.P. Campbell, G.F. Witcombe, and A.M. Roberts. 1986. Low nutrition of ewes in early pregnancy and the residual effect on the offspring. *J. Agric. Sci.* **106:**81–87.
97. Paus, R., E.M.J. Peters, S. Eichmüller, and V.A. Botchkarev. 1997. Neural mechanisms of hair growth control. *J. Invest. Dermatol. Symp. Proc.* **2:**61–68.
98. Plowman, J.E., W.G. Bryson, and T.W. Jordan. 2000. Application of proteomics for determining protein markers for wool quality traits. *Electrophoresis* **21:**1899–1906.
99. Powell, B.C. and G.E. Rogers. 1997. The role of keratin proteins and their genes in the growth, structure and properties of hair, pp. 59–148. *In* P. Jollès, H. Zahn, and H. Höcker, (eds.), Formation and Structure of Human Hair, Birkhauser Verlag, Basel, Switzerland.
100. Powell, B.C., E.A. Passmore, A. Nesci, and S.M. Dunn. 1998. The Notch signalling pathway in hair growth. *Mech. Dev.* **78:**189–192.
101. Pugh, C.W. and P.J. Ratcliffe. 2003. Regulation of angiogenesis by hypoxia: role of the HIF system. *Nat. Med.* **9:**677–684.
102. Quigley, S.P., D.O. Kleemann, M.A. Kakar, J.A. Owens, G.S. Nattrass, S. Maddocks, and S.K. Walker. 2005. Myogenesis in sheep is altered by maternal feed intake during the periconception period. *Anim. Reprod. Sci.* **87:**241–251.
103. Reddy, S., T. Andl, A. Bagasra, M.M. Lu, D.J. Epstein, E.E. Morrisey, and S.E. Millar. 2001. Characterisation of Wnt gene expression in developing and postnatal hair follicles and identification of Wnt5a as a target of Sonic hedgehog in hair follicle morphogenesis. *Mech. Dev.* **107:**69–82.
104. Revell, D.K., S.T. Morris, Y.H. Cottam, J.E. Hanna, D.G. Thomas, S. Brown, and S.N. McCutcheon. 2002. Shearing ewes at mid-pregnancy is associated with changes in fetal growth and development. *Aust. J. Agric. Res.* **53:**697–705.
105. Schinckel, P.G. 1953. Follicle development in the Australian Merino. *Nature* **171:**310–311.
106. Schinckel, P.G. 1955. The relationship of skin follicle development to growth rate in sheep. *Aust. J. Agric. Res.* **6:**308–323.
107. Schinckel, P.G. and B.F. Short. 1961. The influence of nutritional level during pre-natal and early post-natal life on adult fleece and body characteristics. *Aust. J. Agric. Res.* **12:**176–202.
108. Schmidt-Ullrich, R. and R. Paus. 2005. Molecular principles of hair follicle induction and morphogenesis. *BioEssays* **27:**247–261.
109. Shelton, M. 1964. Relation of environmental temperature during gestation to birth weight and mortality of lambs. *J. Anim. Sci.* **23:**360–364.
110. Sherlock R.G., P.R. Kenyon, and S.T. Morris. 2002. Does mid-pregnancy shearing affect lamb fleece characteristics ? *Proc. N.Z. Soc. Anim. Prod.* **62:**57–60.
111. Short, B.F. 1955. Development of the secondary follicle population in sheep. *Aust. J. Agric. Res.* **6:**62–67.

112. Short, B.F. 1955. Developmental modification of fleece structure by adverse maternal nutrition. *Aust. J. Agric. Res.* **6:**863–872.
113. Stephenson, S.K. 1958. Wool follicle development in the New Zealand Romney and N-type sheep. II Follicle population density during fetal development. *Aust. J. Agric. Res.* **8:** 138–160.
114. St-Jacques, B., H.R. Dassule, I. Karavanova, V.A. Botchkarev, J. Li, P.S. Danielian, J.A. McMahon, P.M. Lewis, R. Paus, and A.P. McMahon. 1998. Sonic hedgehog signaling is essential for hair development. *Curr. Biol.* **8:**1058–1068
115. Symonds, M.E., M.J. Bryant, and M.A. Lomax. 1989. Lipid metabolism in shorn and unshorn pregnant sheep. *Br. J. Nutr.* **62:**35–49.
116. Tao, H., Y. Yoshimoto, H. Yoshioka, T. Nohno, S. Noji, and H. Ohuchi. 2002. FGF10 is a mesenchymally derived stimulator for epidermal development in the chick embryonic skin. *Mech. Dev.* **116:**39–49.
117. Taplin, D.E. and G.C. Everitt. 1964. The influence of prenatal nutrition on postnatal performance of merino lambs. *Proc. Aust. Soc. Anim. Prod.* **5:**72–81.
118. Thompson, A.N. and C.M. Oldham. 2004. Lifetime Wool. 1. Project Overview. *Anim. Prod. Aust.* **25:**326.
119. Viallet, J.P., F. Prin, I. Olivera-Martinez, E. Hirsinger, O. Pourquie, and D. Dhouailly. 1998. Chick Delta-1 gene expression and the formation of the feather primordia. *Mech. Dev.* **72:**159–168.
120. Walker, S.K., T.M. Heard, C.A. Bee, A.B. Frensham, D.M. Warnes, and R.F. Seamark. 1992. Culture of embryos of farm animals, pp. 77–92. *In* A. Lauria, and F. Gandolphi, (eds.), Embryonic Development and Manipulation in Animal Production, Portland Press Ltd., London.
121. Wallace, A.L.C. 1979. The effect of hormones on wool growth, pp. 257–268. *In* J.L. Black, and P.J. Reis, (eds.), Physiological and environmental limitations to wool growth, University of New England Publishing Unit, Armidale, N.S.W., Australia.
122. Wallace, C.E., M.W. Simpson-Morgan, and P. McCullagh. 1994. Retarded and excessive development of skin appendages in fetal lambs in response to thyroidectomy before wool follicle appearance. *J. Comp. Pathol.* **110:**275–286.
123. Wilson, N., P.I. Hynd, and B.C. Powell. 1999. The role of BMP-2 and BMP-4 in follicle initiation and the murine hair cycle. *Exp. Dermatol.* **8:**367–368.
124. Yeates, N.T.M. 1956. The effect of high air temperature on pregnancy and birth weight in Merino sheep. *Aust. J. Agric. Res.* 7:435–439.

Chapter 6
Mechanistic Aspects of Fetal Development Relating to Postnatal Health and Metabolism in Pigs

Matthew E. Wilson and Lloyd L. Anderson

Introduction

The productivity of the pig ranks high when compared with the other classes of farm mammals. Reproductive potential is the most important factor contributing to total meat production from this species. The high rate of productivity in the pig is dependent upon early sexual maturity, a comparatively high ovulation rate, relatively short periods of gestation and lactation, as well as the capability of repeating the pregnancy cycle soon after weaning a litter. Some important limitations affecting this production potential are ovulation rate, embryonic death, and perinatal and neonatal losses. This review focuses on (1) the roles of maternal nutrition and ovarian function on fetal development, (2) genotype of the dam affecting fetal development and litter size, (3) maternal circulating hormone concentrations during different stages of gestation as related to fetal development and litter size, (4) effects of specific nutrients and hormones for the dam on fetal development, and (5) maternal reproductive diseases affecting fetal development.

During a gestation period of approximately 115 days, the sow loses about half the potential litter size because of embryonic and fetal death. Most of this loss occurs during the first third of pregnancy. Thus, in sows which ovulate 15–20 ova at estrus, more than 95% of those eggs are fertilised by interaction with spermatozoa, but approximately 7–10 of the developing embryos perish at some stage during the pregnancy. At parturition, litter size often is reduced to approximately 8–11 or 12 living pigs. Thus, these losses may result primarily from a uterine environment that is less than optimal at the time of implantation and during early embryonic development.

After mating, the ovulated eggs are fertilised by spermatozoa migrating to upper regions of the oviducts, and the embryos are usually in the 4-cell stage when they enter the uterus. Within the lumen of the uterus, the embryos develop rapidly into

M.E. Wilson (✉)
Division of Animal & Veterinary Sciences, Davis College of Agriculture, Forestry, and Consumer Sciences, West Virginia University, Morgantown, WV 26506-6108, USA
e-mail: matt.wilson@mail.wvu.edu

P.L. Greenwood et al. (eds.), *Managing the Prenatal Environment to Enhance Livestock Productivity*, DOI 10.1007/978-90-481-3135-8_6,

blastocysts and then form extremely elongated (e.g., 30–100 cm) filamentous membranes. Each of these embryos is arranged end-to-end with no overlap of the membranes. This process results in an even distribution of the embryos throughout the uterine horns; intrauterine spacing of them is virtually complete by day 13 with implantation occurring later (day 14–18). The conceptus (embryo and its placental membranes and fluids) develops rapidly after implantation, but uterine capacity does not become a major limiting factor to embryonic or fetal survival until late stages of the gestation. Successful completion of the pregnancy depends upon functional corpora lutea. Ovariectomy as late as day 110 results in abortion within a few hours. Pregnancy in swine depends upon continued secretion of adequate circulating concentrations of ovarian progestin and oestrogen. During late stages of the pregnancy, the corpora lutea produce increasing amounts of a protein hormone, relaxin, which may play a role in preparation for the process of parturition. Our previous investigations indicated that pregnancy and a high embryo survival rate were sustained after ovariectomising pigs by providing optimal daily levels of exogenous sources of progesterone (80 mg/100 kg body wt) and small amounts of oestradiol benzoate (500 μg/100 kg body wt).

6.1 Effects of Maternal Dietary Restriction and Inanition on Fetal Development

Although pigs lose approximately 40% of their embryos during a gestation period of about 115 days [41, 88, 156] a role for maternal diet intake on survival and development of the conceptuses is less clear. Under conditions of optimal food intake, pregnancy is an anabolic process, benefiting the development of the conceptuses, as well as the dam [3, 10, 178, 216, 217]. The roles of maternal nutrition and ovarian hormones in fetal development were investigated in Yorkshire gilts (5–7 months old and approximately 125 kg body wt) [5]. Pigs were laparotomised to confirm pregnancy and ovulation rate (number of corpora lutea) at day 30. A bilateral ovariectomy (Ov) was performed in half the animals at this time and progesterone (80 mg/100 kg body wt) and oestradiol benzoate (500 μg/100 kg body wt) were given daily by IM injection thereafter. Laparotomised controls were given injections of sesame oil. The pigs were hysterectomised or ovariohysterectomised at day 70 or 110. All animals were fed a ration of 1.82 kg/day (20.95 MJ/day) until laparotomy at day 30 and then either remained on this level of intake as diet controls (1.82 kg/day) or were given a restricted diet (0.80 kg/day; 9.24 MJ/day). The calculated analysis of the diet was: protein 15.92%; calcium 0.99%; phosphorus 0.72%; lysine 0.84%; methionine 0.26%; cystine 0.26%; tryptophan 0.20%; mineral premix; vitamin premix; and metabolisable energy 11.55 MJ/kg DM. The animals were fed individually once daily in small pens in the postoperative room. Abbreviations were designated for pigs laparotomised for the control groups as (Lp control), and for those ovariectomised and given daily intra-muscular (IM) injections of progesterone and oestradiol benzoate as (Ov + P & OE). Pregnancy was confirmed at laparotomy in the

Table 6.1 Fetal survival, and uterus and conceptus growth during dietary restriction in pig

	Days after mating	MJ/day	No. of pregnant pigs	No. of fetuses Living	Degenerating	Fetal survival (%)
Lp control	70	20.95	5	10 ± 1.1	1 ± 0.5	73
Lp control	70	9.24	7	10 ± 1.6	1 ± 0.6	62
Ov + P & OE	70	20.95	6	9 ± 1.5	1 ± 0.6	57
Ov + P & OE	70	9.24	7	8 ± 1.4	3 ± 1.0	47
Lp control	110	20.95	6	10 ± 1.1	2 ± 0.9	62
Lp control	110	9.24	5	7 ± 1.4	3 ± 2.6	45
Ov + P & OE	110	20.95	5	10 ± 0.9	<1 ± 0.2	67
Ov + P & OE	110	9.24	6	8 ± 1.2	1 ± 0.7	54

Values are means ± SE.; Lp control = Laparotomy control, Ov + P & OE = Ovariectomy + Progesterone & Oestradiol Benzoate.

51 pigs at day 30 after mating. These animals were allotted to eight experimental groups of 6 or 7 each, and 47 (92%) of them remained pregnant with one or more living fetuses to day 70 or 110 (Table 6.1). Pregnancy failed in only one pig by day 70 and in three others by day 110. Ovulation rate, as indicated by the number of corpora lutea on day 30, was similar (mean of 14–16 per dam; P>0.05) among groups.

Pigs on a restricted diet weighed less at day 70 (P<0.001) and day 110 (P<0.001) than those receiving 1.82 kg/day. Body weight declined from day 30 to 110 in Ov + P & OE (−3.50 kg) and Lp control (−6.97 kg) pigs given the low level of dietary intake, whereas animals given the control diet gained 24.2 kg after Ov + P & OE and 32.0 kg after laparotomy. Maternal weight gain or loss was accentuated after eliminating the weight of the uterus and developing conceptuses. The mean number of fetuses surviving to day 70 remained similar (P>0.05) in laparotomised controls and in the Ov + P & OE animals given either level of diet (Table 6.1). The percentage of fetal survival was greater (P<0.01) at this time in the Lp control as compared to the Ov + P & OE animals, regardless of level of dietary intake. By day 110, the number of living fetuses was reduced (P<0.001) as well as the percentage survival (P<0.01) in those dams given only 0.80 kg feed daily as compared with full-fed controls. The number of degenerating fetuses (1.2 ± 0.59) was similar (P>0.05) for all groups at day 70 and 110. There was an approximately fourfold increase in fetal growth between day 70 and 110. Fetal wet weight was reduced in both Lp controls and Ov + P & OE, receiving a restricted diet, at both day 70 (P<0.01) and 110 (P<0.01) (Table 6.1). Placental wet weight increased less than twofold between day 70 and 110, regardless of level of diet or gonadal steroid status.

6.1.1 Fetal Development

Brain growth was unaffected by level of diet or hormone treatment to day 70, but by day 110 it was reduced (P<0.05) in Ov + P & OE dams given a restricted diet. The proportion of brain per unit fetus was greater in dams given a restricted diet to day 70 or 110, indicating preferential sparing of the brain as compared with other organs and muscles during adverse conditions for the dam. Thymus weight increased approximately sevenfold between day 70 and 110. There was no effect of maternal diet or gonadal steroid status upon thymus development to day 70; by day 110, however, the thymus was smaller ($P<0.001$) in fetuses from Ov + P & OE dams when effects of diet were held constant. Thyroid gland weight increased approximately threefold from day 70 to 110. Lung weight was unaffected by level of diet or hormone treatment to day 70, but by day 110, lungs were smaller in Ov + P & OE pigs ($P < 0.001$) as well as in animals on a restricted diet ($P<0.001$). Dietary restriction resulted in reduced fetal heart weight at day 70 ($P<0.001$) and 110 ($P<0.001$). Fetal liver weight was reduced markedly by restricting dietary intake of the dams to day 70 ($P<0.001$) or 110 ($P<0.001$) as compared with controls. Maternal dietary restriction resulted in smaller *gastrocnemius* at day 70 ($P<0.02$) as well as at day 110 ($P<0.001$). There were no dietary or hormonal effects on the proportion of *gastrocnemius* per unit of fetus at day 70, but by day 110 fetal muscle development was decreased ($P<0.01$) in all dams given a restricted diet. Similar relationships were found in growth of the *psoas major*. Growth of ovaries, uterus, testes, and epididymides was expressed as a proportion of fetal weight. Ovarian weight declined, whereas uterine weight increased from day 70 to 110. Testes and epididymides increased from day 70 to 110. It was only at day 110 that a restricted maternal diet resulted in a proportionate increase in uterine ($P<0.01$) and epididymal ($P<0.001$) development.

Fetal brain development was adversely affected only after prolonged dietary restriction of 80 days (day 30–110). Similar detrimental effects of maternal diet on brain development have been found in the fetal pig [57], fetal rat [216], and fetal lamb [1] and during postnatal growth in the pig [39, 89, 92]. Under these adverse dietary conditions, the brain was least affected as compared with other fetal organs, endocrine glands, and muscles. The development of fetal thymus and thyroid glands was limited only after day 80, and exogenous progesterone and oestradiol benzoate did not improve their growth. Although the proportions of heart and lung to fetal weight remained consistent regardless of level of diet or hormone treatment, those of liver declined by late pregnancy.

6.1.2 Maternal Nutrition and Litter Size

Although litter size increases by about two piglets between the first and third gestation, this is unrelated to nutritional status of the dam. Restricting metabolisable energy intake (e.g., <20.92 MJ/day) throughout gestation has no detrimental effect on fetal survival rate, and only a slight decrease occurs in birth weight. Numerous experiments have indicated that high energy diets (e.g., >31.38 MJ/day) given throughout gestation reduce litter size whereas piglet birth weight increases [10, 161, 219]. High energy intake stimulates ovulation rate but it also increases fetal mortality; thus, fetal survival rates remain equivalent to those in dams fed normal levels of caloric intake.

Fetal survival rates remain unaltered by marked shifts in caloric or protein intake of the dam at all stages of gestation in this species. Severe protein restriction of the dam, however, reduces both birth weight and postnatal development [130, 162]. Although there is considerable information concerning quantitative protein and energy requirements for pregnant gilts and sows, little is known about the long-term effects of prenatal malnutrition on the development of the progeny.

6.1.3 Embryonic Development During First Third of Pregnancy

Experiments were designed to determine whether gilts can provide an adequate uterine environment for the survival and growth of embryos when the food intake of the dam is limited, and whether supplemental ovarian steroids ensure continued embryonic and placental development under these conditions [2]. Yorkshire gilts were given water only (0 MJ/day, inanition) beginning 10 days before mating and continuing for varying periods after mating. The uterus and conceptuses were recovered by hysterectomy in the experimental animals and full-diet (29.40 MJ/day controls). In one group of gilts subjected to inanition, the ovaries were surgically removed on day 22 and these animals were given daily IM injections of progesterone (0.8 mg/kg body wt) and oestradiol benzoate (5 μg/kg body wt) from day 22 to 34 of inanition. In unmated pigs subjected to inanition, oestrous cycles ceased and ovarian follicles were completely regressed by day 34 (Table 6.2).

Table 6.2 Effect of inanition on embryonic survival in the pig

Days after mating	Days of inanition	No. of pigs		Mean no. of living embryos	Embryonic survival (%)
		Mated	Remaining pregnant		
			Full fed		
14	0	6	6	8.7	82
18	0	6	6	11.1	91
22	0	6	6	11.0	86
26	0	6	6	9.8	87
30	0	6	6	10.8	90
34	0	6	6	10.5	96
			Inanition		
14	25	6	6	6.5	81
18	29	6	5	8.6	92
22	33	6	4	9.3	96
26	37	6	4	9.0	89
30	41	6	2	10.0	94
34	45	6	1	10.0	83
		Inanition plus progesterone and oestradiol benzoate			
34	45	6	6	9.0	87

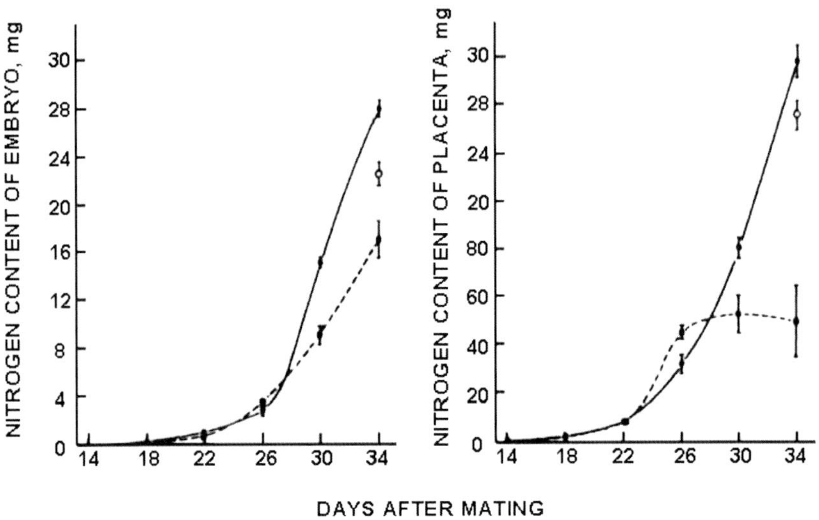

Fig. 6.1 Nitrogen content of embryos and placental membranes in pigs given a full diet (•, solid line), in animals subjected to inanition 25–45 day (■, dashed line), and in pigs given exogenous progesterone and oestradiol benzoate during 45 day inanition (○) [2]. Number of conceptuses in pigs given a full diet was 52, 67, 66, 59, 65, and 63 at days 14, 18, 22, 26, 30, and 34, respectively; in animals subjected to inanition, number of conceptuses was 39, 43, 37, 36, 20, and 10 at days 14, 18, 22, 26, 30, and 34, respectively; and in those given exogenous progesterone and oestradiol benzoate, number of conceptuses was 54 at day 34. Values are means ± SE

Embryonic and placental development were determined on the basis of nitrogen content (Fig. 6.1). Nitrogen content of the embryo increased exponentially between day 14 and 34 in gilts on a full diet and in those subjected to inanition. By day 30, nitrogen content was less ($P<0.05$) in embryos from nutritionally deprived gilts than in embryos from the control animals; this difference increased ($P<0.01$) by day 34. During prolonged inanition, exogenous progesterone and oestrogen from day 22 to 34 increased ($P<0.01$) nitrogen content in embryos; this increase was similar ($P>0.05$) to that found in control dams.

Nitrogen content of the placenta increased in a similar pattern ($P>0.05$) between day 14 and 30 in animals on a full diet and in those on inanition (Fig. 6.1). Continuation of inanition to day 34 resulted in a severe limitation ($P<0.01$) in placental nitrogen deposition as compared with that in controls. Exogenous progesterone and oestrogen given between day 22 and 34 resulted in placental nitrogen content similar ($P>0.05$) to that found in gilts on a full diet. These results indicate that with adequate progesterone and oestrogen, maternal components are made available for development of the conceptuses in spite of nutrient deprivation of dams during early pregnancy. When corpora lutea fail under these adverse conditions for the dam, the pregnancy is lost quickly. Furthermore, an adequate uterine environment permitting transfer of maternal nutrients across the placentas to the embryos for their continued normal development can be maintained by experimental treatment with progesterone and oestrogen in nutritionally deprived gilts during early stages of embryonic development.

6.1.4 Effect of Nutrient Deprivation During Middle Third and Last Third of Pregnancy on Fetal Survival

Yorkshire gilts were subjected to caloric restriction (0 kcal/day; water only) during either the middle third or last third of pregnancy, and fetoplacental development was compared with that in full-fed controls (29.40 MJ/day; 2.72 kg ration/day) [7]. Pregnancies were maintained in 81% of the gilts subjected to inanition during either the middle third or last third of gestation as compared with 100% in full-fed controls (Table 6.3). The average number of living fetuses in nutritionally deprived gilts was similar to that in full-fed controls. Fetal survival rates also were similar in experimental and control animals. Fetal weight increased four-fold between days 70 and 110 (Table 6.3). Fetal growth was limited ($P<0.05$) by prolonged inanition to day 70 compared with the growth in full-diet controls. In starved dams given oestradiol benzoate, fetal development was further reduced ($P<0.01$) in comparison with controls. At day 110, fetal growth in dams subjected to inanition was similar ($P>0.05$) to that found in controls; in those starved gilts given oestradiol benzoate; however, the fetuses were limited severely ($P<0.01$). Placental insufficiency was the primary cause of reduced fetal growth and resulted in abortion in only a few of the dams. Progesterone concentrations in peripheral blood serum of the gilts subjected to inanition during either the middle third or last third of pregnancy were maintained at levels (18–25 ng/mL) similar ($P>0.05$) to those found in full-fed controls.

Table 6.3 Fetal survival during prolonged inanition in the pig

Energy MJ/day	Inanition days of pregnancy	No. mated	No. pregnant	%	Mean no. of living fetuses	Fetal survival (%)	Mean fetal wt, g
0	30–70	4	3	75	9.3	64	215
0	30–70	4	3	75	11.7	70	211
0	30–70	4	3	75	7.3	47	177
29.40		7	7	100	10.9	72	234
0	70–90	4	4	100	11.3	79	542
0	70–90	4	3	75	12.7	68	516
29.40		4	4	100	9.0	59	526
0	70–110	4	2	50	10.0	83	913
0	70–110	4	3	75	8.0	46	717
0	90–110	4	4	100	11.0	68	998
0	90–110	4	4	100	8.3	58	879
29.40	0	8	8	100	8.9	58	984

6.1.5 Effect of Nutrition Deprivation During Middle or Late Pregnancy on Litter Size and Neonatal Growth

In a subsequent series of experiments, Yorkshire gilts were subjected to caloric restriction either during the middle third (days 30–70) or last third (days 50–90) of pregnancy and then gradually realimentated to a full diet and allowed to advance to parturition [89–92]. The full-diet controls received 29.40 MJ/day throughout gestation and those animals subjected to inanition (0 MJ/day) received water only. Pregnancies were maintained in 83% of the gilts calorically restricted in the middle third of pregnancy and in 64% of those dams restricted in the last third as compared with 100% in the controls (Table 6.4). At parturition, average litter size of living pigs in starved dams (9.4) was similar ($P>0.05$) to that found in full-fed controls (8.0).

Maternal blood serum levels of progesterone in starved gilts were maintained at concentrations similar to those in full-fed controls. Progesterone concentrations in the controls remained at a level of 18–25 ng/mL from day 30 until approximately day 105, when the concentrations began to decline, reaching basal levels at parturition and remaining at these levels through day 130 after mating. The progesterone concentrations of the dams starved during the middle third of pregnancy were similar ($P>0.05$) to those in full-fed dams during the 40 day inanition period ending day 70. From day 72 to 99, however, progesterone levels of these dams were reduced ($P<0.01$) to a mean value of 14.5 ng/mL. This reduction in serum steroid levels occurred after the starvation period and during gradual realimentation and full-feeding. By day 100, progesterone levels in this group were again similar to the levels in controls and remained so through day 113. Progesterone levels of dams starved during the last third of pregnancy remained similar to those of full-fed controls throughout gestation. Thus, progesterone concentrations in both experimental groups remained at levels sufficient to sustain pregnancy. These experimental results indicated that maternal nutrient deprivation during the middle third or last third of gestation has little effect on ovarian progesterone secretion and is not a major limitation to fetal survival in this litter-bearing species. Prolonged starvation in these gilts markedly reduced birth weight of living neonates (Table 6.4). Growth of the

Table 6.4 Reproduction in pigs subjected to inanition during middle or late pregnancy on litter size, birth weight and neonatal growth

Energy MJ/day	No. of pigs mated/delivery*	Litter size at birth	Piglet birth wt, kg	Litter wt at birth, kg	Body wt at 80 day of age, kg
29.40	6/6	8±1.5†	1.6±0.05†	12.4±1.9†	27±1.7§
0 [day 30–70]	12/10	9±0.9†	1.2±0.03†	9.5±0.9†	20±1.7†
0 [day 50–90]	11/7	10±0.8†	1.0±0.03§	10.7±0.8†	20±1.7†

Values are means ± SE. Means within column with different superscript are different at $P<0.05$ level of significance.

*Chi-square = 3.317, df = 2, ($P<0.05$).

†a significant difference at $p<0.005$.

§a significant difference at $p<0.005$.

neonates also was reduced to 80 days of age, when they weighed 7 kg less, on average, than the offspring of the control dams. Body weight at 150 days of age, however, was similar (P>0.05) in all experimental and control groups. Litter weight of live piglets at birth was similar among all treatments, primarily as a result of the larger litters born to the starved dams.

6.1.6 Reproduction in Yorkshire Gilts Born to Dams Starved During Pregnancy

After weaning, gilts born to dams that had been starved during the middle third or last third of gestation and gilts born to full-fed control dams were raised in the same outdoor lot. At approximately 150 days of age these gilts were checked for oestrous behavior by a fertile boar and allowed to mate at the third confirmed oestrus. Reproductive performance through the first two gestations was evaluated in a random selection of gilts born to each experimental and control group [92]. Age at pubertal oestrus was reduced (P<0.01) in gilts born to dams starved during pregnancy. Control gilts exhibited first oestrus at an average age of 250 days whereas gilts born to dams starved during the middle or last third of pregnancy displayed pubertal estrus at 208 and 219 days, respectively. Fifteen of 18 gilts born to starved dams from both experimental groups, mated at the third post-pubertal oestrus, remained pregnant to term. First-parity birth weight of living neonates from gilts born to dams starved during pregnancy was similar to birth weights from gilts born to control dams.

Luteolytic capacity is defined as the ability of corpora lutea (CL) to undergo luteolysis after prostaglandin (PG) $F_{2\alpha}$ treatment [56]. Luteolysis in pigs is initiated by secretion of $PGF_{2\alpha}$ from the uterus. In this species a single treatment with $PGF_{2\alpha}$ causes luteolysis in pregnant, hysterectomised, or cycling animals after day 13 but not before this time [142]. Luteolysis results in a rapid decrease in progesterone production by the CL and programmed luteal cell death [21–22]. In vivo treatment with $PGF_{2\alpha}$ decreases protein for the luteinising hormone (LH) receptor in porcine CL [26]. The finding that hypophysectomy leads to a rapid decrease in progesterone secretion from CL is consistent with the idea that inhibition of the LH stimulatory pathway may be important for changes in progesterone production during luteolysis [6]. Progesterone biosynthesis is regulated not only by $PGF_{2\alpha}$ but also inhibition of the LH stimulatory pathway, steroidogenic enzymes, and cholesterol transport [56]. Luteolytic capacity involves a critical change in responsiveness of transcriptional factor (DAX-1), steroidogenic acute regulatory protein (StAR), and LH receptor to $PGF_{2\alpha}$ that results in differential inhibition of luteal progesterone biosynthesis by day 9 and 17 porcine luteal cells in culture [44, 56]. Prolactin is luteotropic during later pregnancy in this species. Prolactin maintains progesterone and relaxin secretion by aging corpora lutea after hypophyseal stalk transection or hypophysectomy in the pig [119]. Peripherial plasma concentrations of prolactin remained elevated in stalk-transected gilts from day 110 to 120 and maintained both progesterone and peak relaxin secretion. In hypophysectomised animals daily IV administration of purified porcine prolactin maintained both progesterone and relaxin secretion whereas in hypophysectomised animals given phosphate buffer saline circulating concentrations of these two hormones rapidly decreased to undetectable levels by day 120.

6.2 Maternal Energy Intake and Fetal Development and Survival

Leptin is produced by adipose tissue and is implicated in reproductive and immune function, as well as in feed intake regulation [24–25]. In gilts fed a high- (28.79 MJ/day) or lower- (21.84 MJ/day) energy diet beginning at day 45 and continuing throughout pregnancy, serum concentrations of leptin remained similar; however, during lactation, both serum and milk concentrations of leptin were greater in animals on the high- versus lower-energy diet [68].

A systematic restricted section of uterine length available to each potential pig embryo was examined at day 20, 25 or 50 [220]. When embryos were restricted to 5 cm, the proportion of surviving fetuses at day 20, 25 and 50 was 61, 12 and 8%, respectively whereas in combined nonrestricted section controls was 82%. Each fetus surviving at day 50 in restricted sections was associated with 36 cm of initial uterine length.

It is clear that the gonadotropin releasing hormone (GnRH) pulse generator system and gonadotropin release are inhibited by undernutrition [209], thus compromising gonadotropin support of folliculogenesis. Metabolic mediators (i.e., hormones, metabolites) also may act at the ovarian level and amplify the gonadotropin-mediated effects of nutrition [166]. The period corresponding to the initiation of preovulatory growth and differentiation seems to be the most sensitive to the influence of undernutrition, and thus a decrease in ovulation rate or delay in ovulation is often observed in prepubertal gilts and weaned sows [167]. Thus, metabolic hormones such as insulin, growth hormone (GH) and IGF-I can play a role in controlling activity of follicular cells of the ovary in a positive way under conditions of optimal nutrition or in a detrimental way during undernutrition in the pig.

LH is essential for maintenance of CL and early pregnancy in the pig [6, 110]. Thus, any factor, including nutrition which affects LH secretion during early pregnancy, may have implications for embryonic survival and maintenance of early pregnancy [152–153]. Environmental factors involved in seasonal infertility and level of reproductive activity in summer-autumn include housing, high ambient temperature, social stress, and boar exposure. In pigs, the European wild boar is a true seasonal breeder with a breeding season occurring in early winter with piglets born once a year in late spring [45, 128]. An important cue affecting onset of sexual activity in the European wild boar is availability of feed [155]. Although domesticated pigs are not classified as seasonal breeders, seasonal infertility may be manifested by reduced farrowing rate, delayed puberty of gilts, prolonged weaning to oestrus interval, and reduced behavioral sexual activity in sows and boars during late summer and autumn resulting in decreased litter size [154]. The nocturnal rise in circulating melatonin concentration in European wild boars is more robust and predictable as compared with that seen in domesticated pigs [154, 190], which do not appear to exhibit a nocturnal rise in melatonin [33]. An important aspect in optimising embryonic survival is to generate an increased number of embryos by a high-plane of nutrition before mating, followed by restricted feeding during the post-mating period that further supports developing embryos [10, 17–18]. To exert nutritional effects on embryonic survival, the progesterone-mediated mechanism is thought to be effective for 3–4 days after mating [76]. Insulin, glucose, free fatty acids (FFA), and amino acids provide metabolic signals to the brain that influence food intake, energy balance and body weight regulation [9, 72]. Feed-restricted pregnant gilts tend to show higher preprandial FFA concentrations, lower preprandial insulin concentrations, and a greater pre- vs. post-prandial difference than in ad libitum fed animals [152]. Seasonally reduced LH secretion in summer and autumn, a time of reduced fertility, is further suppressed by applying feed restriction after mating. Regarding the influence of lactation length (12 vs. 21 days) and feed intake (high, 54.72 MJ/day vs. low, 25.99 MJ/day) on reproductive performance in primiparous sows, greater circulating concentrations of insulin prior to weaning were associated only with level of feed intake treatment [109]. Greater feed intake during lactation was associated with greater circulating concentrations of insulin and glucose, greater LH pulse frequency before weaning and a shorter farrowing-to-oestrous interval in both lactation length treatments. Thus, greater feed intake during lactation improves farrowing-to-oestrous interval through LH release regardless of duration of lactation.

Food deprivation in sows around implantation induced an increase in maternal circulating plasma concentrations of cortisol, $PGF_2\alpha$ metabolite and progesterone [193]. A focus on the effect of 48 h food deprivation of inseminated sows soon after ovulation revealed a decrease in cleavage rate of their embryos, decreased sperm numbers in the oviduct reservoir, as reflected by the number of sperm in the zona pellucida [129]. The decreased sperm numbers in the zona pellucida in food deprived animals may result from reduced sperm viability or their elimination through the oviduct to the peritoneum. This brief period of food deprivation also resulted in increased circulating concentrations of progesterone, cortisol and $PGF_2\alpha$ metabolite as compared with that seen in the control sows. Animals experiencing stress express many behavioral and physiological responses, such as anorexia, aggression, altered motor activity, immunosuppression, and activation of the hypothalamic-pituitary-adrenal axis [115, 191]. Corticotropin-releasing hormone (CRH) has been implicated as an important mediator of behavior, immune, and neuroendocrine systems in animals experiencing stress. In pigs, intracerebroventricular injection of CRH induces immediate dose-dependent behavioral and physiological responses [98]. In addition, intracerebroventricular administration of CRH dose-dependently increased plasma ACTH concentration and elicited hyperactivity and vocalisation, and immune suppression. These acute responses to central administration of CRH showed differences in onset, maintenance and cessation of these behavioral and physiological responses indicate that they are differentially regulated. In sows given repeated ACTH-stimulation during early pregnancy (day 2) circulating concentrations of cortisol and progesterone were significantly increased while $PGF_2\alpha$ metabolite levels decreased [170]. These hormonal changes had no significant effect on ova transport rate, but a negative impact on the oviductal milieu as revealed by fewer spermatozoa attached to the zona pellucida and a decrease in the cleavage-rate of embryos compared with that seen in control animals.

Sows losing excessive bodyweight during lactation can experience extended weaning-to-oestrus intervals and an increase in duration of anoestrus [62]. Feed intake and tissue losses during lactation have a greater effect on weaning-to-oestrus intervals in primiparous compared with multiparous sows [77]. The weaning-to-oestrus intervals in primiparous and multiparous (parities 4–7) Camborough sows in relation to bodyweight, changes to back-fat thickness and circulating serum hormone concentrations (insulin, GH, and cortisol) were examined at the end of gestation and during lactation [87]. Primiparous sows lost a greater amount of back-fat thickness than multiparous sows (–20 and –9%, respectively) during this time. The weaning-to-oestrus intervals also were greater (5.5 and 4.2 days, respectively) in the primiparous compared with multiparous sows. Changes in serum GH concentrations appeared to be a better hormone imbalance indicator to explain differences in weaning-to-oestrus intervals between primiparous and multiparous sows than change in body condition.

Fetal mortality in healthy gilts of a segregating F2 cross of Large White and Meishan pigs was examined on the basis of the length distribution of mummified fetuses and the frequency of non-fresh stillborn piglets to establish whether critical periods of fetal mortality exist [205]. Average fetal mortality rate per gilt was 8.7%. Mummified fetuses averaged 0.84 per litter and ranged in length from 0.4 to 33.0 cm thus indicating a range in fetal age at death of approximately 35–100 days. Non-fresh stillborn piglets averaged 0.30 per litter. Fetal mortality in this population occurred throughout the fetal stage of gestation but seemed to be clustered in three distinct periods which coincided with periods of evident changes in porcine placental growth.

A major limitation for increasing litter size in swine is embryonic loss that occurs during the initial 20 days of gestation. This early embryonic loss can occur during pre-elongation development, trophoblastic elongation and placental attachment [3, 83–85]. Competitive acquisition of adequate uterine space among littermate embryos is established during conceptus elongation on day 12, a brief period of rapid morphological change. Progressive changes in the uterine microenvironment between days 10–16 play a major role in embryonic survival following trophoblast elongation and placental attachment [85].

6.3 Genotype of Dam Affecting Fetal Development and Litter Size

It is believed that the majority of the pig breeds descended from the Eurasian Wild Boar (*Sus scrofa scrofa*). Through natural selection for favorable genotypic and phenotypic traits, more than 70 breeds currently exist worldwide [149]. This does not

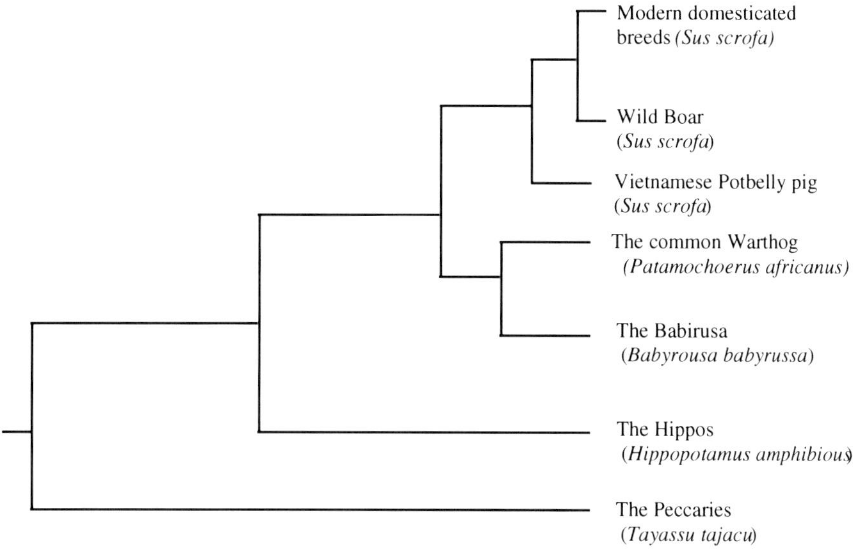

Fig. 6.2 Phylogenetic relationships of *MC4R* variation in the suborder Suiformes drawn with the UPGMA method. The length of branches is not necessarily proportional to real time and genetic distances between taxons

include commercial pig companies that produce lines with traits focused on reproductive efficiency and meat quality. A phylogenetic analysis of the melanocortin-4 receptor gene (*MC4R*) in diverse families of Suiformes from the UPGMA tree showed that all Suidae families cluster together and the *H. amphibious* showed a closer relationship to the Suidae species than the *P. tajacu* (Fig. 6.2) [104]. Maintaining high levels of reproductive efficiency is essential to modern pork production. Litter size is a critical component to efficient pork production. The structure of the swine industry has changed dramatically in the past 10 years. Tools to efficiently identify important genetic variation for litter size and fetal survival are not yet sufficiently developed to maximise the use of the vast information available from human and mouse genome sequencing efforts [171, 172].

Several reproductive traits such as ovulation rate, embryo survival, number born, number weaned and age at puberty in the female and testicular size and libido in the male are important [174]. Litter size is a major determinant of efficiency in swine production. Folate is a vitamin that is important for protein and DNA synthesis in cells, especially for red blood cells and for the appropriate development of the porcine fetus [69, 202]. Folate is delivered to the developing pig fetus bound to an intrauterine secreted folate binding protein (*sFBP*). DNA sequence variation in porcine *sFBP* was associated with differences in uterine capacity, number of ova shed per reproductive cycle and the percentage of embryos that die before day 11 of pregnancy. Because sequence variation in the *sFBP* affected multiple factors influencing litter size, the overall effect of variation in the sequence of the *sFBP* on litter size was negligible.

The use of the prolific Meishan breed to investigate the control of litter size in the pig has led to a greater understanding of the influence of maternal genotype on conceptus growth and development [73, 74]. Meishan gilts give birth to 3–5 more pigs per litter than do Yorkshire gilts of a similar reproductive age, with no difference in ovulation rate or uterine length [27, 30, 216]. From a number of studies (for review see [215]) the overwhelming evidence has led to the suggestion that the Meishan embryo is smaller prior to and after elongation and as a result develops a smaller, but more vascular placenta. Reciprocal embryo transfer between Meishan and Yorkshire females leads to a breed specific pattern of fetal development (Yorkshire fetuses grow at an accelerated rate), but the genotype of the uterine environment has a significant impact with either Meishan or Yorkshire fetuses weighing considerably less when gestated in a Meishan recipient than if they had been gestated in a Yorkshire recipient [73]. However, when Meishan and Yorkshire embryos were co-transferred to either a Meishan or Yorkshire recipient, the littermate pigs were born at the same weight, although overall lighter if gestated in a Meishan uterine environment than in a Yorkshire uterine environment [216].

The role of oestrogen and the oestrogen receptor (ESR) in porcine reproduction has been well studied. Studies using the prolific Meishan breed and Meishan crosses with the Large White breed revealed a positive additive effect on total number born and number born alive [176, 182]. They showed a positive association between a favorable ESR locus and a specific allele (B allele) and reproductive traits that ranged from 1.25 pigs/litter in Meishan crosses to 0.4–0.6 pigs/litter in Large White and Large White crosses. That the ESR gene influences traits such as litter size and other reproductive components was demonstrated by examining the ESR genotype in >300 Yorkshire, Large White and crossbreds as AA, AB or BB genotype [95]. Adding copies of the ESR B allele appears to increase fetal survival, total number of fetuses, total fetal weight, and number of fetuses per uterine horn with the overall effect of the B allele being an increase in reproductive performance. However, others have failed to detect any advantage of the B allele [206].

Uterine capacity in pigs is influenced by both the uterine environment and the ability of the fetuses to survive a crowded uterine environment [203]. A single base change was discovered in the DNA sequence of the erythropoietin receptor (*EPOR)* gene (*EPOR)* which controls red blood cell development in pigs. The difference in the sequence of the *EPOR* affected the uterine capacity, and was associated with a 2.5 pig difference in litter size. The effect on litter size was due to *EPOR* sequence variation in the piglet rather than the dam. This genetic marker based on the difference in the DNA sequence of the pig *EPOR* could be useful for marker assisted selection of gilts for litter size.

Certain swine major histocompatibility complex genotypes have been associated with increased or decreased testicular size and hormone concentrations with 1.5–5% of the phenotypic variation of these traits explained by the pig major histocompatibility complex [210]. Major histocompatibility complex in the female has been linked to ovulation rate, litter size, number born alive and birth and weaning weight [75, 197, 210]. Mean heritability estimates (h^2) for reproductive traits in boars are: testis size 0.40; sperm quantity 0.30; and libido 0.20 [174]. In sows these

h^2 estimates are: age at puberty 0.32; ovulation rate 0.39; embryo survival 0.30; litter size 0.10; litter birth weight 0.29; number born alive 0.07; number weaned 0.06; piglet survival to weaning 0.05; weaning-to-oestrus interval 0.23; and farrowing interval 0.20. Selection for ovulation rate is high but low for litter size. Litter size in the U.S. national pig herd has increased at the rate of approximately 0.052 pigs per year from 1980 to 2000 [96, 157]. The total change is approximately 13%. Improved feed efficiency of growing pigs during this period is evident by a decrease at the rate of approximately –24 g of feed per kg of live weight marketed. Reproductive rate of the U.S. national pig herd during this 20-year period increased 0.0085 in litters per sow per year and 0.052 pigs per litter resulting in an increase of 0.165 pigs per sow per year, a total change of approximately 25% [96].

A selection experiment on pigs randomly split into a "control" group and a "select" or index line originating from a base genetic line traced back to 50% PIC Large Whites and 50% Landrace was carried out over 14 generations at the University of Nebraska [97]. Line I was selected on an index of ovulation rate and embryonic survival for 11 generations, and then it was selected for increased litter size at birth. Line C is the control line. Response in Line I was an increase of 0.19 pigs per generation and the total response after 19 generations is approximately 3.6 pigs per litter [96, 157]. In later generations of this long-term selection experiment, litter size averaged between 13 and 14 pigs or an increase of 0.29 pigs per generation. This response was only 40–50% of the increase in ovulation rate. Embryonic survival and ovulation rate were determined by laparotomy at day 50 of gestation, but found to be an inadequate measure of uterine capacity. The ovulation rate/uterine capacity model of litter size appears to be very effective, but not easily implemented in livestock producer breeder herds. From these results, h^2 of litter size in all selection lines is 0.17, higher than the value of 0.10 reported for most populations [96].

A selection experiment on litter size in the pig was carried on for 17 generations in an INRA (Institut national de la recherché Agronomique) herd in France [32]. The founder population consisted of 10 males and 120 females from the Large White breed. The h^2 estimates for ovulation rate were 0.43 and for number of live embryos at day 30, 0.19, respectively. Total genetic gain was approximately 1.4 piglets at birth per litter. Estimation of genetic parameters for litter size (i.e., number of piglets born alive and number of piglets born) was analyzed from a total of 46,960 litters of Swedish Landrace and Large White breeds [125]. Estimates for h^2 were 0.14 and 0.12, and for repeatability in number born and number born alive were 0.20 and 0.18, respectively. Additive maternal crossbreeding parameter estimates were 0.09 and 0.36 piglets for number born and number born alive, respectively. Estimated maternal heterosis was 0.7 piglets for both traits.

Application of porcine genetic maps in appropriate segregating populations allows for identification of markers linked to quantitative trait loci (QTL) that control components of growth, muscle quality, reproduction and health [63, 223]. Markers closely linked to the QTL can be used in combination with phenotypic information through marker-assisted selection to enhance rates of genetic improvement. QTL discoveries for meat quality and litter size are being used in the pork industry. Progress has been made in identifying markers for genes with major effects on production traits such as back fat [11, 225] and litter size [175].

Selected litter size (H) and a control line (C) were used to estimate correlated responses to litter size in carcass, meat and fat quality traits [67]. No differences were found between the two lines in carcass measurements except back fat depth, which was significantly higher in line H (0.69 + 0.28 mm). Selection for litter size reduced the lean content in the carcass but meat quality traits were not affected. Effect of birth litter size, birth parity number, growth rate, back fat thickness and age at first mating was determined in Swedish Landrace and Swedish Yorkshire gilts and their subsequent reproductive performance in parities 2–5 that included 20,712 farrowing records [195]. Each additional piglet in the litter in which the gilt was born was associated with an increase of her own litter size of between 0.07 and 0.10 piglets per litter ($P<0.001$). Gilts with a higher growth rate to 100 kg body weight for parities 1–5 had a larger litter size ($P<0.05$), shorter weaning-to-first-service interval ($P<0.05$) and higher farrowing rate ($P<0.05$) than gilts with a lower growth rate. A 10 day increase in age at first mating resulted in an increase in litter size of approximately 0.1 piglet for primiparous sows ($P<0.001$) and a decrease ($P<0.05$) for parities 4 and 5. The association between birth litter size, birth parity number, growth traits and subsequent reproductive performance in sows implies that selection of good replacement females to improve one trait may be associated with changes in multiple traits as well as subsequent reproductive performance. Data were analyzed on 7,040 sows about the association between repeat breeding in gilts/sows and their subsequent reproductive performance [196]. Repeat breeding rate is an important parameter in the measurement of efficiency of mating management. Fertilisation failure and early embryonic loss are primary biological components causing repeat breeding. Percentages of litters as a result of repeat breeding in sow parities 1–3 were 6.1, 12.0 and 6.3% for Swedish Landrace and 6.7, 13.1 and 7.4% for Swedish Yorkshire. Among the sows removed from the herds, about 24% of Swedish Landrace and 28% of Swedish Yorkshire were culled due to reproductive problems.

The physiological system governing energy storage in the form of fat might be the most fundamentally adaptive system in humans and other animal species. Mapping of obesity-related genes on porcine chromosomes can be useful in breeding programs for marker assisted selection of those performance traits. Evolutionarily, the pig genome is more closely related to the human genome than laboratory animal species. The discovery of a polymorphism identified in the porcine *MC4R* gene revealed a missense mutation that replaces aspartic acid with asparagine at amino acid position 298 of the *MC4R* protein [102–104]. The Asn298 mutation occurs within a highly conserved motif, NPLIY, of all members of G protein-coupled receptors; whereas, Asp298 is conserved in all five melanocortin receptor subtypes. Functional and phylogenetic analyses of the *MC4R* mutation in domestic pigs demonstrated that Asp298 is required for normal *MC4R* signaling to adenylyl cyclase [104]. Sequencing of the *MC4R* gene in seven diverse genera within the Suiformes revealed 62 nucleotide variations in *MC4R*.

Pituitary hormones including GH and prolactin (PRL) are important in controlling animal growth and reproduction, yet the control of expression of these hormones is poorly understood in pigs. GH is secreted from somatotrophs of the anterior pituitary and has been implicated in regulation of porcine growth, lactation, reproduction [179, 184], and the immune system [48]. PRL is secreted from lactotrophs and mammosomatotrophs of the anterior pituitary [79]. Prolactin has been implicated in a diverse range of physiological processes, including osmoregulation, reproduction, lactation growth [40], as well as immunoregulation [177]. PRL blood concentrations increase markedly in Yorkshire and Meishan gilts during late gestation and lactation [58, 118] and PRL administration maintains progesterone secretion by aging corpora lutea in hypophysectomised pigs [118, 119]. GH plasma concentrations increase prepartum and during early lactation in Meishan sows [58]. POU1F1 (also previously known as PIT-1 or GHF1) is a pituitary-specific factor that binds elements required for pituitary-specific expression of the GH and PRL genes [194]. The POU1F1 protein is also involved in regulating pituitary hormonal gene expression of the TSHβ and GH-releasing hormone receptor (GHRHR) genes [121, 122]. Multiple mRNAs are produced from the POU1F1 gene through alternative splicing. Genetic studies have shown a requirement of POU1F1 for normal GH, PRL, and TSHβ gene expression. Polymorphisms at POU1F1 have been associated with several performance traits in the pig [185, 225]. Pig POU1F1 protein has also been shown to specifically recognise rat GH and PRL promoter sequences required for transactivation [224]. Chinese Meishan pigs develop rapidly with onset of puberty at less than 100 days of age, and have a smaller placental size and larger litter size as compared with British/Continental breeds. Meishan (MS) pigs segregating a *MspI* POU1F1 polymorphism were used to determine differences of GH and PRL at both mRNA and circulating hormone concentrations. Animals from nine litters were used to collect pituitary ($n = 60$) and/or blood samples ($n = 80$) at days 0, 15, and 30 after birth, and all animals were genotyped (CC, CD, DD) for the *MspI* POU1F1 polymorphism. There was a significant effect of POU1F1 genotype on circulating levels of both GH and PRL at birth, but not thereafter. The DD neonates had lower levels of GH, but higher levels of PRL, than other genotypes. POU1F1-α mRNA decreased ($P < 0.05$) from day 0 to 30, which paralleled decreases ($P < 0.05$) in GH mRNA as well as PRL and GH plasma levels over the same period. POU1F1-β mRNA levels did not significantly change over this period. Correlations were significant between POU1F1-α mRNA and both GH mRNA and GH plasma concentration levels, as well as between the two POU1F1 mRNA isoforms. This was the first investigation of GH and PRL at both mRNA and circulating hormone levels in Chinese Meishan pigs segregating the POU1F1 CD polymorphism [188]. The main finding was that a significant POU1F1 genotype effect on both GH and PRL plasma levels was observed, suggesting an association of POU1F1 genotypes with gene expression of these two important pituitary hormones in the pig.

GH released from pituitary under direct control of hypothalamic releasing (i.e., GH-releasing hormone [GHRH]) and inhibiting (i.e., sst or SRIF) hormones is an anabolic hormone that regulates metabolism of proteins, fats, sugars and minerals in mammals [8–9]. The discoveries of synthetic peptide and nonpeptide GH-secretagogues (GHSs) that stimulate GH release, as well as a receptor(s) unique from GHRH receptor revealed that GHRH and GHSs operate through distinct

G protein-coupled receptors to release GH. Isolation and characterisation of ghrelin, the natural ligand for GHS receptor, has opened a new era of understanding the physiology of anabolism, feeding behavior, nutritional homeostasis for GH secretion and gastrointestinal motility through gut-brain interactions. Amino acid sequence of rat ghrelin-28, human ghrelin-28, and porcine ghrelin differs by 2 amino acid changes in position 11 and 12 for human ghrelin and in positions 12, 22 and 26 for porcine ghrelin compared with rat ghrelin. Few studies are available on ghrelin affecting muscle accretion and long-term growth [9]. Myogenic expression vectors containing the full length cDNA of swine ghrelin-28 (pGEM-wt-s*Ghln*) and truncated variant (pGEM-tmt-s*Ghln*) consist of the first seven residues of ghrelin (including Ser^3 substituted with Trp^3) with addition of a basic amino acid Lys (k) at the C-terminus [221]. Secretion of GH is stimulated by GHRH and ghrelin (acting via the GH secretagogue (GSH) receptor (GHSR), and inhibited by somatostatin (SRIF)) [8]. Other peptides/proteins influence GH secretion, at least in some species. The cellular mechanism by which the releasing hormones affect GH secretion by the somatotrope requires specific signal transduction systems (cAMP and/or Ca^{2+} influx and/or mobilisation of intracellular Ca^{2+}) and/or tyrosine kinase(s) and/or nitric oxide (NO/cGMP). At the subcellular level, GH release in response to GHS is accomplished by the following. The GH-containing secretory granules are moved close to the cell surface. There is then transient fusion of the secretory granules with the fusion pores in multiple secretory pits in the somatrope cell surface [4, 8, 43, 112, 113].

6.4 Maternal Circulating Hormone Concentrations During Different Stages of Gestation as Related to Fetal Development and Litter Size

Fetal growth and development is dramatically influenced by the number of conceptuses in the litter. Therefore, factors which influence fetal growth and development can act directly on the fetus altering growth and development or indirectly altering the survival of conceptuses that then leads to a more, or less, crowded uterine environment which then alters growth and development. Historically, utilising a variety of experimental approaches, there has been strong agreement that 30–40% of the embryos in a litter will not survive to day 30 of gestation. At least a portion of this embryonic loss is thought to result from the very dramatic changes in the hormonal and histotrophic milieu to which the embryo is exposed [204]. In the first days following fertilisation there is a dramatic increase in serum concentrations of progesterone. This progesterone, of luteal origin, is thought to regulate the production of histotrophic components by the endometrium. Vallet et al. [201] demonstrated that advancing the timing of the increase in progesterone via exogenous administration of progesterone on day 2 and 3 moderately increased total histotrophic protein secretion between days 10 and 15 of gestation. This same treatment regimen resulted in a 9% increase in fetal weight at 105 day of gestation; however, this treatment also reduced litter size by 19% potentially accounting for the difference in fetal weight near term of gestation. Conversely, Vallet and Christenson [200] treated gilts with the progesterone receptor antagonist mifepristone on day 2 of gestation, but they were unable to ameliorate the negative impact of early treatment with progesterone. The treatment with mifepristone on day 2 of gestation decreased litter size by 36% at 105 day and consequently increased fetal weight by 6%. Furthermore, providing exogenous progesterone from days 1.5 to 4 of gestation in an attempt to decrease the negative impact of a low plane of nutrition prior to weaning also decreased litter size by 35% [127]. Clearly, altering the pattern of progesterone by either advancing or delaying the normal increase can impact the growth and development of the fetus, but the effect appears to be entirely a result of decreasing conceptus survival and thereby increasing fetal weight late in gestation.

There is evidence that there is a moderate ($r = 0.21$) correlation between the amount of progesterone removed from the uterine arterial circulation and the weight of the fetus in that region of the uterus [101]. Supplementation with progesterone from days 4 to 30 of gestation had no effect on the weight of fetuses harvested on day 30, at the end of treatment [17]. Exogenous administration of progesterone and oestradiol-17β daily from days 4 to 15 of gestation decreased the stature and weight of fetuses by day 49, while at the same time increasing the length and weight of the placenta. The authors claimed that these alterations were a result of progesterone treatment, but included no progesterone only treatment group to delineate the steroid was causing the effect [108]. In a follow up study, the administration of progesterone and oestrone had no effect on fetal development when administered from days 4 to 20 of gestation, but increased fetal weight by 10% at day 50 when administered from days 20 to 30 of gestation. Ironically, when these authors administered the same progesterone and oestrone treatment from days 20 to 25 of gestation it led to a 12% reduction in fetal weight at day 50 [51]. Treatment of sows with a combination of progesterone and oestradiol–17β on days 14–20 of gestation had no effect on embryo survival by days 28–32 [29]. Taken together it appears that administration of supplemental progesterone after day 4 of gestation has no consistent effect on the growth and development of the fetus.

On approximately days 11–12 of gestation the embryo will begin synthesising and secreting vast quantities of oestradiol-17β that serves two main functions. First, it serves as the signal for maternal recognition of pregnancy by increasing the synthesis and secretion of prostaglandins from the endometrium, particularly prostaglandin E_2, and by increasing blood flow to the gravid uterine horn 400%, both of which result in a decrease in the concentration of prostaglandin $F_2\alpha$ and an increase in the ratio of prostaglandin E_2 to $F_2\alpha$ leaving via the uterine drainage. Second, embryonic oestradiol-17β triggers the secretion of a large number of the histotrophic components that have been produced under progesterone stimulation. This results in a very rapid and dramatic change in the uterine environment that is believed to be detrimental to the least developed embryos in the litter. In the Meishan pig, which exhibits a reduced embryonic mortality, the embryo is smaller at the time of oestradiol-17β production and triggers a less dramatic alteration in the uterine environment. The Meishan embryo also elongates to a reduced length and has a smaller, more efficient placenta later in gestation. To test whether the reduced size of the Meishan conceptus was related to the difference in the magnitude of the oestradiol-17β signal triggering histotrophic secretion, we treated Meishan gilts with exogenous oestradiol-17β at the time that the embryo would be synthesising and secreting oestradiol-17β. Supplementing with additional exogenous oestradiol-17β at the time of maternal recognition of pregnancy and embryonic elongation resulted in a 40% increase in placental size at day 112 of gestation and a concomitant decrease in placental efficiency. Recently, we have demonstrated that the increase in growth of the placenta associated with exogenous oestradiol-17β supplementation at the time of embryonic elongation appears to be mediated by a near doubling of the proliferation rate of the trophectoderm immediately following the supplementation [204, 211–214].

Implanting gilts with oestrone containing implants designed to release 5 mg/oestrone/day on day 30 of gestation resulted in nearly an order of magnitude increase in plasma oestrone [198]. However, this dramatic increase in circulating concentrations of oestrone only resulted in a 75% increase in endometrial content in endometrial tissue and had no effect on fetal or placental weights at day 45 [198]. Although not reported by the authors, as a result of the lack of effect on fetal weight and a reduction in placental weight there was a 15% increase in placental efficiency as a result of the oestrone treatment. Implanting gilts with oestrone or oestradiol–17β containing implants each designed to release 5 mg/day on day 30 again resulted in nearly an order of magnitude increase in plasma oestrone and oestradiol-17β [199]. In this experiment the authors left the gilts implanted until day 60 of gestation, but not surprisingly, neither the oestrone nor the oestradiol-17β treatment had any effect on the weight of the fetus or the placenta at day 60 [199]. It would appear that although increasing the oestradiol-17β exposure to the peri-implantation embryo can have dramatic effects on the conceptus later in gestation, there appears to be no effect of mid-gestational administration of exogenous oestrogens on the growth and development of the conceptus.

Exogenous treatment of pregnant gilts with GH (5 mg/day) from days 30 to 43 of gestation resulted in a 220% increase in maternal serum IGF-I while at the same time a 30% decrease in maternal serum IGF-II [187]. Furthermore, this 13 day regimen of GH supplementation resulted in a 10% increase in fetal weight and a 22% increase in placental weight, with no increase in fetal or implantation site length. In a similar experiment in which gilts were treated with exogenous GH (approximately 4 mg/day) from days 28 to 40 of gestation, there was no affect of GH treatment on fetal weight at day 41 of gestation, but the length of the fetus on day 41 was 20% greater [100]. This treatment with GH from days 28 to 40 had no affect on birth weight, but at market (102 kg) pigs born to GH treated gilts had slightly (less than 2%) longer carcasses, but no difference in *longissimus* muscle cross sectional area or backfat thickness [100]. Exogenous administration of GH (4 or 8 mg/day) to gilts facing mild nutrient restriction altered a number of nutrient and metabolic hormone concentrations in maternal serum, but had no affect on the number of pigs born per litter, average birth weight, or a number of measures of muscle fiber development [81]. Administration of exogenous GH (5 mg/day) to gilts early (days 0–30 of gestation) reduced fetal weight at day 65 of gestation by 15% compared to administration later (days 30–64 of gestation), with pigs born to untreated controls weighing intermediate between the two [186]. There is some evidence that the duration of GH treatment influences the impact of fetal growth. Exogenous administration of GH (2 mg/day) from days 25 to 100 of gestation resulted in a 17% increase in birth weight and 4% increase in body length, while treatment from days 25 to 50 of gestation had no effect on birth weight or body length [80]. Overall, there does not seem to be a consistent affect of exogenous treatment during the first half of gestation, but a substantial effect if the treatment continues beyond the middle of gestation, on the growth or the development of the fetus.

6.5 Effects of Specific Nutrients and Hormones for the Dam on Fetal Development

Historically, authors have surmised from the available literature that the sow has a tremendous ability to draw on her body reserves to "buffer" the conceptuses from dietary restrictions of either protein or energy [161]. However, very severe restrictions (i.e., inanition experiments described above) can have detrimental impacts on birth weight of the progeny.

6.5.1 Energy

Restriction in energy intake clearly reduces pig birth weight (for review see [161]). However, reducing energy intake by 50% from that of controls, while maintaining the same intake of protein, vitamins and minerals from the day after breeding until parturition, did not alter the number of pigs born per litter or the birth weights of those pigs [14–15]. Reducing energy intake during the first two-thirds of gestation to one-third (8.37 MJ DE/day) of NRC recommendations (25.10 MJ DE/day) did not reduce the number of pigs per litter nor the birth weight of pigs born and does not appear to have a negative impact on postnatal growth performance [163]. Not all sows respond similarly to energy restriction. Pond et al. [164] utilised three different genetic lines to examine the interaction between energy restriction and genotype. The lines of pigs included a lean growth line, an obese line and a contemporary line. Gilts were assigned at days 30–40 of gestation to be fed either a control (25.10 MJ DE/day) or restricted (8.37 MJ DE/day) diet for the remainder of gestation. The authors indicate that there were effects of genotype and genotype by diet interactions, but do not report a means separation. However, in simply looking at the means reported, pigs born to both obese and contemporary gilts were lighter when the gilts had a restricted energy intake during the last 70% of gestation compared to control fed, whereas, pigs born to lean gilts were actually heavier when they were born to gilts that were energy restricted compared to those that were control fed. It is important to note that unlike a number of other studies where energy restriction during gestation has been investigated, there were dramatic differences in litter size observed among the six different genotype by diet combinations [164].

Reproductive performance in the sow is highly dependent on management during previous pregnancies [59, 147] and therefore, recent work has focused on the impact of additional energy during gestation on reproductive performance, especially over several parities. It is worthy to note that approximately 5.02 MJ is deposited into the gravid uterus during pregnancy [145, 146]. In a report from the southern regional project on nutritional systems for swine to increase reproductive performance [47], the authors cite Baker et al. [23] as evidence that pig birth weights increase with increases in dietary energy. However, Baker et al. [23] fed varying levels (0.9–3.0 kg/day) of a common diet containing 16% crude protein and 13.84 MJ ME/kg, and so the effects observed can not be specifically attributed to variation in energy intake in the diets. Nevertheless, increasing the energy content of the gestation diet from 24.68 MJ/day to 30.96 MJ/day, while maintaining protein intake, resulted in a small (3%), but detectable increase in birth weight with no difference in the number of pigs born per litter [47]. The pigs born to sows fed the higher energy diet during gestation did gain 6% more weight by day 21 than those born to control sows [46]. Sows fed isonitrogenous diets containing 18.41, 30.96 or 43.51 MJ DE/day for the first 50 day of gestation followed by 30.96 MJ DE/day for the remainder of gestation gave birth to progeny with similar birth weight, weaning weight, muscle fiber area, or muscle fiber distribution [28]. The authors did report that the progeny of sows fed 43.51 MJ DE/day for the first 50 day of gestation had 6% greater carcass adiposity at slaughter [28]. When compared to sows limit-fed throughout gestation, sows that were fed ad libitum either from day 25 to 50 or day 25–70 of gestation gave birth to pigs with the similar birth weights [144]. Furthermore, pigs born to sows from all three groups had similar weaning weights, carcass weights, percent lean in the carcass, total carcass muscle mass, drip loss and *semitendinosus* fiber number, secondary to primary fiber ratio and fiber area [144]. However, increasing intake 36% (simply by feeding more of a common diet) has been reported to increase the density of secondary muscle fibers and the ratio of secondary to primary fibres at 61 day of age, in spite of a lack of effect on birth weight [81]. Overall, though there is some ability to modify the growth of the fetus by altering simply the energy content of the diet, the magnitude of the effect appears to be small.

6.5.2 Protein

Extreme restriction of protein intake during gestation appears to have a much greater detrimental impact on fetal development than energy restriction and this is seen not only as a decrease in birth weight, but negatively impacts postnatal growth [161]. In a series of experiments in which "protein-free" (approximately 9 g protein consumed/day) diets were fed during different stages of gestation, only in those gilts that were fed a protein-free diet throughout gestation or for all but days 16–20 of gestation gave birth to pigs with reduced birth weights compared to gilts fed a control diet throughout gestation or from day 24 of gestation until parturition [162].

Progeny of sows that were fed the protein-free diet exhibited reduced daily gains to slaughter, regardless of whether the first 5 weeks are included or not [162]. In a similar study conducted in second parity sows and continued through two pregnancies (i.e., not during the intervening lactation), feeding sows a similar "protein-free" diet also resulted in a reduced birth weight and daily weight gain compared to offspring from control fed sows over both parities [163]. Clearly, the reduction in protein intake, which leads to a very dramatic decrease in blood urea and serum protein [14], not only limits the growth of the fetus, but alters its growth potential.

Much like energy intake during gestation, protein intake during gestation can have profound effects on the subsequent performance of the sow, particularly during lactation [158]. Efforts to delineate benefits of feeding high protein diets during gestation have only had impacts on the milk production of the sow during lactation [111]. Increasing dietary protein by 34% in an isocaloric diet had no effect on birth weight, crown-rump length, abdominal circumference or skull width [80].

6.5.3 Vitamins

In 1967 Selke et al. published the vitamin A requirements of gestating swine. In their report the authors indicate that over a wide range of vitamin A inclusion in the diet (0–23,510 I.U. per kg) there were no differences in the number of pigs born per litter or the weight of those pigs [181]. The pregnant sow has a considerable capacity to homeostatically maintain plasma vitamin A concentrations, even in the face of dietary depletion or repletion [34, 126]. Therefore, efforts to investigate the impact of additional vitamin A (and/or β-carotene, a precursor of vitamin A) have relied on injection therapy [34, 46, 168]. It was reported that injection of β-carotene (32.6 mg) with or without vitamin A (12,300 I.U.) increased litter size in deficient animals without a negative impact on pig birth weight [34]. In a follow up experiment, in which sows were either injected with β-carotene (0, 50, 100 or 200 mg) at weaning in a sustained release formulation or injected with either β-carotene (200 mg) or vitamin A (50,000 I.U.) at weaning, breeding and 7 days after breeding, the authors reported an increase in the number of live pigs per litter of approximately 0.6 pigs with no detrimental affect on birth weight [46]. In a larger study (approximately 1,000 litters) designed to identify the optimum time during gestation for the administration of vitamin A (1,000,000 I.U.), the authors report that they were unable to detect any effect of vitamin A administration, regardless of when it was given during gestation [168]. Others have failed to detect any effect of feeding sows 1 g/day of vitamin C from day 108 of gestation until parturition on birth weight or litter size [222]. Supplementation of sows through three parities with biotin (600–750 μg/day) resulted in an increase in litter size over those sows that were not supplemented, but had no measurable affect on the size of the pigs born to supplemented sows [117].

Folic acid supplementation has received considerable attention, in part because it has been demonstrated to be required for normal embryo development in rodents [189]. Furthermore, folic acid is required for nucleic acid and some amino acid synthesis, which both occur at a very high rate in the developing conceptus. However, it is quite difficult to induce a folic acid deficiency in practical pig diets (for review see [123]). Dietary supplementation of gilts with 1 ppm in the diet increased the number of pigs born per litter by 10%, but had no effect on birth weight of those pigs [124]. When the folic acid supplementation was increased to 1.65 or 6.62 ppm there was a quadratic increase in litter size with increasing folic acid inclusion, again with no effect on pig birth weight [192]. However, no effect of folic acid supplementation at 15 ppm was observed during the first 25 days of gestation [86]. Supplementation with a variety of folates at 2.1 ppm inclusion rate had no effect on total pigs born, number born alive and birth weight [93].

6.5.4 Other Nutrients

The addition of copper to sow diets (15, 30, or 60 ppm) for four consecutive gestation-lactation cycles has been reported to increase pig birth weight in a linear fashion [120]. More recently, the addition of 250 ppm of copper to the sow diet resulted in an 8% increase in live pig birth weight and when adjusted for litter size the increase approached 9% [50]. Inclusion of 55 ppm (on an as fed basis) of L-carnitine from mating until day 112 of gestation resulted in a 7% increase in birth weight and a 50% decrease in the number of stillborn pigs [143]. However, in a more in depth characterisation, supplementation of the sow diet with L-carnitine did not alter the number, area, diameter or type of muscle fibers compared to offspring born to sows that received no additional L-carnitine [169]. Maternal intake of long chain *n*–3 polyunsaturated fatty acids can change tissue composition of the piglets at birth [173].

Intrauterine growth retardation (IUGR), defined as impaired growth and development of the mammalian embryo/fetus or its organs during pregnancy, is a major concern in domestic animal production [226]. Knowledge of the underlying mechanisms has important implications for the prevention of IUGR and is crucial for enhancing the efficiency of livestock production and animal health. Fetal growth within the uterus is a complex biological event influenced by genetic, epigenetic, and environmental factors, as well as maternal maturity. These factors impact on the size and functional capacity of the placenta, uteroplacental blood flows, transfer of nutrients and oxygen from mother to fetus, conceptus nutrient availability, the endocrine milieu, and metabolic pathways. Impaired placental syntheses of nitric oxide (a major vasodilator and angiogenic factor) and polyamines (key regulators of DNA and protein synthesis) may provide a unified explanation for the etiology of IUGR in response to maternal undernutrition and overnutrition. There is growing evidence that maternal nutritional status can alter the epigenetic state (stable alterations of gene expression through DNA methylation and histone modifications) of the fetal genome. This may provide a molecular mechanism for the role of maternal nutrition on fetal programming and genomic imprinting.

6.6 Maternal Reproductive Diseases Affecting Fetal Development

6.6.1 Porcine Parvovirus (PPV)

PPV is a member of the family Parvoiridae, genus *Parvovirus*. Parvovirus virions are nonenveloped, 25 nm in diameter, and have icosahedral symmetry. The structure of baculovirus-expressed PPV capsids was solved using x-ray crystallography and found to be similar to the related canine CPV and minute virus of mice (MVM) [183]. The PPV capsid protein has 57% and 49% amino acid sequence identity with CPV and MVM, respectively. Although PPV is distinguishable from parvoviruses of all other species, it is antigenically related to some [135]. PPV is ubiquitous in swine throughout the world [133]. PPV is associated with reproductive problems, including abortion, small litters, stillbirths, neonatal deaths and weak piglets. There is no clinically apparent disease in non-pregnant pigs. Disease occurs when sero-negative dams are infected in the first half of gestation and the virus crosses the placenta. At the time of farrowing most gilts and sows are immune and impart a high level of PPV antibody to their offspring via colostrum [133–136]. This passively acquired antibody persists at progressively lower levels for 4–6 months, during which time pigs are relatively refractory to infection [151]. If gilts or sows are infected any time during the second half of gestation there is still likely to be transplacental infection. Horizontal transmission of PPV is thought to occur directly by contact among acutely infected and naïve pigs, and indirectly by naïve pigs ingesting or inhaling virus-laden secretions and excretions [133]. PPV-induced reproductive failure is typically signaled by an unusually large number of mummified fetuses delivered at or near term. Infection early in gestation results in litters being fewer in number, embryonic death and resorption [134]. PPV-induced reproductive failure can be prevented by ensuring that all females have developed an active immunity before they conceive for the first time. Because infection is endemic in most herds, immunity is often the result of natural exposure. To ensure immunity it is common practice to vaccinate gilts once or twice before conception and at least once annually thereafter. Inactivated vaccines are both safe and effective [132].

6.6.2 Porcine Respiratory and Reproductive Syndrome (PRRS)

PRRS is a virus-induced disease that is characterised clinically by reproductive failure of gilts and sows and respiratory tract illness severe in young pigs [13]. PRRS is classified in the family Arteriviridae, genus *Arterivirus*; viruses of similar genus are lactate dehydrogenase-elvating virus of mice, simian hemorrhagic fever virus, and equine viral arteritis virus [138]. An infectious virion of PRRS is 50–65 nm in diameter with a lipid-containing outer envelope and six structural proteins, four glycoproteins (GP2, GP3, GP4 and GP5), a membrane protein and a nucleocapid protein [139] – major structural proteins thought to include primary determinants for virus neutralisation. An epidemic of PRRS-induced reproductive failure is presented as a broad spectrum of clinical features including abortions, late-term dead fetuses, stillborn pigs and weakborn pigs [133]. Abortions account for only some of the reproductive losses after exposure to PRRS; fetuses are infected transplacentally but can go to term or longer. Diagnosis of PRRS includes virus isolation and PCR before virus neutralisation that occurs within 7–10 days after initial infection with PRRS. There is a genetic variation in susceptibility to PRRS that exists in cross-bred commercial pig populations [207]. Genetic variability in susceptibility between distinct lines of pigs, as assessed by an in vitro flow cytometry assay indicated that the Large White line had significantly higher percentage positive monocyte-derived macrophages over a Duroc-Pietrain synthetic line [208]. A combination of PRRS and *B. bronchiseptica* leaves pigs more susceptible to pulmonary infection with opportunistic bacteria such as a *Pasteuralla multocida* [35]. Attenuated-virus vaccines of PRRS are more efficacious than inactivated vaccines but the virus mutates easily and thus it is advisable to separate vaccinated pigs from naïve pigs for several weeks [133, 136].

6.6.3 Classical Swine Fever (Hog Cholera, Peste du Porc, Cholera Porcine, Virusschweinepest)

Pigs and wild boar are the natural reservoir of this disease. Classical swine fever is a serious and highly contagious viral disease (genus *Pestivirus*, family Flaviviridae) of pigs [16, 42]. Only one serotype has been found. Acute or chronic infections occur, and both are usually fatal. Virus transmission is mainly oral and often spread by feeding uncooked contaminated garbage. Incubation periods range from 2 to 14 days. Herds infected with less virulent isolates may show only symptoms of poor reproductive performance, stillbirths or mummification, or a failure to thrive. Virulent strains can cause abortions or death of piglets soon after birth. Some piglets are born with a congenital tremor or congenital malformations of visceral organs and central nervous system. Other pigs are asymptomatic but persistently infected. These animals remain viremic and become clinically ill after several months [60]. Differential diagnosis includes African swine fever, porcine dermatitis and nephropathy syndrome, erysipelas, epery throzoonosis, salmonellosis, actinobacillosis, Glasser's disease (*Haemophilus para suis* infection), thrombocytopenic purpura, Aujeszky's disease, heavy metal poisoning and salt poisoning [218]. Pigs congenitally infected with bovine virus diarrhea (BVD) virus may look very similar to pigs with classical swine fever virus. This virus is moderately fragile in the environment being sensitive to drying and ultraviolet light and is rapidly inactivated by a pH<3 [60]. Sodium hypochlorite and phenolic compounds are effective disinfectants [31]. Vaccines can protect animals from clinical disease, but do not prevent infections. Good vaccination programs can eventually eliminate infections in herds [42]. Classical swine fever does not affect humans.

6.6.4 *Aujeszky's Disease-Pseudorabies Virus (AD or PRV)*

AD is caused by porcine herpesvirus-1, which belongs to the Alphaherpesvirinae subfamily, Herpesviridae family. Herpesvirus virions are enveloped, approximately 150 nm in diameter, and contain an icosahedral nucleocapsid approximately 100 nm in diameter, composed of 162 capsomers [137, 180]. Replication occurs in the nucleus with sequential transcription of immediate early (α), early (β) and late (γ) genes producing α, β and γ proteins; the earlier genes and their products regulate the transcription of later genes. DNA replication and encapsidation occur in the nuclear envelope. Pigs (*Sus scrofa domesticus* and *Sus scrofa scrofa*) are the main host; secondary hosts include cattle, sheep, goats, dogs, cats and many feral species. Humans are refractory. The virus infects the central nervous system and other organs such as the respiratory tract. AD spreads rapidly through newly infected herds. In pigs <2 weeks old, prostration and death occurs within hours. In older piglets fever, loss of appetite and depression, vomiting, respiratory difficulty, incoordinate locomotion, drowsiness, muscular twitching, involuntary eye movements and paralysis can be seen with a resulting mortality of 20–100%. Grower and finisher pigs show respiratory distress with mortality rates of 1–10%, whereas in adult pigs the disease is often mild or unapparent. The incubation period is approximately 30 hours. In young piglets the course is typically 8 days but may be as short as 4 days. AD frequently goes into latency with infected animals showing no symptoms yet shed virus for a long time. The virus infection can be reactivated from latency, causing re-shedding of virus and new outbreaks [20]. AD is highly contagious and is principally spread via the respiratory route, oral and nasal secretions being a potent source of virus. AD virus can be transmitted via semen, vaginal secretions and transplacental infection, colostrum or milk, and via contaminated veterinary instruments and equipment. Attenuated, inactivated (PRV) and gene-deleted vaccines have been developed. Vaccines protect pigs from clinical disease, reduce the amount and duration of virus excretion, but do not prevent latent infections [82, 137, 180].

6.6.5 *Porcine Brucellosis (*Brucella suis*)*

Brucellosis is an infectious and contagious disease caused by the bacteria *Brucella suis* mainly in pigs but it can affect other domestic livestock species and humans [36, 37, 131, 160]. After an initial bacteraemia causes chronic inflammatory lesions in reproductive organs of both sexes, it also can localise with lesions in other tissues [150]. *Brucella suis* consists of five biovars, but the infection in pigs is caused by biovars 1, 2, or 3. Biovar 2 is rarely pathogenic for humans, whereas biovars 1 and 3 are highly pathogenic, causing severe disease. Porcine brucellosis primarily occurs in adults, but it does not always cause symptoms and therefore clinical diagnosis is difficult. When endemic, common signs are: non-specific infertility, reduced farrowing rate, irregular oestrous cycles, initial fever, testicular pain, reluctance to mate and lameness. When recently introduced to a herd, dramatic signs are: increase in returns to service, abortions and stillbirths, weak piglets resulting in increased pre-weaning mortality rate, abscesses, lameness caused by arthritis and posterior paralysis. Porcine brucellosis is transmitted venereally and by ingestion. Bacteria are excreted in semen, boar urine and in uterine discharges and milk and can survive up to 6 weeks. There are no effective vaccines. Control of the disease requires basic hygiene such as disposal of contaminated material (i.e., placentas), thorough cleaning and disinfection, and quarantine.

6.6.6 Porcine Enterovirus Infection, Swine Vesicular Disease (SVD)

SVD typically is a transient disease of pigs with vesicular lesions appearing in the mouth and on the feet. SVD virus is a single-stranded RNA genome enclosed in a capsid of icosohedral symmetry, and is a porcine enterovirus in the Picornaviridae family [55, 61]. The virus can survive for long periods in the environment being resistant to heat up to 69°C, and pH ranging from 2.5 to 12. SVD virus infection does not cause severe production losses, but it is of major economic importance because it is difficult to distinguish from foot-and-mouth disease. The disease has not occurred in North America or Australia, and as of 2003 SVD is found only in Italy and Portugal. SVD is moderately contagious and morbidity is lower and lesions are less severe than seen in foot-and-mouth disease. Pigs are most easily infected through damaged skin or an ulcerated mucous membrane with an incubation period of 2–7 days. Pigs can secrete the virus from nose or mouth, and excrete in feces about 48 h before clinical signs are evident. Pigs are the only species that are naturally infected, but the virus may be present in sheep or cattle. Differential diagnosis for SVD diseases include foot-and-mouth disease, vesicular stomatitis, vesicular exanthema of swine and chemical or thermal burns. Cell mediated responses in a porcine enterovirus infection in pathogen-free piglets with infection are weak and localised and not associated with significant antiviral activity [38]. This virus is not fatal to humans. Strict quarantine by state or federal authorities should be imposed in farms or areas suspected of having SVD. There are no inactivated vaccines against SVD virus, none are commercially available.

6.6.7 Japanese Encephalitis Virus (JEV)

JEV, potentially severe viral disease that is spread by infected mosquitoes (arbovirus), is a flaviviral (single-stranded RNA) neurologic infection closely related to St. Louis encephalitis and West Nile virus [99]. The virus can infect humans, most domestic animals, birds, bats, snakes and frogs. JEV has a complex life cycle involving domestic pigs and a specific type of mosquito, *Culex tritaemorhynchus*, that lives in rural rice-growing and pig-farming regions. There is no specific treatment for Japanese encephalitis [19].

6.6.8 *Vesicular Stomatitis (VS)*

VS is a viral disease characterised by fever, vesicles, and subsequent erosions in the mouth and epithelium on the teats and feet. Horses, cattle and pigs are naturally susceptible; sheep and goats are rarely affected. The vesicular stomatitis virus is a genus *Vesiculovirus* in the family Rhaboviridae; major subtypes: New Jersey, Indiana [148]. Animals are infected with the virus by eating or coming in contact with contaminated saliva or fluid from lesions of infected animals, and can be transmitted to humans. The incubation period to clinical signs ranges from 2 to 8 days, and animals generally recover completely in 3–4 days. There is no specific treatment. Antibiotics may avoid secondary infection of abraised tissues.

6.6.9 *Foot-and-Mouth Disease (FMD)*

FMD is a severe, highly communicable viral disease of cattle and swine. It also affects sheep, goats, deer, and other cloven-hoofed ruminants [12]. The virus survives in lymph nodes and bone marrow at neutral pH but is destroyed in muscle at pH <6.0. There are at least seven separate types and many subtypes of FMD virus. Immunity to one type does not protect an animal against other types. FMD can be confused with several similar, but less harmful, diseases such as vesicular stomatitis, bluetongue, bovine viral diarrhea, and foot rot in cattle, vesicular exanthema of swine, and swine vesicular disease. FMD is one of the most difficult animal infections to control; livestock animals are highly susceptible to FMD viruses. If FMD appears in animals, report immediately to begin an effective state and federal eradication program.

6.6.10 *Menangle*

Menangle is a newly emerged disease of swine, currently limited to one outbreak in Menangle, New South Wales, Australia [105–107, 159]. This viral disease causes mummified and stillborn piglets, reduced farrowing rates and reduced litter number and size, as well as occasional abortions. The disease appears to be maintained and spread by fruit bats (*Pteropus sp.*); however the route of transmission in swine is currently unknown. Menangle is one of several recently discovered RNA viruses in the family Paramyxoviridae. Molecular characterisation of the virus places it in the genus *Rubulavirus*. The mode of transmission of Menangle virus from bats to pigs is unknown, but fecal-oral or urinary-oral transmission is suspected.

*6.6.11 Porcine Erysipelas Polyarthritis (***Erysipelothrix rhusiopathiae***)*

Erysipelothrix rhusiopathiae is a natural pathogen of swine, causing a self-sustaining, chronic polyarthritis, comparable in many ways to human arthritis [49, 114]. An outbreak of acute Erysipelas was diagnosed in sows housed in a single gestation barn on a commercial 1000 sow farrow-to-finish farm. Erysipelas is a continuing threat to intensively reared swine housed in entirely environmentally-controlled housing. Clinical examination revealed polygonal dark red to purple, sometimes raised lesions on the skin of the dorsum and hams. All animals were reluctant to walk and favored one or more limbs when walking. In addition to Erysipelas, diamond skin lesions have also been reported with septicemia caused by Actinobacillussuis. The source of *E. rhusiopathiae* in outbreaks of Erysipelas was previously thought to be persistence in the soil; however, recent research has shown that it can remain viable in soil for only 35 days under optimal conditions. In healthy swine, estimates of 30 to 50% are tonsillar carriers of *E. rhusiopathiae*. The organism can be shed in oronasal secretions, urine and feces. Rodents and birds can also serve as reservoirs. Spread of infection in a swine heard is by close contact or by contaminated water, feed or bedding. Chronic polyarthritis was induced in pigs by infection with *Erysipelothrix rhusiopathiae* (serovar 2, strain T28) and viable bacteria could be isolated >5 months later from synovial fluid and from isolated chondrocytes [78]. The number of viable bacteria could be increased by hypotonic shock of the chondrocytes indicating a substantial intracellular amount of bacteria; neither viable bacteria nor bacterial antigen were detected in unaffected joints. In similarly infected pigs the in vivo activation of chondrocytes by cytokines was investigated in affected joints by immunocytochemistry [53]. The presence of interleukin 1 in the inflammatory cells of the synovium was confirmed by major histocompatibility complex (MHC) class II antigens that were detected as a marker of synovial activation. In contrast, cartilage removed from an unaffected joint in the same animal showed no chondrocyte activation. In vivo expression and distribution of the porcine homologues of the intercellular adhesion molecule-1 (ICAM-1) and MHC Class II as markers of chondrocyte activation was examined in experimentally induced infection via intra-articular injection of *Erysipelothrix rhusiopathiae* [52]. ICAM-1 was found to be strongly expressed in vivo on chondrocytes and synovial cells in arthritic joints but not in cartilage from unaffected joints. As the disease progressed 5 months post-infection, infiltration of $CD4^{+}$ lymphocytes into damaged cartilage was also apparent. Although ICAM-1 and MHC Class II are not expressed on porcine chondrocytes they appear to be induced as arthritis progresses and their detection serve as markers in arthritic pigs. Other tests include erysipelas serum titration with sheep red blood cells passively sensitised with a cellular extract [165]. Diagnostic differential hemagglutination titrations may then be made with porcine serum for erysipelas antibody. Vaccination of healthy breeding swine as an aid in preventing reproductive failure caused by PPV, erysipelas caused by *Erysipelothrix rhusiopathiae*, and leptospirosis antigens chemically inactivated provides a 26-week duration of immunity against erysipelas.

6.6.12 *Porcine Leptospirosis (Leptospira Pomona, Leptospira Interrogans)*

Leptospirosis is an important bacterial disease of humans and many animal species. It is of economic significance in swine worldwide, not only because of abortions and stillbirths, but also because of the high death rates caused by the *icterohaemorrhagiae* serovar [66, 94]. Losses caused by leptospirosis result from localisation of the bacteria in the host that, during the leptospiremia period, diffuse throughout the animal organism, including the genital tract [65, 66]. During this phase, leptospira can be found in liver, lungs, eyeballs and central nervous system. Leptospires are tightly coiled helical spirochetes, usually 0.1 μm by 6–0.1 by 20 μm, but occasional cultures may contain much longer cells [70, 116]. The genus *Leptospira* was divided into two species *L. interrogans* with >200 serovars comprising all pathogenic strains, and *L. biflexa* with >60 serovars containing the saprophytic strains isolated from the environment [71]. In porcine kidney sections immunohistochemically stained with polyvalent antisera (serovars *canicola*, *grippotyphosa*, *hardjo*, *copenhageni*, and *pomona*) numerous spiroid bacteria are seen within tubular lumens resulting in interstitial nephritis and pyelonephritis [140–141]. The concept that the kidney is the main location of persistent leptospira and the primary maintainer of the chronic disease state was questioned by new findings in swine infected by *bratislava* serovar, which was isolated from the oviduct and uterus of sows that aborted and from the reproductive tract of boars [54, 64]. The natural reservoir of pathogenic leptospires is the proximal convoluted tubule of the kidney and in certain maintenance hosts, the genital tract. Transmission can be direct through urine splashing, in post-abortion discharges, venereally, through milk or transplacentally. Indirect transmission is through contamination of the environment with infected urine. Leptospires penetrate exposed mucous membranes or through abraded or water-softened skin and then disseminate throughout the body. The principal clinically significant aspects of the disease in swine are abortion and birth of weak piglets. A usual pattern in sows is delivery 1–3 weeks prematurely, with some mummified fetuses, some recently dead, and others born alive only to die soon afterward. *Leptospiria interrogans* serovar *pomona* type kennewicki and serovar *grippotyphosa* are the most frequent isolates in cases of porcine leptospirosis. Disease prevention includes vaccination of breeding stock (including boars) twice, 4–6 weeks apart, and then at 6-month intervals. Vaccination of sows about to farrow or with very young litters, should be delayed a week. It is safe to vaccinate pregnant gilts or sows.

Acknowledgements This project was supported by National Research Initiative Competitive Grant no. 2003-35206-12817 from the USDA Cooperative State Research, Education, and Extension Research, and from the Iowa Agriculture and Home Economics Experiment Station, Ames, supported by Hatch and State of Iowa funds.

References

1. Alexander, G. 1974. Birth weight of lambs: influences and consequences, pp. 215–245. *In* K. Elliott and J. Knight (eds.), Size at Birth, Amsterdam, Elsevier.

2. Anderson, L.L. 1975. Embryonic and placental development during prolonged inanition in the pig. *Am. J. Physiol.* **229**:1687–1694.
3. Anderson, L.L. 1978. Growth, protein content, and distribution of early pig embryos. *Anat. Rec.* **190**:143–154.
4. Anderson, L.L. 2004. Discovery of a new cellular structure-the porosome: elucidation of the molecular mechanism of secretion. *Cell Biol. Int.* **28**:3–5.
5. Anderson, L.L. and D.W. Dunseth. 1978. Dietary restriction and ovarian steroids on fetal development in the pig. *Am. J. Physiol.* **234**:E190–E196.
6. Anderson, L.L., G.W. Dyck, H. Mori, D.M. Henricks, and R.M. Melampy. 1967. Ovarian function in pigs following hypophysial stalk transection or hypophysectomy. *Am. J. Physiol.* **212**:1188–1194.
7. Anderson, L.L., D.L. Hard, and L.P. Kertiles. 1979. Progesterone secretion and fetal development during prolonged starvation in the pig. *Am. J. Physiol.* **236**:E335–E341.
8. Anderson, L.L., C.G. Scanes, and S. Jeftinija. 2004. Invited MINIREVIEW titled: Growth hormone secretion: Molecular and cellular mechanisms and *in vivo* approaches. *Exp. Biol. Med.* **229**:291–302.
9. Anderson, L.L., S. Jeftinija, C.G. Scanes, M.H. Stromer, J.-S. Lee, K. Jeftinija, and A. Glavaski-Joksimovic. 2005. Physiology of ghrelin and related peptides. Proceedings of the 5th International Conference on Farm Animal Endocrinology, July 4–6, 2004, Budapest, Hungary. *Domest. Anim. Endocrinol.* **29**:111–144.
10. Anderson, L.L. and R.M. Melampy. 1972. Factors affecting ovulation rate in the pig, pp. 329–383. *In* D.J.A. Cole (ed.), Pig Production, Proceedings of the 18th Easter School In Agricultural Science, University of Nottingham. Butterworths, London.
11. Andersson, L., C.S. Haley, H. Ellegren, S.A. Knott, M. Johansson, K. Anderson, L. Anderson-Eklund, I. Edfors-Lilja, M. Fredholm, I. Hansson, J. Hakansson, and K. Lundstrom. 1994. Genetic mapping of quantitative trait loci for growth and fatness in pigs. *Science* **263**:1771–1774.
12. http://www.aphis.usda.gov/lpa/pubs/fsheet_faq_notice/fs_ahfmd.html. 2002. Foot-and mouth disease.
13. http: //ars.usda.gov/research/projects/projects.htm?ACCN_NO=405351&fy=2002. Virus and Prion Diseases of Livestock. 2002 Annual Report.
14. Antinmo, T., W.G. Pond, and R.H. Barnes. 1974a. Effect of dietary energy vs. protein restriction on blood constituents and reproductive performance in swine. *J. Nutr.* **104**:1033–1040.
15. Antinmo, T., W.G. Pond, and R.H. Barnes. 1974b. Effect of maternal energy vs. protein restriction on growth and development of progeny in swine. *J. Anim. Sci.* **39**:703–711.
16. Artois, M., K.R. Depner, V. Guberti, J. Hars, S. Rossi, and D. Rutili. 2002. Classical swine fever (hog cholera) in wild boar in Europe. *Rev. sci. tech. Off. int. Epiz.* **21**:287–303.
17. Ashworth, C.J. 1991. Effect of pre-mating nutritional status and post-mating progesterone supplementation on embryo survival and conceptus growth in gilts. *Anim. Reprod. Sci.* **26**:311–321.
18. Ashworth, C.J. 1998. Advances in embryo mortality research, pp. 231–237. *In* Proceedings of the 15th IPVS Congress Birmingham, England.
19. http://www.astdhpphe.org/infect/jpenceph.html. 2006. Japanese encephalitis.
20. Aujeszky's Disease Virus (ADV). 2006. http://www.multiplex-eu.org/adv.php
21. Auletta, F.J. and A.P. Flint. 1988. Mechanisms controlling corpus luteum function in sheep, cows, nonhuman primates, and women especially in relation to the time of luteolysis. *Endocr. Rev.* **9**:88–105.
22. Bacci, M.L., A.M. Barazzoni, M. Forni, and G.L. Costerbosa. 1996. In situ detection of apoptosis in regressing corpus luteum of pregnant sow: evidence of an early presence of DNA fragmentation. *Domest. Anim. Endocrinol.* **13**:361–372.
23. Baker, D.H., D.E. Becker, H.W. Norton, C.E. Sasse, A.H. Jensen, and B.G. Harmon. 1969. Reproductive performance and progeny development in swine as influenced by feeding intake during pregnancy. *J. Nutr.* **97**:489–495.

24. Barb, C.R. 1999. The brain-pituitary-adipocyte axis: role of leptin in modulating neuroendocrine function. *J. Anim. Sci.* **77**:1249–1257.
25. Barb, C.R, J.B. Barrett, R.R. Kraeling, and G.B. Rampacek. 2001. Serum leptin concentrations, luteinizing hormone and growth hormone secretion during feed and metabolic fuel restriction in the prepubertal gilt. *Domest. Anim. Endocrinol.* **20**:47–63.
26. Barb, C.R., R.R. Kraeling, G.B. Rampacek, and C.A. Pinkert. 1984. Luteinizing hormone receptors and progesterone content in porcine corpora lutea after prostaglandin F2 alpha. *Biol. Reprod.* **31**:913–919.
27. Bazer, F.W., W.W. Thatcher, F. Martinat-Botte, and M. Terqui. 1988. Sexual maturation and morphological development of the reproductive tract in Large White and prolific Chinese Meishan pigs. *J. Reprod. Fert.* **83**:723–728.
28. Bee, G. 2004. Effect of early gestation feeding, birth weight, and gender of progeny on muscle fiber characteristics of pigs at slaughter. *J. Anim. Sci.* **82**:826–836.
29. Belstra, B.A., M.A. Diekman, B.T. Richert, and W.L. Singleton. 2002. Effects of lactation length and an exogenous progesterone and estradiol-17β regimen during embryo attachment on endogenous steroid concentrations and embryo survival in sows. *Theriogenology* **57**:2063–2081
30. Biensen, N.J., M.E. Wilson, and S.P. Ford. 1998. Meishan and Yorkshire fetal and placental development in either a Meishan or Yorkshire uterus to day 70, 90 and 110 of gestation. *J. Anim. Sci.* **76**:2169–2176.
31. Blackwell, J.H. 1998. Cleaning and disinfection, pp. 445–448. *In* Foreign Animal Diseases, Richmond, VA, United States Animal Health Association.
32. Bolet, G., J.P. Bidanel, and L. Ollivier. 2001. Selection for litter size in pigs. II. Efficiency of closed and open selection lines. *Genet. Select. Evol.* **33**:515–528.
33. Bollinger, A.L., M.E. Wilson, A.E. Pusateri, M.L. Green, T.G. Martin, and M.A. Diekman. 1997. Lack of a nocturnal rise in serum concentrations of melatonin as gilts attain puberty. *J. Anim. Sci.* **75**:1885–1892.
34. Brief, S. and B.P. Chew. 1985. Effects of vitamin A and β-carotene on reproductive performance in gilts. *J. Anim. Sci.* **60**:998–1004.
35. Brockmeier, S., K. Lager, and M. Palmer. 2002. Interactions among porcine reproductive and respiratory system virus (Prrsv), *Bordetella bronchiseptica* and *Haemophilus parasuis* in swine respiratory disease. Pig Veterinary Society International Congress Proceedings.
36. 1998. Brucellosis in pigs, pp. 1001–1002. *In* National Publishing Inc. Eight (ed.), Merck Veterinary Manual, Philadelphia.
37. 1997. Brucellosis caused by *Brucella suis*, pp. 807–810. *In* Saunders, Eight (ed.), Veterinary Medicine. London.
38. Brundage, L.J., J.B. Derbyshire, and B.N. Wilkie. 1980. Cell mediated responses in a porcine enterovirus infection in piglets. *Can. J. Comp. Med.* **44**:61–69.
39. Buitrago, J.A., E.F. Walker, Jr., W.I. Snyder, and W.G. Pond. 1974. Blood and tissue traits in pigs at birth and at 3 weeks from gilts fed low or high energy diets during gestation. *J. Anim. Sci.* **38**:766–771.
40. Buntin, J.D. 1993. Prolactin-brain interactions and reproductive function. *Amer. Zool.* **33**:229–243.
41. Casida, L.E. 1953. Fertilization failure and embryonic death in domestic animals, pp. 27–37. *In* E.T. Engle (ed.), Pregnancy Wastage. Thomas, Springfield, Ill.
42. Center for Food Security and Public Health, College of Veterinary Medicine, Iowa State University. Classical Swine Fever. August 2, 2005, pp. 1–3.
43. Cho, S.-J., K. Jeftinija, A. Glavaski, S. Jeftinija, B.P. Jena, and L. L. Anderson. 2002. Structure and dynamics of the fusion pores in live GH-secreting cells revealed using atomic force microscopy. *Endocrinology* **143**:1144–1148.
44. Clark, B.J., J. Wells, S.R. King, and D.M. Stocco. 1994. The purification, cloning, and expression of a novel luteinizing hormone-induced mitochondrial protein in MA-10 mouse Leydig tumor cells. Characterization of the steroidogenic acute regulatory protein (StAR). *J. Biol. Chem.* **269**:28314–28322.

45. Claus, R. and U. Weiler. 1985. Influence of light and photoperiodicity on pig prolificacy. *J. Reprod. Fertil. Suppl.* **33**:199–208.
46. Coffey, M.T. and J.H. Britt. 1993. Enhancement of sow reproductive performance by β-carotene or vitamin A. *J. Anim. Sci.* **71**:1198–1202.
47. Coffey, M.T., B.G. Diggs, D.L. Handlin, D.A. Knabe, C.V. Maxwell, Jr., P.R. Noland, T.J. Prince, and G.L. Cromwell. 1994. Effects of dietary energy during gestation and lactation on reproductive performance of sows: A cooperative study. *J. Anim. Sci.* **72**:4–9.
48. Conley, L.K., M.C. Aubert, A. Giustina, and W.W. Wehrenberg. 1996. Role of GRA and GH in the onset of puberty and during corticoid-altered growth, pp. 185–209. Proceedings of the Serono Symposium on Growth Hormone Secretagogues, New York.
49. Coombs, R.R.A. 1989. Viewpoint of a general immunologist. *Vet. Rec.* **124**:553–557.
50. Cromwell, G.L., H.J. Monegue, and T.S. Stahly. 1993. Long-term effects of feeding a high copper diet to sows during gestation and lactation. *J. Anim. Sci.* **71**:2996–3002.
51. Dalton, D.L. and J.W. Knight. 1983. Effects of exogenous progesterone and estrone on conceptus development in swine. *J. Anim. Sci.* **56**:1354–1361.
52. Davies, M.E., A. Horner, and B. Franz. 1994. Intercellular adhesion molecule-1 (ICAM-1) and MHC Class II on chondrocytes in arthritic joints from pigs experimentally infected with *Erysipelothrix rhusiopathiae. FEMS Immunol. Med. Microbiol.* **9**:265–272.
53. Davies, M.E., A. Horner, B. Franz, and H.J. Schuberth. 1992. Detection of cytokine activated chondrocytes in arthritic joints from pigs infected with *Erysipelothrix rhusiopathiae. Ann. Rheum. Dis.* **51**:978–982.
54. Delbem, A.C.B., J.C. de Freitas, A.P.F.R.L. Bracarense, E.E. Müller, and R.C. de Oliveira. 2002. Leptospirosis in slaughtered sows: serological and histopathological investigation. *Braz. J. Microbiol.* **33**:174–177.
55. Derbyshire, J.B., M.C. Clarke, and D.M. Jessett. 1969. The isolation of adenoviruses and enteroviruses from pig tissues. *J. Comp. Pathol.* **79**:97–100.
56. Diaz, F.J. and M.C. Wiltbank. 2005. Acquisition of luteolytic capacity involves differential regulation by prostaglandin F2α of genes involved in progesterone biosynthesis in the porcine corpus luteum. *Domest. Anim. Endocrinol.* **28**:172–189.
57. Dickerson, J.W.T. and J. Dobbing. 1967. Prenatal and postnatal growth and development of the central nervous system of the pig. *Proc. R. Soc. Lond. [Biol.]* **166**:384–395.
58. Dlamini, B.J., Y. Li, J. Klindt, and L.L. Anderson. 1995. Acute shifts in relaxin, progesterone, prolactin and growth hormone secretion in Chinese Meishan gilts during late pregnancy and after hysterectomy. *J. Anim. Sci.* **73**:3732–3742.
59. Dourmad, J.Y. 1991. Effect of feeding level in the gilt during pregnancy on voluntary feed intake during lactation and changes in body composition during gestation and lactation. *Livest. Prod. Sci.* **27**:309–315.
60. Dulac, G.C. 1998. Hog cholera, pp. 273–282. *In* Foreign Animal Diseases, Richmond, VA, United States Animal Health Association.
61. Dunne, H.W., J.T. Wang, and E.H. Ammerman. 1971. Classification of North American porcine enteroviruses: a comparison with European and Japanese strains. *Infect. Immun.* **4**:619–631.
62. Einarsson, S. and T. Rojkittikhun. 1993. Effects of nutrition on pregnant and lactating sows. *J. Reprod. Fertil.* **48**:229–239.
63. Ellegren, H., B.P. Chonvdhary, M. Johansson, L. Marklund, M. Fredholm, I. Gustavsson, and L. Anderson. 1994. A primary linkage map of the porcine genome reveals a low rate of genetic recombination. *Genetics* **137**:1089–1100.
64. Ellis, W.A., P.J. McParland, D.G. Bryson, A.B. Thiermann, and J. Montgomery. 1986. Isolation of leptospires from the genital tract and kidneys of aborted sows. *Vet. Rec.* **118**:294–295.
65. Ellis, W.A., J.J. O'Brien, J.A. Cassells, S.D. Neill, and J. Hanna. 1985. Excretion of *Leptospira interrogans* serovar *hardjo* following calving or abortion. *Res. Vet. Sci.* **39**:296–298.
66. Ellis, W.A. and A.B. Thiermann. 1986. Isolation of *Leptospira interrogans* serovar *bratislava* from sows in Iowa. *Am. J. Vet. Res.* **47**:1458–1460.

67. Estany, J., D. Villalba, M. Tor, D. Cubiló, and J.L. Noguera. 2002. Correlated response to selection for litter size in pigs: II. Carcass, meat, and fat quality traits. *J. Anim. Sci.* **80**:2566–2573.
68. Estienne, M.J., A.F. Harper, D.M. Kozink, and J.W. Knight. 2003. Serum and milk concentrations of leptin in gilts fed a high- or low-energy diet during gestation. *Anim. Reprod. Sci.* **75**:95–105.
69. Fahrenkrug, S.C., T.P. Smith, B.A. Freking, J. Cho, J. White, J.L. Vallet, T.H. Wise, G.A. Rohrer, and J.W. Keele. 2002. Porcine gene-discovery by normalized CDNA-library sequencing and EST cluster assembly, pp. 475–478. *Mamm. Genome*, v. 13.
70. Faine, S., B. Adler, C. Bolin, and P. Perolat. 1999. *Leptospira* and leptospirosis, 2nd ed. MedSci, Melbourne, Australia.
71. Faine, S. and N.D. Stallman. 1982. Amended descriptions of the genus *Leptospira* Noguchi 1917 and the species *L. interrogans* (Stimson 1907) Wenyon 1926 and *L. biflexa* (Wolbach and Binger 1914) Noguchi 1918. *Int. J. Syst. Bacteriol.* **32**:461–463.
72. Fernstrom, J.D. 1983. Role of precursor availability in control of monoamine biosynthesis in brain. *Physiol. Rev.* **63**:484–546.
73. Ford, S.P. 1997. Embryonic and fetal development in different genotypes in pigs. *J. Reprod. Fertil. Suppl.* **52**:165–176.
74. Ford, S.P. and R.K. Christenson. 1979. Blood flow to uteri of sows during the estrous cycle and early pregnancy: Local effect of the conceptus on the uterine blood supply. *Biol. Reprod.* **21**:617–624.
75. Ford, S.P., N.K. Schwartz, M.F. Rothschild, A.J. Conley, and C.M. Warner. 1988. Influence of SLA haplotype on preimplantation embryonic growth rate in miniature swine. *J. Reprod. Fertil.* **84**:90–104.
76. Foxcroft, G.R. 1997. Mechanisms mediating nutritional effects on embryonic survival in pigs. *J. Reprod. Fertil. Suppl.* **52**:47–61.
77. Foxcroft, G.R., F.X. Aherne, E.J. Clowes, H. Miller, and L. Zak. 1995. Sow fertility: The role of suckling inhibition and metabolic status, pp. 377–393. *In* M. Ivan (ed.), Animal Science Research and Development-Moving Towards a New Century. Agriculture and Food Canada, Ottawa, Ont., Canada.
78. Franz, B., M.E. Davies, and A. Horner. 1995. Localization of viable bacteria and bacterial antigens in arthritic joints of *Erysipelothrix rhusiopathiae*-infected pigs. *FEMS Immunol. Med. Microbiol.* **12**:137–142.
79. Frawley, L.S. and F.R. Boockfor. 1991. Mammosomatotropes: presence and functions in normal and neoplastic tissue. *Endocr. Rev.* **12**:337–355.
80. Gatford, K.L., J.M. Boyce, K. Blackmore, R.J. Smits, R.G. Campbell, and P.C. Owens. 2004. Long-term, but not short-term, treatment with somatotropin during pregnancy in underfed pigs increases the body size of progeny at birth. *J. Anim. Sci.* **82**:93–101.
81. Gatford, K.L., J.E. Ekert, K. Blackmore, M.J. De Blasio, J.M. Boyce, J.A. Owens, R.G. Campbell, and P.C. Owens. 2003. Variable maternal nutrition and growth hormone treatment in the second quarter of pregnancy in pigs alter semitendinosus muscle in adolescent progeny. *Br. J. Nutr.* **90**:283–293.
82. Geering, W.A., A.J. Forman, and M.J. Nunn. 1995. Exotic diseases of animals, pp. 440. *In* Aust Gov Publishing Service, Canberra.
83. Geisert, R.D., J.W. Brookbank, R.M. Roberts, and F.W. Bazer. 1982. Establishment of pregnancy in the pig: II. Cellular remodeling of the porcine blastocyst during elongation on day 12 of pregnancy. *Biol. Reprod.* **27**:941–955.
84. Geisert, R.D., R.H. Renegar, W.W. Thatcher, R.M. Roberts, and F.W. Bazer. 1982. Establishment of pregnancy in the pig: I. Interrelationships between peri-implantation development of the pig blastocysts and the uterine endometrial secretions. *Biol. Reprod.* **27**: 925–939.
85. Geisert, R.D. and R.A.M. Schmitt. 2002. Early embryonic survival in the pig: Can it be improved? *J. Anim. Sci.* (E. Suppl. 1) **80**:E54–E65.

86. Guay, F., J.J. Matte, C.L. Girard, M.-F. Palin, A. Giguère, and J.-P. Laforest. 2002. Effect of folic acid and glycine supplementation on embryo development and folate metabolism during early pregnancy in pigs. *J. Anim. Sci.* **80**:2134–2143.
87. Guedes, R.M.C. and R.H.G. Nogueira. 2001. The influence of parity order and body condition and serum hormones on weaning-to-estrus interval of sows. *Anim. Reprod. Sci.* **67**:91–99.
88. Hanly, S. 1961. Prenatal mortality in farm animals. *J. Reprod. Fertil.* **2**:182–194.
89. Hard, D.L. and L.L. Anderson. 1979. Maternal starvation and progesterone secretion, litter size, and growth in the pig. *Am. J. Physiol.* **237**:E273–E278.
90. Hard, D.L. and L.L. Anderson. 1982a. Interaction of maternal blood volume and uterine blood flow with porcine fetal development. *Biol. Reprod.* **27**:79–90.
91. Hard, D.L. and L.L. Anderson. 1982b. Ornithine decarboxylase activity in maternal and fetal tissues during inanition in pigs. *Biol. Reprod.* **27**:91–98.
92. Hard, D.L. and L.L. Anderson. 1982c. Growth, puberty, and reproduction in gilts born to nutrient-deprived dams during pregnancy. *J. Anim. Sci.* **54**:1227–1234.
93. Harper, A.F., J.W. Knight, E. Kokue, and J.L. Ursy. 2003. Plasma reduced folates, reproductive performance, and conceptus development in sows in response to supplementation with oxidized and reduced sources of folic acid. *J. Anim. Sci.* **81**:735–744.
94. Hathaway, S.C. and T.W.A. Little. 1981. Prevalence and clinical significance of leptospiral antibodies in pigs England. *Vet. Rec.* **108**:224–227.
95. Isler, B.J., K.M. Irvin, S.M. Neal, S.J. Moeller, M.E. Davis, and D.L. Meeker. 2006. Examination of the relationship between the estrogen receptor gene and reproductive tract components in swine. Ohio State University Research and Reviews: Poultry and Swine Special Circular 171–00. http://ohioline.osu.edu/sc171_8.html
96. Johnson, R. 2006. History of litter size selection. pp. 1–6. http://mark.asci.ncsu.edu/nsif/00proc/johnson.htm
97. Johnson, R.K., M.K. Nielsen, and D.S. Casey. 1999. Responses in ovulation rate, embryonic survival and litter traits in swine from 14 generations of selection to increase litter size. *J. Anim. Sci.* **77**:541–557.
98. Johnson, R.W., E.H. von Borell, L.L. Anderson, L.D. Kojic, and J.E. Cunnick. 1994. Intracerebroventricular injection of corticotropin-releasing hormone in the pig: acute effects on behavior, adrenocorticotropin secretion, and immune suppression. *Endocrinology* **135**:642–648.
99. Kallen, A.J. 2005. Japanese encephalitis. http://www.emedicine.com/med/topic3158.htm
100. Kelley, R.L., S.B. Jungst, T.E. Spencer, W.F. Owsley, C.H. Rahe, and D.R. Mulvaney. 1995. Maternal treatment with somatotropin alters embryonic development and early postnatal growth of pigs. *Domest. Anim. Endocrinol.* **12**:83–94.
101. Kephart, K.B., D.R. Hagen, L.C. Griel, Jr., and M.M. Mashaly. 1981. Relationship between uterine progesterone and fetal development in pigs. *Biol. Reprod.* **25**:349–352.
102. Kim, K.-S., L.L. Anderson, J.M. Reecy, N.T. Nguyen, G.S. Plastow, and M.F. Rothschild. 2003. Molecular genetic studies of porcine genes for obesity, pp. 269–271. *In* G. Medeiros-Neto, A. Halpern, and C. Bouchard (eds.), Progress in Obesity Research, John Libbey Eurotext Ltd.
103. Kim, K.-S., N. Larsen, T. Short, G.S. Plastow, and M.F. Rothschild. 2000. A missense variant of the melanocortin 4 receptor gene is associated with fatness, growth and feed intake traits. *Mamm. Genome* **11**:131–135.
104. Kim, K.-S., J.M. Reecy, W.H. Hsu, L.L. Anderson, and M.F. Rothschild. 2004. Functional and phylogenetic analyses of a melanocortin-4 receptor mutation in domestic pigs. *Domest. Anim. Endocrinol.* **26**:75–86.
105. Kirkland, P.D., P.W. Daniels, M.N. Mohd Nor, R.J. Love, A.W. Philbey, and A.D. Ross. 2002. Menangle and Nipah virus infections of pigs. *Vet. Clin. N. Amer. Prac. Food Anim.* **18**:557–571.
106. Kirkland, P.D., R.J. Love, A.W. Philbey, A.D. Ross, R.J. Davis, and K.G. Hart. 2001a. Epidemiology and control of Menangle virus in pigs. *Aust. Vet. J.* **79**:199–206.

107. Kirkland, P.D., R.J. Love, A.W. Philbey, R.J. Davis, and K.G. Hart. 2001b. Epidemiology and control of Menangle virus in pigs. *Aust. Vet. J.* **79**:199–206.
108. Knight, J.W., F.W. Bazer, and H.D. Wallace. 1974. Effect of progesterone induced increase in uterine secretory activity on development of the porcine conceptus. *J. Anim. Sci.* **39**:743–746.
109. Koketsu, Y., G.D. Dial, J.E. Pettigrew, J.L. Xue, H. Yang, and T. Lucia. 1998. Influence of lactation length and feed intake on reproductive performance and blood concentrations of glucose, insulin and luteinizing hormone in primiparous sows. *Anim. Reprod. Sci.* **52**:153–163.
110. Kraeling, R.R., C.R. Barb, and G.B. Rampacek. 1992. Prolactin and luteinizing hormone secretion in the pregnant pig. *J. Anim. Sci.* **70**:3521–3527.
111. Kusina J., J.E., Pettigrew, A.F. Sower, M.E. White, B.A. Crooker, and M.R. Hathaway. 1999. The effect of protein intake during gestation and lactation on lactational performance of the primiparous sow. *J. Anim. Sci.* **77**:931–941.
112. Lee, J.-S., K. Jeftinija, S. Jeftinija, M.H. Stromer, and L.L. Anderson. 2004. Immunocytochemical distribution of somatotrophs in porcine anterior pituitary. *Histochem. Cell Biol.* **122**:571–577.
113. Lee, J.-S., M.S. Mayes, M.H. Stromer, C.G. Scanes, S. Jeftinija, and L.L. Anderson. 2004. Number of secretory vesicles in growth hormone cells of the pituitary remains unchanged after secretion. *Exp. Biol. Med.* **229**:632–639.
114. Leibold, W. 1981. Cellular immune reactions in Erysipelas polyarthritis in pigs, pp. 173–185. *In* H. Deicher, and L.C. Schulz (eds.), Arthritis: Models and Mechanisms, Springer-Verlag, Berlin.
115. Lenz, H.J., A. Raedler, H. Greten, and M.R. Brown. 1987. CRF initiates biological actions within the brain that are observed in response to stress. *Am. J. Physiol.* **252**:R34–R39.
116. Levett, P.N. 1999. Leptospirosis: re-emerging or re-discovered disease? *J. Med. Microbiol.* **48**:417–418.
117. Lewis, A.J., G.L. Cromwell, and J.E. Pettigrew. 1991. Effects of supplemental biotin during gestation and lactation on reproductive performance of sows: a cooperative study. *J. Anim. Sci.* **69**:207–214.
118. Li, Y., C.J. Huang, J. Klindt, and L.L. Anderson. 1991. Divergent effects of antiprogesterone, RU 486, on progesterone, relaxin and prolactin secretion in pregnant and hysterectomized pigs with aging corpora lutea. *Endocrinology* **129**:2907–2914.
119. Li, Y., J.R. Molina, J. Klindt, D.J. Bolt, and L.L. Anderson. 1989. Prolactin maintains relaxin and progesterone secretion by aging corpora lutea after hypophysial stalk transection or hypophysectomy in the pig. *Endocrinology* **124**:1294–1304.
120. Lillie, R.J. and L.T. Frobish. 1978. Effect of copper and iron supplements on performance and hematology of confined sows and their progeny through four reproductive cycles. *J. Anim. Sci.* **46**:678–682.
121. Lin, C., S.C. Lin, C.P. Chang, and M.G. Rosenfeld. 1992. Pit-1 dependent expression of the receptor for growth hormone releasing factor mediates pituitary cell growth. *Nature* **360**:765–768.
122. Lin, S.C., S. Lin, D.W. Drolet, and M.G. Rosenfeld. 1994. Pituitary ontogeny of the Snell dwarf mouse reveals pit-1 independent and pit-1 dependent origins of the thyrotrope. *Development* **120**:515–522.
123. Lindemann, M.D. 1993. Supplemental folic acid: A requirement for optimizing swine reproduction. *J. Anim. Sci.* **71**:239–246.
124. Lindemann, M.D. and E.T. Kornegay. 1989. Folic acid supplementation to diets of gestating-lactating swine over multiple parities. *J. Anim. Sci.* **67**:459–464.
125. Logar, B., M. Kovač, and Š. Malovrh. 1999. Estimation of genetic parameters for litter size in pigs from different genetic groups. *Acta Agraria Kaposváriensis* **3**:135–143.
126. Mahan, D.C. and J.L. Vallet. 1997. Vitamin and mineral transfer during fetal development and the early postnatal period in pigs. *J. Anim. Sci.* **75**:2731–2738.

127. Mao, J. and G.R. Foxcroft. 1998. Progesterone therapy during early pregnancy and embryonal survival in primiparous weaned sows. *J. Anim. Sci.* **76**:1922–1928.
128. Mauget, R. 1982. Seasonality of reproduction in the wild boar, pp. 509–526. *In* D.J.A. Cole, and G.R. Foxcroft (eds.), Control of Pig Production, Butterworths, London.
129. Mburu, J.N., S. Einarsson, H. Kindahl, A. Madej, and H. Rodriquez-Martinez. 1998. Effects of post-ovulatory food deprivation on oviductal sperm concentration, embryo development and hormonal profiles in the pig. *Anim. Reprod. Sci.* **52**:221–234.
130. McCance, R.A. and E.M. Widdowson. 1974. The determinants of growth and form. *Proc. Roy. Soc. London Ser. B* **185**:1–17.
131. McMilliam. 1999. Brucellosis, pp. 385–395. *In* Diseases of Swine, Iowa State University Press, Ames, Iowa, USA.
132. Mengeling, W.L., T.T. Brown, Jr., P.S. Paul, and D.E. Gutekunst. 1979. Efficacy of an inactivated virus vaccine for prevention of porcine parvovirus-induced reproductive failure. *Am. J. Vet. Res.* **40**:204–207.
133. Mengeling, W.L., K.M. Lager, and A.C. Vorwald. 2000. The effect of porcine parvovirus and porcine reproductive and respiratory syndrome virus on porcine reproductive performance. *Anim. Reprod. Sci.* **60–61**:199–210.
134. Mengeling, W.L., P.S. Paul, and T.T. Brown, Jr. 1980. Transplacental infection and embryonic death following maternal exposure to porcine parvovirus near the time of conception. *Arch. Virol.* **65**:55–62.
135. Mengeling, W.L., J.F. Ridpath, and A.C. Vorwald. 1988. Size and antigenic comparisons among the structural proteins of selected autonomous parvoviruses. *J. Gen. Virol.* **69**:825–837.
136. Mengeling, W.L., A.C. Vorwald, K.M. Lager, D.F. Clouser, and R.D. Wesley. 1999. Identification and clinical assessment of suspected vaccine-related field strains of porcine reproductive and respiratory syndrome virus. *Am. J. Vet. Res.* **60**:334–340.
137. Merck Veterinary Manual. 1998. Pseudorabies, 8th ed., pp. 964–966. National Publishing Inc., Philadelphia, PA
138. Meulenberg, J.J.M., M.M. Hulst, E.J. de Meijer, P.L.J.M. Moonen, A. den Besten, E.P. de Kluyver, G. Wensvoort, and R.J.M. Moorman. 1993. Lelystad virus, the causative agent of porcine epidemic abortion and respiratory syndrome (PEARS), is related to LDV and EAV. *Virology* **192**:62–72.
139. Meulenberg, J.J.M., A. Petersen-den Besten, E.P. de Kluyver, R.J.M. Moorman, W.M.M. Schaaper, and G. Wensvoort. 1995. Characterization of proteins encoded by ORFs 2 to 7 of Lelystad virus. *Virology* **206**:155–163.
140. Miller, D.A., M.A. Wilson, and C.A. Kirkbride. 1989. Evaluation of multivalent leptospira fluorescent antibody conjugates for general diagnostic use. *J. Vet. Diagn. Invest.* **1**:146–149.
141. Miller, D.A., M.A. Wilson, W.J. Owen, and G.W. Beran. 1990. Porcine leptospirosis in Iowa. *J. Vet. Diagn. Invest.* **2**:171–175.
142. Moeljono, M.P., W.W. Thatcher, F.W. Bazer, M. Frank, L.J. Owens, and C.J. Wilcox. 1977. A study of prostaglandin F2alpha as the luteolysin in swine: II Characterization and comparison of prostaglandin F, estrogens and progestin concentrations in utero-ovarian vein plasma of nonpregnant and pregnant gilts. *Prostaglandins* **14**:543–555.
143. Musser, R.E., R.D. Goodband, M.D. Tokach, K.Q. Owen, J.L. Nelssen, S.A. Blum, S.S. Dritz, and C.A. Civis. 1999. Effects of L-carnitine fed during gestation and lactation on sow and litter performance. *J. Anim. Sci.* **77**:3289–3295.
144. Nissen, P.M., V.O. Danielsen, P.F. Jorgensen, and N. Oksbjerg. 2003. Increased maternal nutrition of sows has no beneficial effects on muscle fiber number or postnatal growth and has no impact on the meat quality of the offspring. *J. Anim. Sci.* **81**:3018–3027.
145. Noblet, J., W.H. Close, R.P. Heavans, and D. Brown. 1985. Studies on the energy metabolism of the pregnant sow. Uterus and mammary tissue development. *Br. J. Nutr.* **53**:251–260.
146. Noblet J., J.Y. Dourmad, M. Etienne, and J. Le Dividich. 1997. Energy metabolism in pregnant sows and newborn pigs. *J. Anim. Aci.* **75**:2708–2714.

147. O'Grady, J.F., F.W.H. Elsey, R.M. MacPherson, and I. McDonald. 1975. The response of lactating sows and their litters to different dietary energy allowances. *Anim. Prod.* **17**:65–72.
148. OIE World Organisation for Animal Health. 2005. Vesicular stomatitis. http://www.oie.int/eng/maladies/fiches/A_A020.HTM
149. Oklahoma State University Board of Regents. 1996. *Breeds of Livestock*. http://www.ansi.okstate.edu/breeds/swine/Swine-w.htm
150. Olsen, S.C. 2004. Porcine brucellosis, pp. 777–784. *In* Oie Terrestrial Manual, 5th ed., Paris, France, International Des Epizooties.
151. Paul, P.S., W.L. Mengeling, and E.C. Pirtle. 1982. Duration and biological half-life of passively-acquired colostral antibodies to porcine parvovirus. *Am. J. Vet. Res.* **43**:1376–1379.
152. Peltoniemi, O.A.T., R.J. Love, C. Klupiec, and G. Evans. 1997. Effect of feed restriction and season on LH and prolactin secretion, adrenal response, insulin and FFA in group housed pregnant gilts. *Anim. Reprod. Sci.* **49**:179–190.
153. Peltoniemi, O.A.T., A. Tast, and R.J. Love. 1999. Seasonal, environmental and nutritional effects on gonadotrophins – implications for establishment and maintenance of early pregnancy in the pig. *Domest. Anim. Reprod.* **34**, Suppl 6:96–100.
154. Peltoniemi, O.A.T., A. Tast, and R.J. Love. 2000. Factors effecting reproduction in the pig: seasonal effects and restricted feeding of the pregnant gilt and sow. *Anim. Reprod. Sci.* **60–61**:173–184.
155. Pepin, D. and R. Mauget. 1989. The effect of planes of nutrition on growth and attainment of puberty in female wild boar raised in captivity. *Anim. Reprod. Sci.* **20**:71–77.
156. Perry, J.S. and I.W. Rowlands. 1962. Early pregnancy in the pig. *J. Reprod. Fertil.* **4**:175–188.
157. Petry, D.B., J.W. Holl, and R.K. Johnson. 2004. Responses to 19 generations of litter size selection in the NE Index line. II. Growth and carcass responses estimated in pure line and crossbred litters. *J. Anim. Sci.* **82**:1895–1902.
158. Pettigrew, J.E. and H. Yang. 1997. Protein nutrition of gestating sows. *J. Anim. Sci.* **75**:2723–2730.
159. Philbey, A.W., P.D. Kirkland, A.D. Ross, R.J. Davis, A.B. Gleeson, R.J. Love, P.W. Daniels, A.R. Gould, and A.D. Hyatt. 1998. An apparently new virus (Family Paramyxoviridae) infectious for pigs, humans and fruit bats. *Emerg. Infect. Diseases* **4**:269–271.
160. Plommet, M., R. Diaz, and J.M. Verger. 1998. Brucellosis, pp. 23–35. *In* S.R. Palmer, L. Soulsey, and D.I.H. Simpson (eds.), *Zoonoses*. Oxford University Press, Bath Press, Avon.
161. Pond, W.G. 1973. Influence of maternal protein and energy nutrition during gestation on progeny performance in swine. *J. Anim. Sci.* **36**:175–182.
162. Pond, W.G., D.N. Strachan, Y.N. Sinha, E.F. Walker, Jr., J.A. Dunn, and R.H. Barnes. 1969. Effect of protein deprivation of swine during all or part of gestation on birth weight, postnatal growth rate, and nucleic acid content of brain and muscle of progeny. *J. Nutr.* **99**:61–67.
163. Pond, W.G., J.T. Yen, and H.J. Mersmann. 1987. Effect of severe dietary protein, nonprotein calories or feed restriction during gestation on postnatal growth of progeny in swine. *Growth* **51**:355–371.
164. Pond, W.G., J.T. Yen, H.J. Mersmann, and L.H. Yen. 1983. Effect of gestation diet intake on plasma lipids and progeny birth and weaning weights of genetically lean, obese and contemporary swine. *J. Nutr.* **113**:436–446.
165. Poole, G.M. and F.T. Counter. 1973. Erysipelas serum titration with sheep red blood cells passively sensitized with a cellular extract. *Appl. Environ. Microbiol.* **26**:211–212.
166. Prunier, A. and H. Quesnel. 2000a. Nutritional influences on the hormonal control of reproduction in female pigs. *Livest Prod. Sci.* **63**:1–16.
167. Prunier, A. and H. Quesnel. 2000b. Influence of the nutritional status on ovarian development in female pigs. *Anim. Reprod. Sci.* **60–61**:185–197.
168. Pusateri, A.E., M.A. Diekman, and W.L. Singleton. 1999. Failure of vitamin A to increase litter size in sows receiving injections at various stages of gestation. *J. Anim. Sci.* **77**:1532–1535.

169. Ramanau, A., R. Schmidt, H. Kluge, and K. Eder. 2006. Body composition, muscle fibre characteristics and postnatal growth capacity of pigs born from sows supplemented with L-carnitine. *Arch. Anim. Nutr.* **60**:110–118.
170. Razdan, P., A.M. Mwanza, H. Kindahl, H. Rodriquez-Martinez, F. Hultén, and S. Einarsson. 2002. Effect of repeated ACTH-stimulation on early embryonic development and hormonal profiles in sows. *Anim. Reprod. Sci.* **70**:127–137.
171. Rohrer, G.A., B.A. Freking, and D.J. Nonneman. 2002. Genetic mapping porcine EST sequences using length polymorphisms, p. 130. Proceedings of the 28th International Conference on Animal Genetics. Abstract No. D115.
172. Rohrer, G.A., D.J. Nonneman, B.A. Freking, G.P. Harhay, W.M. Snelling, and J.W. Keele. 2003. Production of a porcine-human comparative map based on the porcine genetic map and human genome sequence, p. 612. Proceedings of the Plant and Animal Genome XI.
173. Rooke, J.A., A.G. Sinclair, and M. Ewen. 2001. Changes in piglet tissue composition at birth in response to increasing maternal intake of long chain *n*-3 polyunsaturated fatty acids are non-linear. *J. Nutr.* **86**:461–470.
174. Rothschild, M.F. 1996. Genetics and reproduction in the pig. *Anim. Reprod. Sci.* **42**:143–151.
175. Rothschild, M.F., C. Jacobsen, D.A. Vase, C.K. Tuggle, T. Short, S. Sasaki, G.R. Eckardt, and D.G. McLaren. 1994. A major gene for litter size in pigs, pp. 225–228. Proceedings of the 5th World Congress on Genetics Applied to Livestock Production **21**.
176. Rothschild, M., C. Jacobson, D. Vaske, C. Tuggle, L. Wang, T. Short, G. Eckardt, S. Sasaki, A. Vincent, D. McLaren, O. Southwood, H. van der Steen, A. Mileham, and G. Plastow. 1996. The estrogen receptor locus is associated with a major gene influencing litter size in pigs. *Proc. Natl. Acad. Sci. USA* **93**:201–205.
177. Sabharwal, P.R., R. Glaser, W. Lafuse, S. Varma, Q. Liu, S. Arkins, R. Kooijman, L. Kutz, K.W. Kelley, and W.B. Malarkey. 1992. Prolactin synthesized and secreted by human peripheral blood mononuclear cells: An autocrine growth factor for lymphoproliferation. *Proc. Natl. Acad. Sci. USA* **89**:7713–7716.
178. Salmon-Legagneur, E., C. Legault, and A. Aumaitre. 1966. Relations entre les variations pondérales de la truie en reproduction et les performances d'élevage. *Ann. Zootech.* **15**:215–229.
179. Samaras, S.E., D.R. Hagen, K.A. Bryan, J.S. Mondschein, S.F. Canning, and J.M. Hammond. 1994. Effects of growth hormone and gonadotropin on the insulin-like growth factor system in the porcine ovary. *Biol. Reprod.* **50**:178–186.
180. Saunders. 1997. Pseudorabies, pp. 1094–1103. *In* Eight (ed.), Veterinary Medicine, London.
181. Selke, M.R., C.E. Barnhart, and C.H. Chaney. 1967. Vitamin A requirement of the gestating and lactating sow. *J. Anim. Sci.* **26**:759–763.
182. Short, T.H., M. Rothschild, O. Southwood, D. McLaren, A. de Vries, H. van der Steen, G. Eckardt, C. Tuggle, J. Helm, D. Vaske, A. Mileham, and G. Plastow. 1997. Effect of the estrogen receptor locus on reproduction and production traits in four commercial pig lines. *J. Anim. Sci.* **75**:3138–3142.
183. Simpson, A.A., B. Hebert, G.M. Sullivan, C.R. Parrish, Z. Zadori, P. Tijssen, and M.G. Rossmann. 2002. The structure of porcine parvovirus: comparison with related viruses. *J. Mol. Biol.* **315**:1189–1198.
184. Spicer, L.J., J. Klindt, F.C. Buonomo, R. Maurer, J.T. Yen, and S.E. Echternkamp. 1992. Effect of porcine somatotropin on number of granulosa cell luteinizing hormone/human chorionic gonadotropin receptors, oocyte viability, and concentrations of steroids and insulin-like growth factors I and II in follicular fluid of lean and obese gilts. *J. Anim. Sci.* **70**:3149–3157.
185. Stancekova, K., D. Vasicek, D. Peskovicova, J. Bulla, and A. Kubek. 1999. Effect of genetic variability of the porcine pituitary-specific transcription factor (PIT-1) on carcass traits in pigs. *Anim. Genet.* **30**:313–315.
186. Sterle, J.A., T.C. Cantley, R.L. Matteri, J.A. Carroll, M.C. Lucy and W.R. Lamberson. Effect of recombinant porcine somatotropin on fetal and placental growth in gilts with reduced uterine capacity. *J. Anim. Sci.* **81**:765–771.

187. Sterle, J.A., T.C. Cantley, W.R. Lamberson, M.C. Lucy, D.E. Gerrard, R.L. Matteri, and B.N. Day. 1995. Effects of recombinant porcine somatotropin on placental size, fetal growth, and IGF-I and IGF-II concentrations in pigs. *J. Anim. Sci.* **73**:2980–2985.
188. Sun, H.S., L.L. Anderson, T.-P. Yu, K.-S. Kim, J. Klindt, and C.K. Tuggle. 2002. Neonatal Meishan pigs show POU1F1 genotype effects on plasma GH and PRL concentration. *Anim. Reprod. Sci.* **69**:223–237.
189. Tagbo, I.F. and D.C. Hill. 1977. Effect of folic acid deficiency in pregnant rats and their offspring. *Can. J. Physiol. Pharmacol.* **55**:427–433.
190. Tast, A., O. Hälli, S. Ahlstrom, H. Andersson, R.J. Love, and O.A.T. Peltoniemi. 2001. Seasonal alterations in circadian melatonin rhythms of the European wild boar and domestic gilt. *J. Pineal Res.* **30**:43–49.
191. Tazi, A., R. Dantzer, M. Le Moal, J. Rivier, W. Vale, and G.F. Koob. 1987. Corticotropin-releasing factor antagonist blocks stress-induced fighting in rats. *Regul. Pept.* **18**:37–42.
192. Thaler, R.C., J.L. Nelssen, R.D. Goodband, and G.L. Allee. 1989. Effect of dietary folic acid supplementation on sow performance through two parities. *J. Anim. Sci.* **67**:3360–3369.
193. Tsuma, V.T., S. Einarsson, A. Madej, H. Kindahl, and N. Lundeheim. 1996. Effect of food deprivation during early pregnancy on endocrine changes in primiparous sows. *Anim. Reprod. Sci.* **41**:267–278.
194. Tuggle, C.K. and A. Trenkle. 1996. Control of growth hormone synthesis. *Domest. Anim. Endocrinol.* **13**:1–33.
195. Tummaruk, P., N. Lundeheim, S. Einarsson, A.-M. Dalin. 2001a. Effect of birth litter size, birth parity number, growth rate, back fat thickness and age at first mating of gilts on their reproductive performance as sows. *Anim. Reprod. Sci.* **66**:225–237.
196. Tummaruk, P., N. Lundeheim, S. Einarsson, A.-M. Dalin. 2001b. Repeat breeding and subsequent reproductive performance in Swedish Landrace and Swedish Yorkshire sows. *Anim. Reprod. Sci.* **66**:267–280.
197. Vaiman, M., Ch. Renard, and N. Bourgeaux. 1988. SLA, the major histocompatibility complex in swine: Its influence on physiological and pathological traits, pp. 23–38. *In* C.M. Warner, M.F. Rothschild, and S.J. Lamont (eds.), The Molecular Biology of the Major Histocompatibility Complex of Domestic Animal Species, Iowa State University Press, Ames, IA.
198. Vallet, J.L. and R.K. Christenson. 1994. Effect of estrone treatment from day 30 to 45 of pregnancy on endometrial protein secretion and uterine capacity. *J. Anim. Sci.* **72**:3188–3195.
199. Vallet, J.L. and R.K. Christenson. 1996. The effect of estrone and estradiol treatment on endometrial total protein, uteroferrin, and retinol-binding protein secretion during midpregnancy or midpseudopregnancy in swine. *J. Anim. Sci.* **74**:2765–2772.
200. Vallet, J.L. and R.K. Christenson. 2004. Effect of progesterone, mifepristone, and estrogen treatment during early pregnancy on conceptus development and uterine capacity in swine. *Biol. Reprod.* **70**:92–98.
201. Vallet, J.L., R.K. Christenson, W.E. Trout, and H.G. Klemcke. 1998. Conceptus, progesterone, and breed effects on uterine protein secretion in swine. *J. Anim. Sci.* **76**:2657–2670.
202. Vallet, J.L., B.A. Freking, K.A. Leymaster, and R. Christenson. 2005. Allelic variation in the secreted folate binding protein gene is associated with uterine capacity in swine. *J. Anim. Sci.* **83**:1860–1867.
203. Vallet, J.L., B.A. Freking, K.A. Leymaster, and R.K. Christenson. 2005b. Allelic variation in the erythropoietin receptor gene is associated with uterine capacity and litter size in swine. *Anim. Genet.* **36**:97–103.
204. Vallet, J.L., H.G. Klemcke, and R.K. Christenson. 2002. Interrelationships among conceptus size, uterine protein secretion, fetal erythropoiesis, and uterine capacity. *J. Anim. Sci.* **80**:729–737.
205. van der Lende, T. and B.T.T.M. van Rens. 2003. Critical periods for foetal mortality in gilts identified by analysing the length distribution of mummified foetuses and frequency of non-fresh stillborn piglets. *Anim. Reprod. Sci.* **75**:141–150.

206. van Rens, B.T.T.M., W. Hazeleger, and T. van der Lende. 2000. Periovulatory hormone profiles and components of litter size in gilts with different estrogen receptor (ESR) genotypes. *Theriogenology* **53**:1375–1387.
207. Vincent, A., B. Thacker, P. Halbur, M. Rothschild, and E. Thacker. 2006. An investigation of susceptibility to porcine reproductive and respiratory syndrome virus between two genetically diverse commercial lines of pigs. *J. Anim. Sci.* **84**:49–57.
208. Vincent, A.L., B.J. Thacker, P.G. Halbur, M.F. Rothschild, and E.L. Thacker. 2005b. The in vitro susceptibility of macrophages to porcine reproductive and respiratory syndrome virus varies between genetically diverse lines of pigs. *Viral Immunol.* **18**:506–512.
209. Wade, G.N., J.E. Schneider, and H.Y. Li. 1996. Control of fertility by metabolic cues. *Am. J. Physiol. Endocrinol. Metab.* **270**:E1–E19.
210. Warner, C.M. and M.F. Rothschild. 1991. The swine major histocompatibility complex (SLA), pp. 368–397. *In* R. Srivastava, B. Ram, and P. Tyle (eds.), Immunogenetics of the Major Histocompatibility Complex, VCH, New York.
211. Wilmoth,T.A., D.L. Smith, J.M. Koch, and M.E. Wilson. 2006. Effects of estradiol on the uterine environment and trophectoderm in the gilt. *J. Anim. Sci.* **84** (In Press).
212. Wilson, M.E. 2005. Influence of the embryo on extending luteal function in the pig. *Havemeyer Foundation Monograph Series* **16**:13–15.
213. Wilson, M.E. and S.P. Ford. 1997. Differences in trophectoderm mitotic rate and P450 17α-hydroxylase expression between the late preimplantation Meishan and Yorkshire conceptus. *Biol. Reprod.* **56**:380–385.
214. Wilson, M.E. and S.P. Ford. 2000. Effect of estradiol-17β administration during the time of conceptus elongation on placental size at term in the Meishan pig. *J. Anim. Sci.* **78**:1047–1052.
215. Wilson, M.E. and S.P. Ford. 2001. Comparative aspects of placental efficiency. *J. Reprod. Fertil. Suppl.* **58**:223–232.
216. Wilson, M.E., N.J. Biensen, C.R. Youngs, and S.P. Ford. 1998. Development of Meishan and Yorkshire littermate conceptuses in either a Meishan or Yorkshire uterine environment to day 90 of gestation and to term. *Biol. Reprod.* **58**:905–910.
217. Winick, M. 1970. Cellular growth in intrauterine malnutrition. *Pediatr. Clin. North Am.* **17**:69–78.
218. World Organization for Animal Health. 2000. Classical swine fever (hog cholera), pp. 199–211. *In* Manual of Standards for Diagnostic Tests and Vaccines, Paris.
219. Wrathall, A.E. 1971. Prenatal Survival in Pigs. Part I. Ovulation Rate and Its Influence on Prenatal Survival and Litter Size in Pigs, pp. 1–108. Farnham Royal, Slough, Eng.: Commonwealth Agricultural Bureaux.
220. Wu, M.C., Z.Y. Chen, V.L. Jarrell, and P.J. Dziuk. 1989. Effect of initial length of uterus per embryo on fetal survival and development in the pig. *J. Anim. Sci.* **67**:1767–1772.
221. Xie, Q.F., C.X. Wu, Q.Y. Meng, and N. Li. 2004. Ghrelin and truncated ghrelin variant plasmid vectors administration into skeletal muscle augments long-term growth in rats. *Domest. Anim. Endocrinol.* **27**:1155–1164.
222. Yen, J.T. and W.G. Pond. 1983. Responses of swine to periparturient vitamin C supplementation. *J. Anim. Sci.* **556**:621–624.
223. Young, L.D. 1995. Review of swine genetics in the U.S. USDA-ARS, U.S. Meat Animal Research Center, Clay Center, Nebraska, pp. 1–9. http://mark.asci.ncsu.edu/nsif/95proc/review.htm
224. Yu, T.P., H.S. Sun, S. Wahls, I. Sanchez-Serrano, M.F. Rothschild, and C.K. Tuggle. 2001. Cloning of the full length pig PIT1 (POU1F1) cDNA and a novel alternative PIT1 transcript, and functional studies of their encoded proteins. *Anim. Biotech.* **12**:1–19.
225. Yu, T.-P., C.K. Tuggle, C.B. Sehmitz, and M.F. Rothschild. 1995. Association of PIT1 polymorphisms with growth and carcass traits in pigs. *J. Anim. Sci.* **73**:1282–1288.
226. Wu, G., F.W. Bazer, J.M. Wallace, and T.E. Spencer. 2006. Intrauterine growth retardation: implications for the animal sciences. *J. Anim. Sci.* **84**:2316–2337.

Chapter 7
Regulatory Aspects of Fetal Growth and Muscle Development Relating to Postnatal Growth and Carcass Quality in Pigs

Charlotte Rehfeldt, Marcus Mau and Klaus Wimmers

Introduction

The phenotype of a newborn piglet is the result of its embryonic and fetal development, which is a very complex and highly integrated process. Fetal growth retardation results in low birth weight, which has detrimental consequences for the piglets' vitality, postnatal growth rate as well as carcass and meat quality [3, 38, 164]. Therefore, understanding mechanisms that control fetal growth are essential to develop successful strategies to reduce the incidence of fetal growth retardation.

Prenatal growth in the pig, as in other mammals, is determined by the genotype of the conceptus, but largely depends on the maternal uterine milieu of hormones, nutrients, and growth factors (Fig. 7.1). Feto-maternal interactions and fetal feedback mechanisms may also play a role. The supply of nutrients to the embryo/fetus and its capacity to utilise the available substrates is of major importance for fetal growth. Nutrient partitioning and utilisation in the feto-maternal unit are under the control of hormones and growth factors although, conversely, nutrition may also influence the hormonal status [11, 12, 170, 187, 197]. Maternal characteristics, as determined by genetic (breed, genotype) and environmental (nutrition, housing, etc.) factors, further modify these interactions. In the pig, about 18% of the phenotypic variability in birth weight result from maternal genetic effects, and only 7% from direct genetic effects (h^2) as estimated from 20,000 Yorkshire piglets [178], with the effects of maternal environment difficult to estimate. Conclusively, numerous other factors contribute to variation in fetal growth and birth weight. In the pig, which is a polytocous mammal, competition among littermates for nutrients in utero is a major constraint of fetal growth, with the placenta being the main factor contributing to maternal variation in nutrient supply [3, 5, 133, 198].

Prenatal skeletal muscle development (myogenesis) in pigs has an irreversible impact on postnatal growth and muscle accretion, which are main constituents of

---

C. Rehfeldt (✉)
Research Institute for the Biology of Farm Animals, 18196 Dummerstorf, Germany
e-mail: rehfeldt@fbn-dummerstorf.de

P.L. Greenwood et al. (eds.), *Managing the Prenatal Environment to Enhance Livestock Productivity*, DOI 10.1007/978-90-481-3135-8_7,

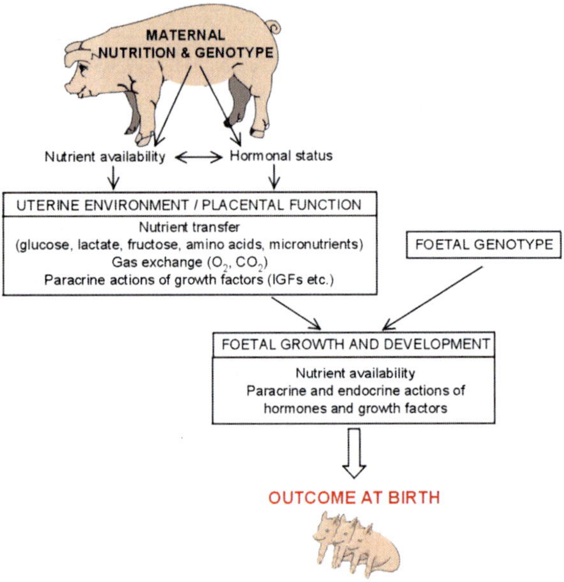

Fig. 7.1 Summary of important influences on porcine fetal growth and development and the resulting phenotype of newborn piglets. The *arrows* indicate the direction of effects. This schematic neither includes potential feedback by fetus to the dam, nor interactions among fetuses

carcass and meat quality [162, 164, 184]. As with fetal growth in general, myogenesis is regulated by nutrient availability and is under the control of hormones and growth factors which can alter metabolism at the level of transcription and translation of regulatory and structurally important genes. Therefore, the purpose of this chapter is to explore aspects of physiological and genomic regulation of porcine fetal growth and myogenesis.

7.1 Prenatal Muscle Development and Relation to Growth and Carcass Quality

Skeletal myogenesis and its control has been the subject of numerous comprehensive reviews [11, 13, 46, 115, 125, 153, 206]. The elementary events during myogenesis are stem cell commitment, proliferation and apoptosis of myoblasts, differentiation and fusion of myoblasts to myotubes and, finally, their maturation into muscle fibres. Myoblasts are precursors of muscle cells and have their origin early in embryogenesis within specific somites. They do not contain any myofibrillar proteins. Determined myoblasts are committed to a myogenic fate. They only differentiate and express a muscle-specific phenotype after they have received appropriate signals from the milieu within which they reside. Once proliferating myoblasts enter the differentiation programme, they withdraw from the cell cycle and form post-mitotic myofibres. This transformation is accompanied by muscle-specific gene expression. Post-myogenic muscle growth is characterised by an increase in length and diameter of myofibres, particularly during postnatal life. Once formed, the multinucleated myofibres commence production of myofibrillar proteins. The myofibre types are distinguished by different isoforms of myofibrillar myosin associated with different contractile and metabolic properties. Throughout life, the myofibres remain in a state of dynamic adaptation in response to hormones, mechanical activity and innervation, which modulate differential gene expression during functional acclimatisation.

Muscle tissue development in the pig, as in other mammals, includes at least two phases. Muscle fibres that form during the initial stages of myoblast fusion into primary myofibres provide a framework for a larger population of smaller secondary fibres [9, 125]. These are formed from fetal myoblasts during a second wave of differentiation. A further population of myoblasts do not form fibres, but locate on the surface of myofibres between the sarcolemma and the basal lamina; these are termed satellite cells and they are able to divide and serve as the source of new myonuclei during postnatal growth [128, 176]. Satellite cells contribute to growth of the myofibres and also participate in regeneration processes after injury or severe nutritional and growth restriction, whereas myonuclei remain mitotically quiescent.

Porcine primary myofibres form between about days 35 and 60, and secondary myofibres between about days 54 and 90 days of gestation [104, 207]. A third generation of very small diameter fibres forms shortly after birth [104], which may explain increases in total fibre number observed in porcine muscle between birth and 5 weeks of age [161]. About 20 secondary myofibres form around each primary fibre in the pig. Consequently, the adult muscle is composed mainly of fibres originating from the secondary myofibres. The primary myotubes in turn play a critical role in the formation of the secondary myofibres [206]. This relationship is reflected by a high correlation between the primary and secondary fibre number in pigs [35, 137]. Thus, even though primary myotubes constitute only a minor percentage of the total fibre number, they significantly influence muscle fibre number and, consequently, muscle size. The importance of the number of both primary myotubes and secondary myofibres for muscle growth is underscored by a lower number of primary fibres and lower secondary to primary fibre ratio in small compared to large pig breeds [185]. Primary myofibres comprise slow myosin heavy chain and exhibit slow myosin ATPase activity from about day 60 of gestation. During late gestation, some secondary fibres located adjacent to single primary fibres convert to the slow type and thereby generate clusters of slow myofibres, which is typical for porcine muscle [9, 62, 104, 163]. However, in the superficial region of some muscles (e.g. *semitendinosus*), initially slow primary myofibres can convert to the fast type, whereas adjacent secondary fibres do not convert to slow type as in limb muscles from other mammalian species [206]. Initially, muscle fibre type differentiation depends on different myoblast lineages, but later during myogenesis becomes more dependent on a series of extrinsic factors. The different stages of porcine skeletal muscle development are shown in Fig. 7.2.

Muscle mass is mainly a result of the number and size of the constituent muscle fibres. Muscle structure and functional properties affect an animal's growth performance and are constitutive factors influencing meat quality traits *post mortem* [100, 106, 107, 161, 181]. Prenatal myogenesis is highly significant for postnatal growth, carcass and meat quality, because it determines the number of skeletal

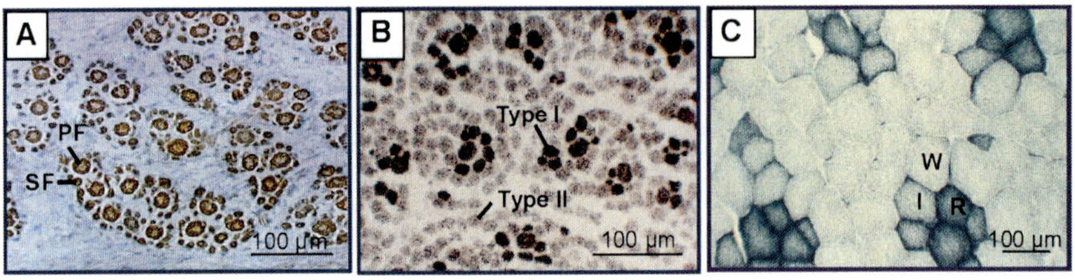

Fig. 7.2 Different stages of porcine muscle development: (**a**) day 62 of gestation, *semitendinosus* muscle with primary (PF) and secondary (SF) muscle fibres; aniline blue/orange G staining; (**b**) day of birth, *semitendinosus* muscle with type I (*dark staining*) and type II (*light staining*) myofibres; myosin ATPase stain with acid pre-incubation at pH 4.2; (**c**) day 240 postnatal, *longissimus* muscle with red (R), intermediate (I), and white (W) fibres; NADH tetrazolium reductase staining

muscle fibres, which remains constant after birth. The postnatal increase in skeletal muscle mass results mainly from an increase in muscle fibre size, which in turn is limited by genetic and physiological factors [161]. Therefore, piglets with more skeletal myofibres exhibit a higher potential for postnatal muscle accretion (Fig. 7.3). In pigs, low birth weight resulting from prenatal undernutrition is related to reduced muscle fibre and myonuclei numbers and lower DNA content [36, 61, 62, 164, 207, 208]. In addition, low birth weight pigs exhibit reduced growth and accretion of lean tissues, but higher levels of fat deposition and tend to develop poor meat quality at market weight [8, 57, 79, 157, 164, 168]. Reduced meat quality in terms of tenderness scores, drip loss, pH value 45 min *post mortem*, and impedance values have been recently reported, whereas intramuscular fat content was highest in low birth weight pigs, consistent with their higher degree of fatness (57, 164, 168).

7.2 Nutritional and Hormonal Regulation

7.2.1 Regulation of Fetal Growth and Myogenesis by Maternal Factors

The maternal diet controls fetal growth directly by providing glucose, amino acids, and other essential nutrients and metabolites for the conceptus [170]. These are transferred across the placenta by passive and active transport mechanisms. Thus, the placenta is central to extrinsic regulation of fetal growth. Hence, maternal factors that influence placental function, such as the contact area with the uterine wall, blood flow and transplacental transport capacity, are of major importance for fetal growth [3, 5].

Porcine fetal and birth weights are inversely correlated with litter size [5, 126, 151, 198], indicating competition for nutrients among littermates in utero. This proposition is further supported by the relationship between fetal size and placental size [4, 124] or placental blood flow [215]. Placental function in the pig has, in turn, been shown to be affected by maternal nutrition and hormonal status and their interactions.

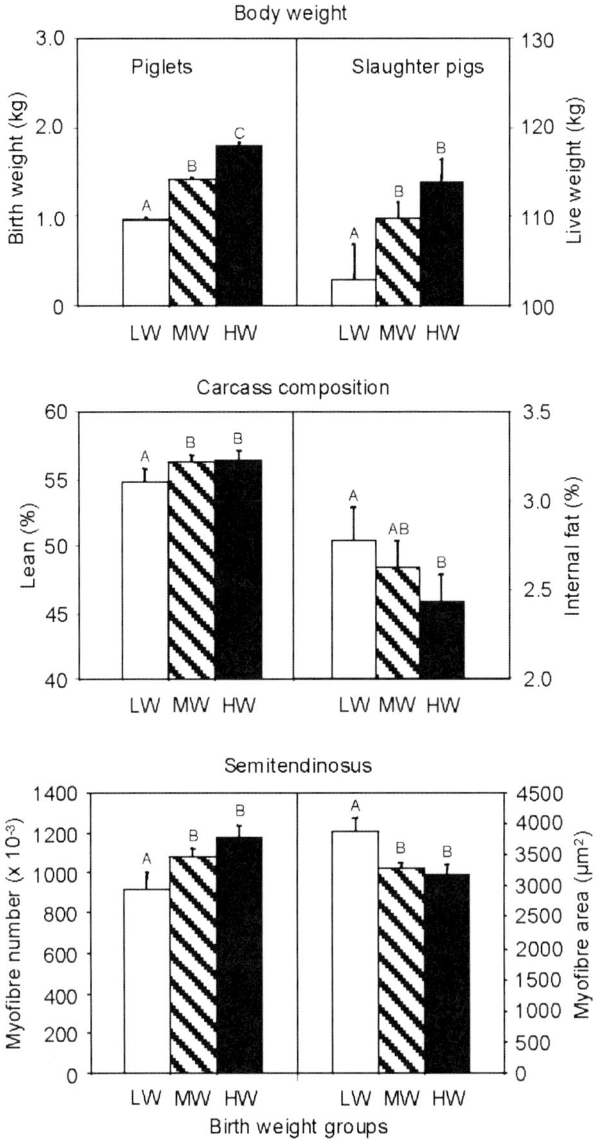

Fig. 7.3 Growth, carcass and associated *semitendinosus* muscle fibre characteristics at slaughter of different birth weight class German Landrace pigs: low (LW), mid (MW) and heavy (HW) birth weight [164]. Significant differences between means are denoted by different letters (all $P \leq 0.09$)

The major substrates reaching the porcine fetus are glucose, placentally-derived lactate and fructose, and amino acids, while transfer of fatty acids by the epitheliochorial placenta is very low in the pig [150]. The primary reason why maternal undernutrition negatively influences fetal growth is the lack of these nutrients reaching the fetus. On the other hand, maternal nutrition also affects hormonal status, especially circulating IGF-I, which is not known to cross the placental barrier but could modify placental function [167].

7.2.1.1 Nutritional Regulatory Effects

Progeny of gilts or sows receiving a diet deficient in energy or protein have reduced fetal and birth weight [15, 154, 155, 169, 174, 175]. Low energy diets are associated with a reduction in skeletal myofibre and myonuclear numbers. For low protein feeding, results for changes of skeletal muscle structure are not yet available. Skeletal muscle protein, RNA, and DNA were reduced in total, while concentrations remained unaffected [154, 156]. Studies on restricted protein feeding have revealed decreases in placental growth [174], amino acid concentration in allantoic fluid and fetal plasma [216], and in placental and endometrial synthesis of nitric oxide and polyamines [217]. These results suggest that protein deficiency may impair placental transport of amino acids from the maternal to the fetal blood. From a recent study, McPherson et al. [122] concluded that nutrient needs of pregnant gilts should be based on dynamic compositional changes in individual fetal tissues during gestation, which occurs at different rates.

Current evidence suggests that nutrition regulates hormones and growth factors that control mammalian growth [5, 11, 12, 170, 187, 197]. An inadequate supply of nutrients leads to consequential changes in fetal hormones. In particular, the regulatory proteins of the growth hormone (GH) – insulin-like growth factor (IGF) axis have been shown to be modulated by nutrition. In young pigs, feed restriction decreased the circulating concentration of IGF-I and increased that of GH [190], changes that are similar to those seen in most mammalian species [187]. Similar changes could be expected in pregnant sows. Thus, the GH peak amplitude was increased in response to maternal feed restriction [155]. With regard to the offspring a lower circulating IGF-I concentration has been observed in newborn piglets in response to maternal protein restriction [23]. There have been several studies clearly showing the negative effects of maternal feed restriction on fetal IGF-I concentrations in sheep [167].

In addition to maternal feed restriction, maternal feed intake above standard requirements also alters the endocrine and metabolic status in the feto-maternal unit (Table 7.1). Feeding pregnant sows ad libitum resulted in significant increases in circulating IGF-I and urea nitrogen concentrations. Maternal glucose concentrations remained unchanged or even decreased, and increases in circulating insulin were not consistently apparent. Thus, IGF-I, which is a pleiotropic growth factor, seems to be a key controller in nutritional regulation of growth. Maternal serum IGF-I may mediate the effects of nutrition on fetal growth by improving placental growth and function, thereby reducing maternal constraints in the pig and other litter-bearing species. Fetuses of sows fed above requirements had greater or equal circulating concentrations of IGF-I, increased urea nitrogen and lactate in umbilical blood, while glucose concentration remained unchanged [167]. Circulating amino acids and fructose concentrations were not measured in fetal blood, thus it remains

Table 7.1 Changes in important endocrine and metabolic factors in maternal and fetal blood in response to sow nutrition above standard nutritional requirements during gestation

Factor	Mother	Fetus	References
IGF-I	↑ (=)	↑ =	Gatford et al. [51], Musser et al. [133], Nissen et al. [138]
Insulin	↑ =	=	Gatford et al. [51], Musser et al. [133], Nissen et al. [138], Père et al. [152]
Glucose	= (↓)	=	Gatford et al. [51], Musser et al. [133], Nissen et al. [138], Père et al. [152]
Lactate	=	↑	Gatford et al. [51], Musser et al. [133], Nissen et al. [138]
Urea N	↑	↑	Gatford et al. [51], Musser et al. [133]

↑ Increase; ↓ Decrease; = No change.

unclear whether transfer of these nutrients to fetuses was affected. As there is some evidence that only small littermates may benefit from improved maternal nutrition, it would be necessary to consider the effects on the individual feto-placental units. Apart from one study where slight increases in birth weight occurred [222], no changes in average birth weight have been observed in response to maternal feeding above requirements in pigs. However, Dwyer et al. [37] found a reduced number of piglets within litters exhibiting extremely low numbers of myofibres in response to feeding pregnant sows 100% above standard requirements. Furthermore, after *ad libitum* feeding of sows during early pregnancy, Musser et al. [133] no longer observed an inverse correlation between fetal number and average fetal weight, as occurred in offspring of control fed sows, and is commonly seen within pig litters. In conclusion, maternal feeding above requirements seems to stimulate growth of fetuses and development of their muscles in what would normally be small, disadvantaged littermates, thereby reducing variation between fetuses within litters.

7.2.1.2 Maternal Hormonal Regulation

There is increasing evidence that placental function and fetal growth are modified by the maternal somatotropic axis, as shown previously in other species [5]. The GH-IGF axis plays a major role in the control of growth and differentiation. Circulating GH acts on the liver and other tissues and stimulates the expression of IGF-I and IGF binding protein (IGFBP) genes (e.g. IGFBP-3), increasing their levels in the circulation [46, 108]. The insulin-like growth factors (IGF-I and IGF-II) are single chain proteins (7.5 kDa) of which about 99% are bound to high affinity IGFBPs (IGFBP 1–6) that circulate in the blood [83]. The mitogenic and anabolic effects of IGF-I and IGF-II on proliferation and differentiation are mediated by the type-1 IGF-receptor (IGF-1R).

Treatment of pregnant sows with porcine growth hormone (pGH) is associated with substantial increases in circulating IGF-I and glucose [52, 165, 173, 183], and stimulates placental growth and alters mRNA expression of regulatory proteins at the fetal-maternal interface. Regulatory proteins within the IGF system, such as GH receptor (GH-R), IGF-1R, IGF-I, IGF-II and IGFBPs are expressed in porcine endometrium and placenta and/or porcine embryos, suggesting autocrine and paracrine actions of IGFs [49, 177, 179, 182]. In this regard, Kelley et al. [95] observed a positive response in IGF-I mRNA expression in uterine endometrium following maternal pGH treatment from day 28 to 40 of gestation. In a further experiment, changes in the maternal (endometrium) and fetal (chorion) components of the placenta were studied in response to pGH treatment at day 28 following treatment from day 10 to day 27, and after cessation of treatment at day 37 and day 62 of gestation [49, 165]. The concentrations of RNA and protein in the endometrium were increased by pGH treatment, suggesting greater capacity for endometrial protein synthesis. During prenatal life, the weight of the fetal placenta was unchanged; however it was increased by 20% at birth and had a 30% increase in protein concentration. Treatment with pGH resulted in down-regulation of endometrial GH-R mRNA and IGF-I mRNA at day 28. However, withdrawal of pGH was followed by an over-compensatory increase at day 62 (Fig. 7.4). On the other hand, higher

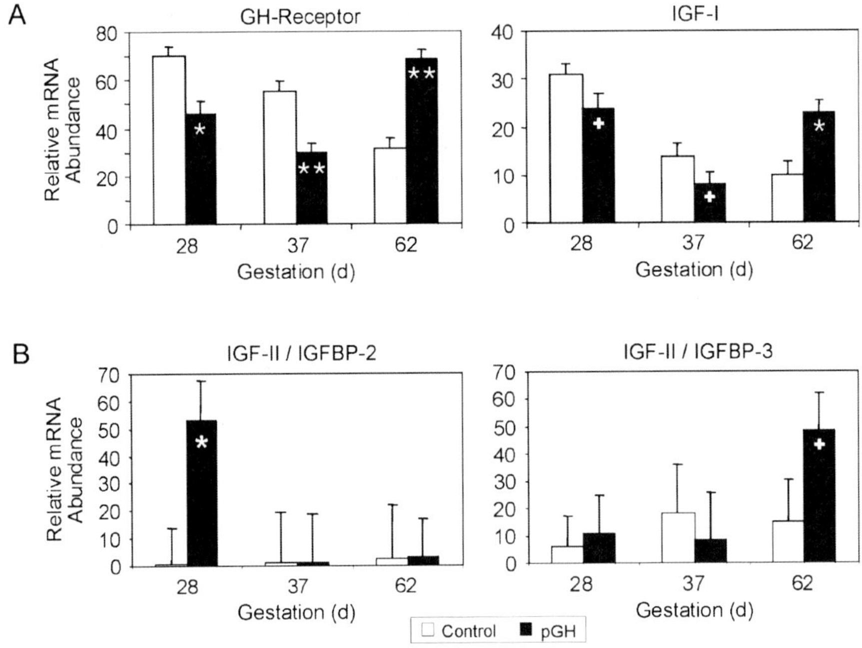

Fig. 7.4 Maternal growth hormone treatment modifies the expression of important regulatory proteins in the endometrium and placenta [49]. Sows were treated daily with 6 mg pGH or a placebo (control) from days (d) 10–27 of gestation. (**a**) mRNA expression of growth hormone receptor (GH-R) and IGF-I in endometrium; (**b**) ratios of the mRNA expression levels of IGF-II to IGFBP-2 and IGF-II to IGFBP-3 in placental chorion determined at days 28, 37, and 62 of gestation. * $P<0.01$; ** $P<0.001$; $^{+}$ $P=0.08$–0.12 for differences between pGH and control pigs

ratios of IGF-II mRNA to IGFBP-2 or IGFBP-3 mRNA concentrations suggested that more free IGF-II may have been available in placental tissue at the end of pGH treatment (day 28) and after complete recovery from pGH withdrawal at day 62. These changes may have positively influenced placental function and, indeed, were related to observed signs of improved embryonic or fetal growth [165, 166].

Treatment with pGH has been found to influence myogenesis. Fetuses of treated sows had greater numbers of primary and secondary myofibres, and this was associated with increased expression of the myogenic regulatory factors MyoD and Myf-5 during mid-gestation (day 62), indicating a higher proportion of proliferating myoblasts [166]. Birth weight and total myofibre number were especially increased by maternal GH treatment in small littermates more commonly disadvantaged by insufficient nutrient supply.

Average fetal weight was only increased after long-term pGH treatment, from 25 to 100 days of gestation, with the effects being greatest in the largest litters, again suggesting that maternal constraints to fetal growth were reduced [50]. By comparison, treatment with pGH or GH releasing factor during late gestation increased birth weight and body lipid content, which was explained by the diabetogenic state of the dam, but did not increase the number of muscle fibres [42, 99, 163].

Other data have also shown the importance of estrogens for placental and fetal growth. Treatment of ovariectomised or cyclic gilts with estrogen increased IGF-I expression in the uterus. Furthermore, research has indicated that estrogens, the pregnancy recognition signal from the pig conceptus, increases uterine epithelial fibroblast growth factor (FGF)-7 expression and, in turn, FGF-7 stimulates proliferation and differentiation of conceptus trophoectoderm [88, 89, 180]. To date, it has not been determined whether fetal myogenesis is influenced by the action of estrogens.

7.2.1.3 Specific Micronutrients

Supplementation of the maternal diet with specific micronutrients may also be significant for fetal growth, given they play a crucial role in the regulation of metabolism, including muscle performance and energy utilisation [29]. Mahan and Vallet [113] reviewed the knowledge on mineral and vitamin transfer to the developing pig and concluded that scientific research on vitamin and mineral nutrition of pregnant pigs to optimise fetal and postnatal growth and development was still in its infancy. Positive effects on porcine reproductive performance have been reported for vitamin E and selenium, both of which exhibit antioxidant properties. A high intake of riboflavin may increase transfer of riboflavin to the conceptus and improve conception rate and or embryonic survival resulting in increased farrowing rates and or litter size [113]. Supplementation of the maternal diet with folic acid, which plays an important role in amino acid and nucleotide metabolism during early gestation has increased litter size, particularly in multiparous sows [110]. However, no benefit was demonstrated when diets for reproducing sows were supplemented with additional vitamin C. Iron deficiency in mothers has been shown to result in lower birth weights in humans and in rodent models [118]. However, there were only marginal effects on the offspring when supplemental Fe was administered to pregnant sows [34, 60, 140]. Fetal development is not greatly influenced by maternal dietary levels of the minerals Ca and P, which suggests that mineral reserves are diverted from maternal bone tissue to meet fetal needs when the sow is inadequately supplied. Other minerals, such as K, Na, Cl, I, Zn, Mg and Cu, are clearly necessary to avoid defects in the developing fetus. It has been suggested that Se and I interact to alter thyroid function in the fetus of sheep, cattle, and rats [170], indicating they may also be important in development of fetal pigs. Studies in humans and rodents also provide evidence that specific micronutrients, such as Mg, K, and Zn influence the IGF system [29]. However, no results on the effects of deficiency or oversupply on fetal myogenesis have been explicitly reported for any of the vitamins and minerals mentioned above.

Another specific nutrient that has more recently become a subject of interest is the vitamin-like molecule L-carnitine, which is involved in protein synthesis [143], glucose homeostasis [26], and ß-oxidation [120]. The supplementation of pregnant sows with L-carnitine results in increased birth weight and/or litter weight [39, 135, 158, 203]. In addition, myogenesis was affected by L-carnitine supplementation, resulting in an increase in the number of muscle fibres in the *semitendinosus* muscle in piglets [134]. This is in contrast to the apparently unchanged numbers of muscle fibres in response to L-carnitine supplementation reported by [159], although this latter study only assessed number of fibres/unit area rather than the total number of myofibres in the entire cross-section of the muscle. Waylan et al. [203] examined the mRNA expression of IGF-I and -II as well as IGFBP-3 and -5 in response to maternal L-carnitine supplementation at mid-gestation. No changes were observed in fetal muscle, liver, uterus, and placenta, although there was reduced IGF-II, myogenin and IGFBP-3 mRNA expression in porcine myoblasts from fetuses of L-carnitine treated sows, following culture for 96 h under standard conditions [203] (Fig. 7.5). The authors interpreted these changes to indicate delayed and prolonged proliferation, with potential for increased muscle fibre number at birth.

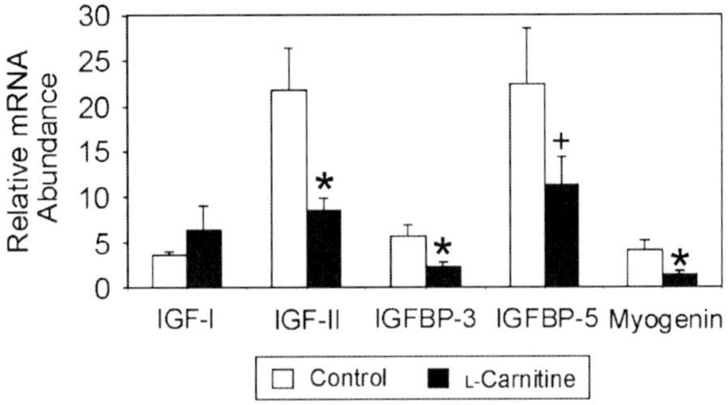

Fig. 7.5 L-carnitine supplemented to gestating sows modifies myoblast characteristics in developing fetuses. The columns represent growth factor mRNA abundance in porcine muscle cells derived from fetuses (days 54–59) of sows supplemented with 100 mg L-carnitine per day or from non-supplemented sows [198]. * $P<0.05$ and $^{+}$ $P=0.13$ for differences between L-carnitine supplemented and control sows

Of the various specific nutrients that may affect placental and fetal development, conjugated linoleic acids (CLA) are also worthy of note. Bee [7] supplemented sow diets with 2% of conjugated linoleic acids (CLA) or linoleic acid during gestation and lactation and observed higher weight gain and carcass weight, as well as larger loin eye areas and *semitendinosus* muscle weights in the progeny of CLA vs. linoleic acid supplemented sows. The mechanisms of action remain to be elucidated.

7.2.2 Endocrine and Paracrine Factors Affecting Fetal Growth

7.2.2.1 Effects on the Whole Animal

Fetal hormones and growth factors are under the control of the uterine environment and fetal genotype, and affect growth and development in utero by altering metabolism and gene expression in fetal tissues [3]. Many hormones and growth factors have been implicated in mammalian fetal and muscle development and gene expression (including thyroid hormones, GH, IGFs, the transforming growth factor ß (TGF-ß) superfamily, glucocorticoids, insulin and steroid hormones), and interactions between hormonal systems are also important [3, 22, 55, 68, 127].

Initially, fetal pituitary hormones (e.g. GH) were thought not to play any role in development of the pig as a result of apparently normal growth and development of decapitated fetuses [188]. Adipose and muscle tissue did not respond to the removal of the pituitary before day 70 of gestation [73, 77]. However, other studies revealed that hypophysectomy at 72–74 days of gestation alters tissue biochemistry and cytochemistry [72, 74, 160] and lowers serum and tissue IGF-I, but not IGF-II concentrations [84, 101]. This, and other evidence [6, 33], suggest that fetal development is increasingly influenced by pituitary hormones in the last third of gestation, whereas during the earlier stages the supply of nutrients is the overriding regulator of fetal growth.

Studies using fetal hypophysectomy in pigs have also shown that the influence of thyroxine in enhancing tissue development may be mediated by increased circulating IGF-I and IGFBPs [76, 96, 101] (Fig. 7.6), and that the stimulating effect of thyroxine is counter-regulated by pituitary GH [69]. Hausman [70] concluded that fetal obesity may be directly associated with elevated thyroid hormone levels and suppressed GH levels, but not with elevated thyroxine levels alone. Other studies have shown that hypophysectomy caused deficiencies in skeletal muscle vascularisation, fibre type transformations and in oxidative capacity, as demonstrated by changes in structure and enzyme histochemistry [72, 74, 78]. Evidence for the importance of thyroid hormones for porcine fetal and muscle development has also been provided by a reduction in the number of thyroid hormone receptors in small for gestational age piglets [22].

It has been strongly suggested, from experimental studies using pigs and other mammals, that various peptide growth factors are important endocrine and/or paracrine regulators of fetal growth. In particular, IGF-I and IGF-II, the IGF-1 receptor (IGF1-R), and IGFBPs seem to play a central role in development of the pig, as

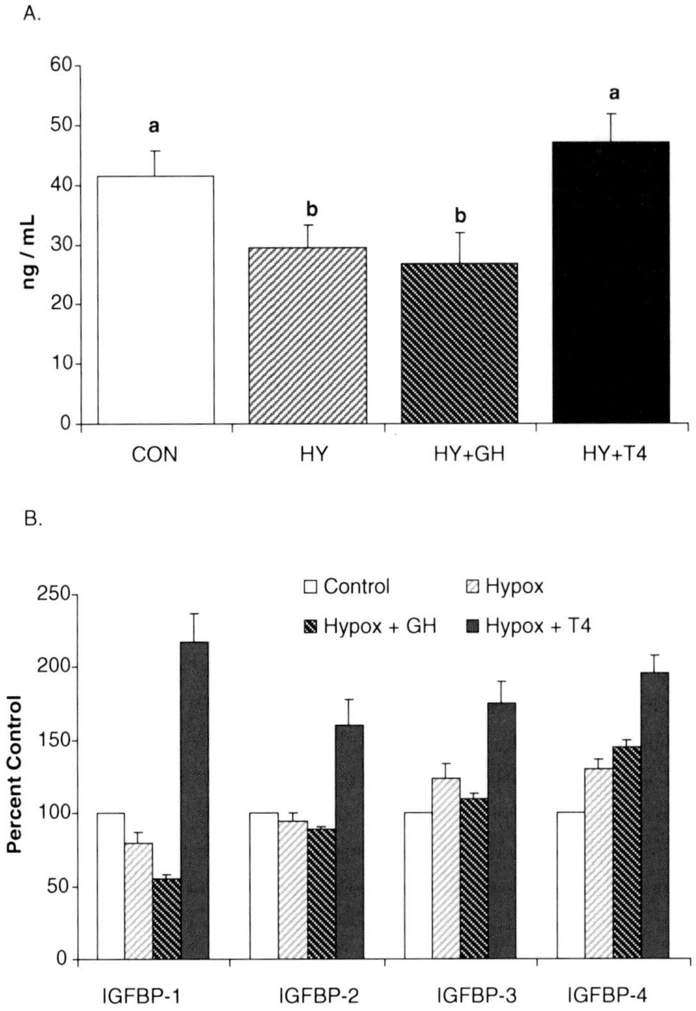

Fig. 7.6 Serum IGF-I (**a**) and IGFBPs (**b**) in porcine fetuses hypohysectomised (Hypox or Hy) on day 70 of gestation, in response to treatment with growth hormone (GH) or thyroxine (T4) from day 90 to 105 of gestation (G.J. Hausman, personal communication: based on [69, 71, 101])

concluded from their presence, ontogeny in serum and/or tissues [53, 102, 111, 149], and changes in their response to various experimental conditions (see above). Less is known of the direct effects of IGFs and IGFBPs on porcine cellular growth and differentiation in vivo or of the importance of other peptide growth factors in fetal pig growth, but inference can be drawn from in vitro studies. Because skeletal muscle is one of the most important tissues that determine postnatal growth and carcass quality in pigs, the following section reviews the current knowledge on the effects of hormones and growth factors on cultured porcine muscle cells.

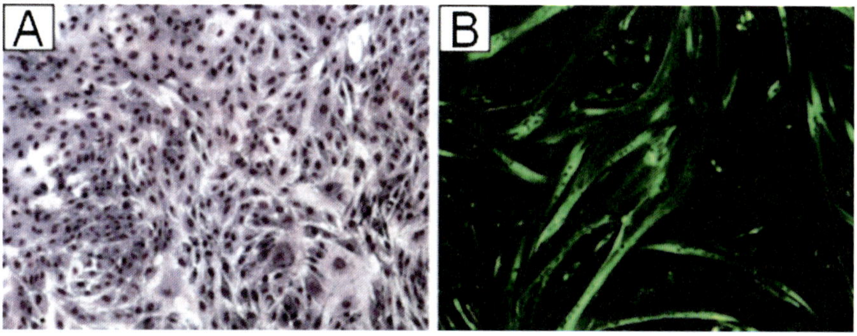

Fig. 7.7 Cell culture of porcine muscle satellite cells isolated from *semitendinosus* muscle of newborn piglets. (**a**) Proliferating myoblasts at day 4 of cultivation in growth medium; (**b**) Immunofluorescent staining of myotubes for desmin at day 7 of cultivation in serum-free differentiation medium

7.2.2.2 Direct Effects on Cultured Porcine Muscle Cells

The control of cultured muscle cell growth and differentiation (Fig. 7.7) is under very tight multi-factorial control, involving various metabolic and reproductive hormones, growth factors and nutrients [11, 27, 63].

Significant effects of reproductive hormones like 17β-estradiol on rodent L_6, C_2C_{12}, Sol_8 [90] and bovine muscle cell cultures [93] have been reported. However, there is little information on their effects on growth and differentiation of porcine satellite cell cultures. Mau et al. [116] observed slight decreases in DNA synthesis rate in proliferating porcine myoblasts after incubation with 17β-estradiol and estrone at supra-physiological concentrations (1 nM and 1 μM), but no changes at physiological concentrations. Both the estrogen receptor α (ERα) and β (ERβ) have recently shown to be expressed in porcine skeletal muscle satellite cells, which, consequently, could be considered a target for estrogens or estrogen-like compounds [91]. At high concentrations ($\geq$10 μM), the isoflavonic phytoestrogens genistein and daidzein were shown to act as toxins and inhibitors of porcine myoblast growth [116], but the mechanisms of action remain to be investigated. On the other hand, physiological concentrations of estrogens and selected concentrations of genistein (0.1 μM) and daidzein (10 μM) were able to reduce protein degradation in differentiating cultures (Mau and Rehfeldt, unpublished).

Testosterone up-regulated expression of androgen receptors (AR) in porcine satellite cells and decreased fusion and subsequent expression of myosin and creatine phosphokinase, markers of muscle differentiation [31, 123]. Although testosterone increased the expression of AR, it did not influence proliferation and protein accretion [1, 31]. This is not consistent with results showing positive effects of testosterone on the mitotic activity of porcine skeletal muscle satellite cells in vivo [129].

Metabolic hormones that are known to regulate muscle growth involve GH, the thyroid hormones thyroxine (T_4) and triiodothyronine (T_3), glucocorticoids, and insulin. Indications that GH might act directly on muscle are generally not supported by cell culture experiments [45]. For instance, Harper et al. [65] could not find any effects of GH on the protein turnover of ovine muscle cells. However, these cell culture studies used muscle cells that correspond more closely to embryonic than to adult muscle, and embryonic growth is relatively insensitive to GH [139]. Studies on the direct effects of GH on cultured porcine muscle cells are lacking. Sera from swine injected with porcine GH stimulated in vitro muscle cell proliferation [97], which may not result solely from GH but, rather, from indirect effects of the hormone. Data on the effects of thyroid hormones on cultured skeletal muscle cells of farm animals is limited to cells derived from chicken. Thyroxine (T_4) increased protein synthesis rate and the amount of protein in cultured chick skeletal muscle cells [201]. In serum-free medium, T_3 suppressed both the differentiation and proteolysis of muscle cells originating from the chicken embryo [28].

Cortisol and related steroids are generally regarded as hormones that result in catabolism of protein in animals. Their effects in muscle cell culture appear to be different from effects on differentiated post-mitotic muscle. Here, glucocorticoids like dexamethasone are often components of serum-free media [18, 45, 67] as they were shown to stimulate both proliferation and differentiation. Maximum stimulation of growth by dexamethasone has been determined in the range of 10^{-8} to 10^{-7} M [48]. The direct effects of glucocorticoids on muscle cell proliferation and differentiation involve increases in the IGF1-R signaling pathway [54] and decreases in IGFBP expression [119]. On the other hand, recent studies imply that the addition of medium containing 80 nM dexamethasone does not increase myotube formation in myoblast cultures derived from porcine *semitendinosus* muscle, when cultivated in laminin-coated dishes [75]. Like other steroid hormones, glucocorticoids are toxic to cultured myoblasts at high concentrations.

Insulin has been shown to increase proliferation and differentiation of various types of muscle cells. At supra-physiological concentrations, insulin caused mitogenic or differentiative effects on muscle cells, reflecting its ability to bind to the IGF-1R [11]. In cultured porcine myotubes, insulin stimulated protein synthesis while it decreased protein degradation [32, 80].

Various peptide growth factors, such as IGF-I and IGF-II, basic FGF, and TGF-β are also potent effectors of satellite cell proliferation and differentiation. Results obtained from cell lines and primary cell cultures show that both IGF-I and IGF-II stimulate cellular growth and differentiation in a time- and dose-dependent manner. IGF-I stimulated proliferation in porcine satellite cell clones [30]. Further, the addition of 15 ng/ml IGF-I to differentiated porcine myogenic cultures increased protein synthesis by 21% and decreased protein degradation by 18% [80]. The responses to IGF-II were almost identical to those of IGF-I, but there was no additive response when the IGFs were combined [30]. An example of the effects of IGF-I in serum-free medium on porcine primary satellite cells in culture is presented in Fig. 7.8.

Effects of IGF-I and -II are controlled and regulated by the IGFBPs, which are believed to protect the IGFs against proteolysis, stimulate or inhibit the biological actions of IGF or, possibly, have IGF-independent actions [17]. It is known that muscle cell cultures produce various amounts of IGFs and IGFBPs depending on the origin (species and muscle type) and developmental state of these cells [141]. The production and secretion of IGFBPs, in turn, is under the control of IGFs and

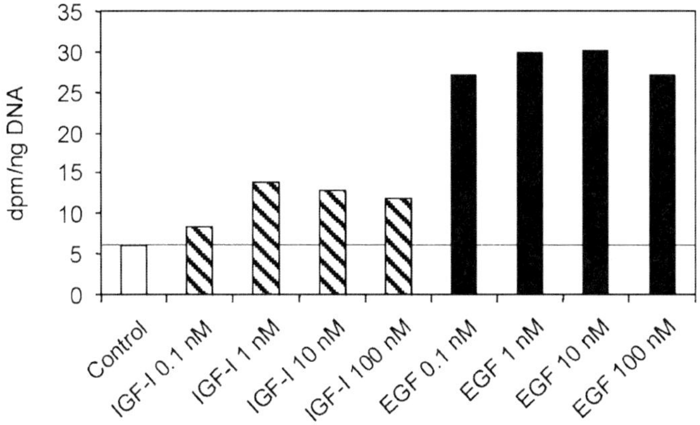

Fig. 7.8 Insulin-like growth factor I (IGF-I) and epidermal growth factor (EGF) stimulate the growth of porcine satellite cells derived from *semitendinosus* muscle (117). The cells were exposed to differing concentrations of IGF-I and EGF in serum-free medium for 26 h. DNA synthesis rate (dpm/ng DNA) was measured during the final 6 h of the incubation period using incorporation of ^{3}H-thymidine. Least squares means are represented by columns; SE = 0.6–0.8 dpm/ng DNA. Each column represents the data from 12 microplate wells from two replicates. Least squares means for all growth factor concentrations are significantly different from the control ($P<0.05$)

other growth factors such as bFGF and TGF-β [81, 221]. In porcine myogenic cell cultures, IGFBPs, such as IGFBP-2 and IGFBP-4, are produced both by satellite cells and fibroblasts [221]. However, in addition to IGFBP-2 and IGFBP-4, porcine satellite cells produce and secrete IGFBP-3 [81, 86, 221] and IGFBP-5 [81, 146, 221]. In proliferating porcine satellite cell cultures, the addition of IGF-I increased the concentrations of IGFBP-3 by 1.7-fold and IGFBP-5 by 2.5-fold in the conditioned medium [221]. Protein and mRNA levels of IGFBP-3 were decreased with differentiation then increased after differentiation concluded [81, 86]. Moreover, IGFBP-2 and IGFBP-3 mRNA levels showed a three-fold increase in extensively differentiated cultures of porcine embryonic myogenic cells after 144 h of incubation indicating that these IGFBPs play a role in differentiation [87]. Furthermore, multiple passaging of cloned porcine satellite cells resulted in increased secretion of IGFBP-2, which was associated with depressed cell proliferation and myotube formation [44]. It is suggested this may have been caused by surplus IGFBP-2 that specifically bound IGF-I and reduced its bioactivity. Overall, these results suggest that the IGFBPs are important regulatory factors during myogenesis in the pig, in particular, in the transition from proliferation to differentiation. With the recent development of recombinant porcine (rp) IGFBPs [145, 146] the examination of direct effects of specific IGFBPs on satellite cell growth has become realisable. The IGF-I-stimulated proliferation of porcine embryonic myogenic cells was suppressed by the addition of rpIGFBP-3 or rpIGFBP-5, both in an IGF-dependent (effective IGF-I-binding limiting availability of IGF-I to the IGF-1R) and IGF-independent manner [145, 146]. This suggests potential for IGFBP-3 and IGFBP-5 to affect porcine muscle cell growth during critical periods of development and to impact on the ultimate muscle mass in the postnatal animal.

Epidermal growth factor (EGF), another peptide growth factor, was found to stimulate growth [10, 64, 171] and differentiation [64] of muscle cells from various species through its interaction with the EGF receptor (EGF-R) which is expressed in porcine skeletal muscle [148]. Porcine primary satellite cell cultures expressing EGF-R increased DNA synthesis rate 4- to 5-fold (Fig. 7.8) when 10 nM EGF were added to a basal serum-free medium [117]. In contrast, EGF did not stimulate growth of clonal porcine satellite cells in serum-free culture medium of identical composition [30].

Fibroblast growth factor (FGF) is one of the most potent mitogens, exerting its effects at the sub-nanomolar concentrations. It stimulates proliferation but inhibits differentiation of cultured muscle cells [11, 47]. The mitogenic effects of FGF have been shown for pig [30], cattle, turkey and sheep satellite cell cultures [27]. The bFGF-mediated increase of proliferation in porcine satellite cell clones was higher with medium that contained 2% FBS (fetal bovine serum) than with basal serum-free medium [30] suggesting that interactions with other serum factors play a role (see below).

Platelet derived growth factor-BB (PDGF-BB) is another potent mitogenic component of serum, and is released from platelets during the clotting of blood. Specific forms of PDGF might also be produced by somatic cells such as muscle cells. PDGF-BB was found to promote proliferation of porcine satellite cells [30].

Finally, TGF-β and myostatin, which are members of the TGF-β superfamily, also affect porcine muscle cell growth. Transforming growth factor-β suppressed proliferation and differentiation of cultured porcine embryonic myoblasts [144] and satellite cells [18]. In basal serum-free medium, TGF-β stimulated porcine satellite cell proliferation by 25%, whereas in serum-containing medium it inhibited growth by 58%. Consequently, TGF-β may affect cell cultures depending on whether serum or other growth factors are present in the medium [30]. However, when added to FGF-containing serum-free medium, TGF-β stimulated the proliferation of porcine satellite cells [18]. Myostatin negatively regulates myogenic cell growth by suppressing both proliferation and differentiation as shown using porcine myogenic cell cultures [92]. Furthermore, treatment with either TGF-β or myostatin increased the production of IGFBP-3 mRNA and protein [92].

Combinations of two or more growth factors typically result in synergistic responses [30]. The combination of bFGF and IGF-I dramatically increases proliferation, and IGF-I has been shown to synergise with EGF to increase proliferation of porcine satellite cell clones. Interactions of IGF-I, EGF and bFGF with PDGF-BB stimulated cell proliferation. Furthermore, a 5-fold increase in DNA occurred in satellite cell clones grown in basal serum-free medium when they were simultaneously exposed to bFGF, PDGF-BB, EGF and IGF-I. Removal of IGF-I, bFGF or PDGF-BB resulted a reduction in proliferation by 40, 20 and 10%, respectively, indicating that all of these mitogens are physiological regulators of porcine satellite cell growth. Moreover, the addition of bFGF in combination with IGF-I and EGF in serum-free medium stimulated cellular proliferation to a level which indicated synergistic effects of these mitogenic growth factors [30].

Although a variety of hormones and growth factors that have not yet been characterised and incorporated into in vitro experiments are likely to exist in vivo, it is only a matter of time until they are described and researchable using muscle cell cultures. According to Dodson et al. [27] the definition of extrinsic factors that regulate satellite cells is vital to the eventual development of practicable strategies for exogenous manipulation of muscle satellite cell activity and muscle growth.

7.3 Genomic Regulation

It is well established that structural and functional properties of muscle fibres are correlated with meat quality traits in pigs [105, 162]. Muscle fibre traits have moderate to high heritabilities (h^2=0.20–0.59), similar to growth traits, although these are generally higher than those for meat quality traits (h^2=0.15–0.32) [43, 100]. Direct selection of pigs on high muscle fibre diameter and proportion of the glycolytic fibre type increased stress susceptibility and had a negative impact on meat quality [205]. Moreover, it has been demonstrated by simulated selection using a large data set from pigs that selection responses in growth, carcass and meat quality traits could be markedly improved if muscle structural traits were included in selection indices [43]. Consequently, genes affecting fetal growth and muscle development impact on postnatal growth as well as carcass and meat quality traits. Hence, functional candidate genes may be derived from genes involved in the prenatal muscle determination and formation of myoblasts, and in the regulation of anabolic and catabolic processes in muscle. Also, genes encoding muscle structural components are functional candidate genes for traits related to muscularity and meat quality. In this regard, expression profiling of muscle tissue at different prenatal stages enables new hypotheses on the genetic basis of variation in growth and muscle traits to be developed, and new functional candidate genes to be identified. Linkage QTL (quantitative trait locus) analysis can deliver positional information on genes affecting growth, carcass, meat quality and muscle structural traits without any prior hypothesis on the physiology of the trait, and complements candidate gene and expression analyses. The following section highlights certain regulatory genes involved in myogenesis, genes encoding components of muscle structure, current progress in expression profiling of prenatal muscle, as well as results of QTL analyses for traits associated with muscle structure.

7.3.1 Intrinsic Regulatory Genes of Myogenesis

Myogenesis depends on the strictly synchronised expression of a number of genes and their interactions. These genes control delamination and migration, proliferation, as well as determination and differentiation. Within this interactive network, myogenic regulatory factors (MRFs) play a key role and depend, themselves, on activating and inhibitory factors. The network of signalling pathways leads through the inhibition or stimulation of the MRFs to the formation of myogenic precursor cells and myoblasts (for review see [13, 115]). An overview of genes whose roles in myogenesis are well established is provided in Table 7.2. However, in addition to these, a number of other genes, including cyclins, integrins and members of the

Table 7.2 Genes encoding major regulators of myogenesis

	Genes	Delamination/ migration	Proliferation	Determination/ differentiation	Specific features
MRFs	MYOG (myogenin)			x	Basic helix-loop-helix transcription factor (bHLH); terminal differentiation of myogenic precursor cells
	MYF6 (myogenic factor 6)			x	Basic helix-loop-helix transcription factor (bHLH); terminal differentiation of myogenic precursor cells and of myotubes
	MYOD1 (myoblast determination protein 1)			x	Basic helix-loop-helix transcription factor (bHLH); induction of cell-cycle arrest and differentiation to myoblasts
	MYF5 (myogenic factor 5)		x		Basic helix-loop-helix transcription factor (bHLH); determination and proliferation of myoblasts
Protein – DNA interaction, other transcription factors	MEOX-2 (homeobox protein Mox-2)	x			Mesoderm induction and regional specification
	LBX1 (ladybird homeobox homolog 1)	x			Key regulator of muscle precursor cell migration
	PAX3 (paired box protein Pax-3)	x	x		Delamination and proliferation; repression of MyoD1, interacts with MSX1, mediates transcription of MET
	MSX1 (homeobox protein MSX-1)		x	x	Down-regulates expression of MyoD1, interacts with PAX3
	MEF2, myocyte specific enhancer factor 2A, 2B, 2C			x	Determination of type of muscle-specific differentiation; interacts with MRF
	FOXK1, FOXO1a (forkhead box protein K1, ~O1a)			x	FOXK1 enhances myoglobin expression; FOXO1a controls rate of myotube fusion
	GATA-2			x	Transcription factor in calcineurin pathway;

Table 7.2 (continued)

	Genes	Delamination/ migration	Proliferation	Determination/ differentiation	Specific features
	NFATC3 (nuclear factor of activated T-cells)			x	Member of NFAT transcription factor family, Ca responsive isoform 3; (IGF1)-calcineurin-NFATC3 signal pathway;
Protein-protein interaction,intracellular signal transduction	HGF (hepatocyte growth factor)	x			Mesemchymal protein that stimulates mitogenesis and cell motility
	SHH (sonic hedgehog)	x	x		Expansion of early myogenic precursor cells; induces PAX3; upregulates MYF5
	MET (hepatocyte growth factor receptor)	x	x		Delamination and migration, proliferation
	ID (inhibitor of DNA bin-ding 1, dominant negative helix-loop-helix protein)		x		Disables heterodimerisation of MRF with E12 and E2A; prolonged proliferation; delayed differentiation
	GDF8 (myostatin, growth/ differentiation factor 8		x	x	Member of bone morphogenetic protein family and transforming growth factor-β superfamily; negative regulator of myogenesis
	Notch		x	x	Determination of cells to become myoblasts; regulation of MYOD1; prolonged proliferation
	WNT (wingless-type MMTV integration site family)		x	x	WNT members 1, 3a and 7a; secreted signalling proteins involved in activation of MYF5 and MYOD1
	laminins (heteropolymers of α, β, and γ chains)		x	x	Enhance proliferation of myoblasts; stimulate myoblast motility; promote myotube formation
	BMP (bone morphogenetic proteins) BMP4, BMP7		x	x	Members of transforming growth factor-β superfamily; inhibit MRF

Table 7.2 (continued)

Genes	Delamination/ migration	Proliferation	Determination/ differentiation	Specific features
NOG (noggin)			x	Transforming growth factor- β superfamily signalling; restriction of proliferation, promotion of differentiation
FST (follistatin)				Inhibition of GDF8
MLP (muscle LIM protein), SLIM (skeletal muscle LIM proteins)			x	LIM proteins interact with MYOD- and MYOG- E12 complexes
GRB2 (growth factor receptor-bound protein 2)			x	Signalling of HGF and MET
PPP3R2 (protein phosphatase 3, regulatory subunit B, β isoform)			x	Calcineurin B, type II; initiation of myogenesis; determination of fibre type, hypertrophy and regeneration

Sources: Krempler and Brenig [98]; Maltin et al. [115]; TePas and Soumillion [194]; Buckingham et al. [14]; Parker et al. [147]; www.neuro.wustl.edu/neuromuscular/mother/myogenesis.html

FGF family, TGF family, and IGFs are known to influence myogenesis. While many of these genes have been identified in the pig, only some have been considered as candidate genes for postnatal growth and carcass quality.

The MRF gene family includes the genes Myogenin (MYOG or MYF4), myogenic determination factor 1 (MYOD1 or MYF3), myogenic factor 5 (MYF5) and myogenic factor 6 (MYF6 or MRF4 or herculin). These are transcription factors that activate stage-specific expression of genes during myogenesis (for review see [194]). The expression of MYOD1 and MYF5 is necessary for initiation of myoblast formation and satellite cell proliferation [142]. MYOG induces the differentiation or fusion of myoblasts, and MYF6 is subsequently involved in the maintenance of mature muscle fibres [56, 66, 142]. Porcine MYOG has been shown to be polymorphic, and association of variant forms with muscle mass and growth rate is proven [195]. Myogenic factor 5 also exhibits a number of polymorphisms [186, 192, 199], however no associations with growth, carcass and meat quality traits have been found. Currently, there are known polymorphisms in MYOD1 and MYF6, although associations with production traits have been mostly negative or not consistent [41, 200, 218, 219].

Myostatin is a member of the TGF-β superfamily and a negative regulator of myofibre formation [121]. Myostatin and its variants have been shown to have large affects on muscle growth in several species. The gene in its non-functional form was shown to be causative of the double muscling phenotype in cattle [58, 59, 94], and a mutation was shown to cause the *compact* hypermuscular mutation in mice [189]. Porcine studies have shown that runt piglets have increased myostatin expression compared to the more heavily muscled control piglets [85]. However, association of variants in the coding region of Myostatin with muscle mass have not been shown to date.

LIM domain proteins are essential regulators of muscle development, both in embryos and adults [2, 112]. For some porcine genes encoding LIM domain proteins, polymorphisms have been found as well as different allele frequencies or differential expression in neonatal pigs of western commercial breeds and some Chinese breeds [109, 202].

Current knowledge of the regulation of myogenesis has enabled various genes to be analysed to assess effects of prenatal developmental events in relation to postnatal growth and carcass quality in pigs.

7.3.2 Genes Encoding Structural Components of Muscle Tissue

Myofibrillar proteins exist as multiple isoforms that derive from multigene (isogene) families. Additional isoforms can be generated from the same gene through alternative splicing or use of alternative promoters. Major components of the thick filaments are the myosin heavy chain and light chain proteins, and actin. Tropomyosin and troponin are compounds of the thin filaments. Myofibrillar protein isogenes are differentially expressed in various muscle and fibre types, but can also be co-expressed within the same fibre. The variable expression of myofibrillar protein isoforms is a major determinant of the contractile properties of skeletal muscle fibres. These isoforms are related to the differences in the parameters of chemo-mechanical-transduction, such as ATP hydrolysis rate and shortening velocity. Expression of members of the myosin heavy chain (MYH) gene family is temporally regulated during myogenesis. The porcine cardiac and skeletal MYH genes are arranged in two clusters in the order MYH slow/β and α on porcine chromosomes 7, and embryonic, 2a, 2x, 2b and perinatal on chromosome 12, respectively, [20, 21, 25]. MYH slow/β and α are the only MYH isoforms expressed in heart muscle. In terms of gene order, intergenic distances, and head-to-tail orientation, these clusters are conserved in human, mouse and pig. In prenatal mammalian muscles, the embryonic, perinatal and slow/β MYH isoforms represent the three dominant skeletal muscle fibre types in the developing fetus. Shortly after birth, the expression of embryonic and perinatal MYH genes is down-regulated. In the pig, the postnatal MYH isoforms (2a, 2x and 2b) are already switched on at least from day 35 of gestation, and postnatally there are four major MYH isoforms: slow/β, 2a, 2x, and 2b. During myogenesis the transcript abundance of these isoforms correlates with their arrangement within the gene cluster (2a>2x>2b) [20, 21]. The four major muscle fibre types of adults are characterised by the expression of the respective MYH isoform. These fibres differ in metabolic, biochemical and biophysical characteristics, which may affect growth and meat quality traits.

By combining real time RT-PCR and microscopic analysis of serial porcine muscle sections to quantify and localise MYH isoforms and their transcripts within its corresponding fibre type, it was shown that the relative fibre type-restricted expression of each postnatal MYH gene showed wide spatial and temporal variation within a given muscle and between muscles, as well as among pig breeds [19]. Significant correlations were found between fibre typing by ATPase staining and quantitative RT-PCR assays of MYH isoforms for types I, IIa and IIx/IIb (r=0.71, 0.67 and 0.52, respectively) [212, 213]. However, MYH isoform transcript abundance and histochemical fibre typing characterise muscle on different levels with the first being a quantitative trait taking into account co-expression of the various isoforms within a muscle fibre, while the second is a categorical trait assigning a single fibre to a particular fibre type. Taking into account (1) the fact that the conventional histochemical fibre typing into types I, IIA and IIB is not well adapted for porcine skeletal muscles, with four fibre types present based on MYH isoforms, i.e. types I, IIa, IIx and IIb [103] and (2) that considerable variation of mRNA expression of each MYH isoform among muscles even within the same animal which is likely to reflect differences in the physiological state of individual muscles [19], the abundance of transcripts of MYH isoforms has been proposed as a new more precise phenotype towards the elucidation of genetic background of variation in traits related to growth and muscularity [212, 213]. In this regard, significant differences in MYH transcript abundance depending on the *m. longissimus* cross-sectional areas were shown [211], as was genomic variation for some MYH genes [24].

Genes coding for sarcomeric proteins may also play a key role in muscle accretion and meat quality. Troponin and tropomyosin isoforms are associated with sensitivity to calcium, whereas titin isoforms dictate the elastic properties of muscle fibres at rest. Myopalladin is a component of the sarcomere, which tethers nebulin in skeletal muscle to alpha-actinin at the Z lines. Myosin and troponin isoforms contribute to the differences in the resistance to fatigue of muscle fibres [172], while myopalladin and titin are associated with carcass traits [211].

7.3.3 Temporal Changes in the Muscle Transcriptome During Prenatal Development

In livestock research, several attempts to use gene expression profiling techniques have been made to elucidate tissue-specific differential gene expression in relation to particular stages of development and/or expression of certain phenotypes. The results will significantly increase and refine the list of candidate genes for economically important traits including growth, carcass composition and meat quality, for which muscle is the major target tissue.

Currently, there are several efforts to apply (micro-)array techniques to identify porcine genes that are differentially expressed in muscle tissue, either among different developmental stages/ages, breeds, or housing and feeding conditions. Using a porcine skeletal muscle cDNA macro-array containing 327 cDNAs derived from whole embryo and skeletal muscle, 48 genes were identified as being differentially expressed in muscle tissue derived from 75- and 105-day fetal hind limb muscles, and in 1 and 7 week-old postnatal *semitendinosus* muscles [223]. Hybridisation of a cDNA microarray of more than 700 clones of normalised skeletal muscle cDNA libraries from hind limb muscle of pigs at 45 and 90 days of gestation, birth, 7 weeks and 1 year of age with targets derived from total RNA of skeletal muscle of pigs at 60 days of gestation and 7 weeks of age revealed 55 clones that were over expressed by at least 2-fold in 60-day fetal skeletal muscle as compared to 7-week postnatal muscle [40, 220]. Furthermore, competitive heterologous hybridisation of human cDNA arrays with *longissimus* muscle mRNA of neonatal pigs revealed higher expression of genes encoding myofibrillar proteins, ribosomal proteins, transcription regulation proteins and glycolytic metabolism enzymes in Duroc compared to Taoyuan pigs [109]. These analyses mark the pathways that reflect genetic control of differences in traits related to muscle growth. Using differential display techniques, genes related to musculoskeletal growth, immune system function, and cellular regulation have also been shown to be differentially expressed at 21, 35, and 45 days of gestation [204].

Prenatal differentially expressed muscle transcripts at seven ages (days 14, 21, 35, 49, 63, 77, 91) corresponding with key developmental stages in the Pietrain and Duroc breeds pigs have been analysed in order to identify genes affecting postnatal growth and postmortem meat quality (EU-Project PorDictor (QLK5_2000_01363) [214]. Several techniques for expression profiling including cDNA-microarrays,

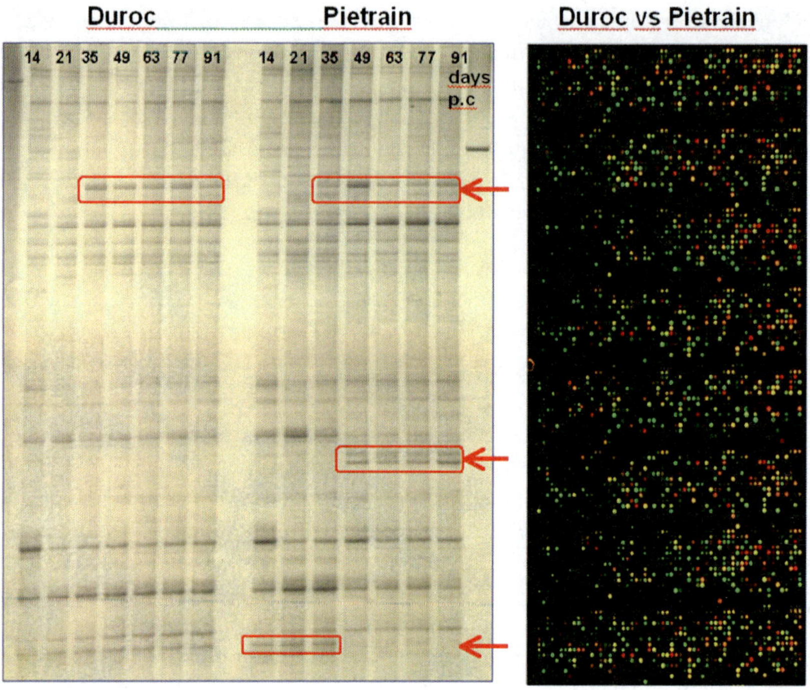

Fig. 7.9 Example of a differential display and an application-specific-microarray used to compare prenatal muscle expression between Duroc and Piétrain breeds of pigs. The differential display represents an open system that allows comparison of multiple samples simultaneously, but requires further analyses to identify differentially expressed genes. Bands representing genes differentially expressed between breeds or expressed at specific developmental stages are marked. In microarray analyses, two samples are competitively hybridized to known probes revealing information on the expression levels that can immediately be used for bioinformatic analysis to elucidate key factors and pathways controlling traits of interest

differential display-RT-PCR, construction of stage-specific muscle cDNA libraries, and subtractive hybridisation were applied (Fig. 7.9). Application specific cDNA microarrays covering more than 500 genes known to be involved in muscle development, growth and structure were used for between breed comparisons at each of the seven prenatal stages, and to provide a temporal expression profile of prenatal muscle in the Duroc breed. Differential expression of genes regulating myogenesis, muscle structure, and energy metabolism suggests not only differential timing of myogenesis between the two breeds, but also that the differences that contribute to varying postnatal growth and carcass and meat quality between the Duroc and Pietrain breeds, commence during early prenatal development [16, 191, 193, 196]. The changes in the muscle transcriptome expression profiles during prenatal muscle development of the Duroc show a profile of waves of expression of: (i) myoblast proliferation stimulating genes followed by; (ii) myoblast proliferation inhibiting and differentiation stimulating genes during primary myotube development, which is repeated at a lower magnitude during secondary myofibre development. Furthermore, expression of energy metabolism genes reaches a nadir when differentiation of myoblasts into myotubes takes place [191, 193, 196]. Differential display RT-PCR expression profiles of each of the seven stages during myogenesis in both Duroc and Pietrain breeds revealed 144 fragments differentially expressed between breeds and 301 fragments differentially expressed between stages. Sequence analysis allowed assignment of the majority of differentially displayed bands to functional groups including myogenesis regulating genes, muscle structural genes, and energy metabolism genes. Furthermore, a number of unknown transcripts and/or genes with unknown function were identified [130, 210]. Thus, differential display RT-PCR complemented microarray analyses as an open system to enable new candidate genes to be identified. The majority of the genes showed one of three expression profiles associated with myogenesis, as shown by quantitative RT-PCR of a number of selected genes: (1) up-regulation at the time points of the formation of both the primary myotubes and the secondary myofibres; (2) up-regulation at the time points of the formation of the primary fibres; and (3) steady up- or down-regulation throughout myogenesis [131]. Together, the various expression profiling approaches revealed more than 500 genes that are regulated during myogenesis. A shortlist of 53 promising candidate genes was established taking into account the significance of results, matching of results obtained with different methodologies, knowledge of gene function, and their mapping position in genomic regions of known QTL for meat quality traits. For a subset of genes, the relationship between expression level and breeding values were analysed. This provided evidence for association of transcript levels and phenotypes. At the same time, some findings of differences between breeds demonstrate the complexity of regulatory mechanisms including cis- and trans-regulation and possible gene-gene and gene-environment interactions. However, further identification of regulated genes and bioinformatic analysis of these genes has revealed pathways involved in the genetic control of muscle development. Finally, association with carcass and meat quality traits has been demonstrated for ten of the genes [132, 210, 211].

7.3.4 Quantitative Trait Loci (QTL) for Muscle Fibre Traits

Genome scans are the most general approach to identifying genomic regions exhibiting QTL without prior hypotheses of the physiological and genetic control of a trait. Within pedigrees the co-segregation of trait phenotypes and marker genotypes is observed. Among the number of markers distributed throughout the genome, at least some will be linked to QTL for the trait of interest. QTL analysis depends on the fact that where such linkage occurs, the marker locus and the QTL will not segregate independently, but that linkage disequilibrium exists within the pedigrees examined. Genome scans were conducted in different experimental and commercial pig populations and revealed QTL for traits related to growth, leanness and meat quality on all 18 autosomes and on chromosome X, as summarised on the Pig Quantitative

Trait Loci (QTL) database (PigQTLdb) (http://www.animalgenome.org/QTLdb/; [82]). The power of QTL analyses depends largely on the size of the experiment in terms of number of animals and markers used and also on the trait analysed, with high heritability being preferable. Disentanglement of complex traits in their constituent phenotypes facilitates the identification of QTL and elucidation of the pleiotropic nature of QTL effects. Muscle fibre traits that largely depend on prenatal events but affect postnatal growth and muscularity are highly heritable. Thus they explicitly allow for powerful QTL analysis. In the pig, QTL for myofibre traits obtained by either electrophoretic discrimination of myosin isoforms or histochemical fibre type differentiation have been shown to be governed by genetic variation at many loci distributed throughout the genome [114, 136, 209]. Quantitative trait loci with genome wide significance are summarised in Table 7.3.

In a Berkshire x Yorkshire crossbred population of pigs a QTL reaching chromosome-wide significance for the proportion of type I (slow/β) myofibres was detected on chromosome 8 (SSC8) [114]. Analyses conducted in Berlin Miniature x Duroc experimental crossbred pigs also revealed a number of QTL at the chromosome-wide level of significance, including those for total myofibre number on SSC2, 5, 11, 12, 14, and 15 and for proportion of giant muscle fibres on SSC1, 3, 4, 12, 15, 16 and 18. Regions with either significant QTL for muscle fibre traits or significant QTL for meat quality, muscularity, or both were detected on SSC1, 2, 3, 4, 5, 13, 14, 15, and 16. Genomic regions affecting the complex traits of muscularity and meat quality, as well as micro-structural properties, may point to QTL that, in the first instance, affect muscle fibre traits and secondarily, meat quality [209]. In particular, QTL for myofibre number and, to a lesser extent, for muscle fibre types are likely to represent genes involved in myogenesis. A number of positional candidate genes for the QTL have been proposed. These include genes of the calcineurin signalling pathway involved in muscle fibre type switching, namely PPP3CC (protein phosphatase 3 catalytic subunit γ isoform, calcineurin A γ), PPP3CB (protein phosphatase 3 catalytic subunit β isoform, calcineurin A β) and NFAM1 (NFAT activation module 1), which all map to SSC14 [136]. Genes of the myostatin axis, including MYOG (myogenin; SSC9), MYOD1 (myogenic determination factor 1; SSC2), MYF5 (myogenic factor 5; SSC5), MYF6 (myogenic factor 6, SSC5) and Myostatin (SSC15) also represent positional candidates. Other established positional candidate genes are MEF2C (MADS box transcription enhancer factor 2, polypeptide C; SSC2), MEF2D (MADS box transcription enhancer factor 2, polypeptide D; SSC4), PPARGC1A, (peroxisome proliferative activated receptor, gamma, coactivator 1, alpha; SSC8) and PPARG (peroxisome proliferator-activated receptor gamma 2; SSC13) [209]

7.3.5 Combining Results of QTL and Expression Analysis

The genome scans provide positional information of regions containing QTL for meat quality that segregate within the resource populations analysed. The expression

Table 7.3 Overview of quantitative trait loci (QTL) for muscle fibre traits detected in the pig

Trait	Position SSC[1]	Close marker	F-value[2]	Population	References
Proportion of number of type I fibres [%]	1 14 X	SW970 SWR925 SW2588	10.3* 10.6* 90.4*	Japanese Wild Boar x Large White	Nii et al. [136]
Proportion of relative area of type I fibres [%]	X	SW2588	10.8**		
Proportion of number of type IIa fibres [%]	2 2 14	SW942 SW1879 SW1027–SW540	10.7** 11.4** 8.9*		
Proportion of relative area of type IIa fibres [%]	2 6	SW942 SW2406	9.3* 12.4**		
Proportion of number of type IIb fibres [%]	2 14 14	SW1879 SW1027 SWR925	10.4* 12.4* 11.7**		
Proportion of relative area of type IIb fibres [%]	6 14 14	SW1329 SW1027–SW540 SW104–SWR925	12.6* 12.0* 11.1**		
Diameter of angular fibres [μm]	1	SW1515	8.4*	Berlin Miniature Pig x Duroc (DUMI)	Wimmers et al.[209]
Diameter of FTG fibres [μm]	2 4	FTH1– SW240 S0214	9.4* 8.6*		
Diameter of giant fibres [μm]	12	S0143	7.4*		
Average fibre diameter [μm]	2	FTH1-SW240	9.5*		
Diameter of white fibres [μm]	14	VCL-SWC27	7.9*		
Number of fibres per mm^2	11	S0386	5.2*		
Proportion of intermediate fibres [%]	2 8	STS2-C3 SW2410	5.8* 6.0*		
Proportion of giant fibres [%]	4 12 15	S0214–S0097 S0143 S0355	6.2* 10.3* 12.7**		

[1]Chromosome
[2]** and *=1% and 5% genome-wide significance levels, respectively

analyses provide functional candidate genes for these traits based on their temporo-spatial and/or phenotype-associated expression. The application of bioinformatic tools and the use and generation of mapping information of differentially expressed genes combined with QTL information reveals segregating functional positional candidate genes. These are highly valuable resources for further association and molecular genetic analyses to provide statistical and functional evidence of their impact on meat quality and carcass traits, and to elucidate prenatal events that affect postnatal growth and meat quality.

References

1. Allen, R.E., K.C. Masak, P.K. McAllister, and R.A. Merkel. 1983. Effect of growth hormone, testosterone and serum concentration on actin synthesis in cultured satellite cells. *J. Anim. Sci.* **56**:833–837.
2. Arber, S., G. Halder, and P. Caroni. 1994. Muscle LIM protein, a novel essential regulator of myogenesis, promotes myogenic differentiation. *Cell* **79**:221–231.
3. Ashworth, C.J., A.M. Finch, K.R. Page, M.O. Nwagwu, and H.J. McArdle. 2001. Causes and consequences of fetal growth retardation in pigs. *Reprod. Suppl.* **58**:233–246.
4. Ashworth, C.J. and H.J. McArdle. 1999. Both placental amino acid uptake and fetal amino acid concentrations differ between small and normally grown *porcine* fetuses. *Early Human Dev.* **54:**90–91.
5. Bauer, M.K., J.E. Harding, N.S. Bassett, B.H. Breier, M.H. Oliver, B.H. Gallaher, P.C. Evans, S.M. Woodall, and P.D. Gluckman. 1998. Fetal growth and placental function. *Mol. Cell Endocrinol.* **140**:115–120.
6. Bauer, M. and N. Parvizi. 1996. Pulsatile and diurnal secretion of GH and IGF-I in the chronically catheterized pig fetus. *J. Endocrinol.* **149**:125–133.
7. Bee, G. 2000. Dietary conjugated linoleic acid consumption during pregnancy and lactation influences growth and tissue composition in weaned pigs. *J. Nutr.* **130**:2981–2989.
8. Bee, G. 2004. Effect of early gestation feeding, birth weight, and gender of progeny on muscle fiber characteristics of pigs at slaughter. *J. Anim. Sci.* **82**:826–836.
9. Beermann, D.H., R.G. Cassens, and G.J. Hausman. 1978. A second look at fibre type differentiation in *porcine* skeletal muscle. *J. Anim. Sci.* **46**:125–132.
10. Blachowski, S., T. Motyl, A. Orzechowski, K. Grzelkowska, B. Interewicz 1993. Comparison of metabolic effects of EGF, TGF-α and TGF-β1 in primary culture of fetal bovine myoblasts and rat L6 myoblasts. *Int. J. Biochem.* **25**:1571–1577.
11. Brameld, J.M., P.A. Weller, J.C. Saunders, P.J. Buttery, and R.S. Gilmour. 1998. Hormonal control of insulin-like growth factor-I and growth hormone receptor mRNA expression by *porcine* hepatocytes in culture. *J. Endocrinol.* **146**:239–245.
12. Breier, B.H. 1999. Regulation of protein and energy metabolism by the somatotropic axis. *Dom. Anim. Endocrinol.* **17**:209–218.
13. Buckingham, M. 2001. Skeletal muscle formation in vertebrates. *Curr. Opin. Genet. Dev.* **11**: 440–448.
14. Buckingham, M., L. Bajard, T. Chang, P. Daubas, J. Hadchouel, S. Meilhac, D. Montarras, D. Rocancourt, and F. Relaix. 2003. The formation of skeletal muscle: from somite to limb. *J. Anat.* **202**:59–68.
15. Buitrago, J.A., E.F. Walker, Jr., W.I. Snyder, and W.G. Pond. 1974. Blood and tissue traits in pigs at birth and at 3 weeks from gilts fed low or high energy diets during gestation. *J. Anim. Sci.* **38**:766–771.
16. Cagnazzo, M., M.F.W. Te Pas, J. Priem, A.A.C. de Wit, M.H. Pool, R. Davoli, and V. Russo. 2006. Comparison of prenatal muscle tissue expression profiles of two pig breeds differing in muscle characteristics. *J. Anim. Sci.* **84**:1–10.
17. Clemmons, D.R. 1998. Role of insulin-like growth factor binding proteins in controlling IGF actions. *Mol. Cell Endocrinol.* **140**:19–24.
18. Cook, D.R., M.E. Doumit, R.A. Merkel. 1993. Transforming growth factor-beta, basic fibroblast growth factor and platelet-derived growth factor-BB interact to affect proliferation of clonally derived *porcine* satellite cells. *J. Cell Physiol.* **157**:307–312.

19. Da Costa, N., R. Blackley, H. Alzuherri, and K.C. Chang. 2002. Quantifying the temporospatial expression of postnatal *porcine* skeletal myosin heavy chain genes. *J. Histochem. Cytochem.* **50**:353–634.
20. Da Costa, N. and K.C. Chang. 2005. Molecular characterisation of the *porcine* skeletal myosin heavy chain cluster and a major candidate regulatory domain. *Arch. Anim. Breed.* **48** (Special Issue):32–39.
21. Da Costa, N., C. McGillivray, and K.C. Chang. 2003. Postnatal myosin heavy chain isoforms in prenatal *porcine* skeletal muscles: insights into temporal regulation. *Anat. Rec.* **273**:731–740.
22. Dauncey, M.J., M. Katsumata, and P. White. 2004. Nutrition, hormone receptor expression and gene interactions: implications for development and disease, pp. 103–124. *In* M.F.W. Te Pas, M.E. Everts, and H.P. Haagsman, (eds.), Muscle Development of Livestock Animals: Physiology, Genetics, and Meat Quality, CAB Int., Wallingford, Oxon, UK.
23. Davis, T.A., M.L. Fiorotto, D.G. Burrin, W.G. Pond, and H.V. Nguyen. 1997. Intrauterine growth restriction does not alter response of protein synthesis to feeding in newborn pigs. *Am. J. Physiol* **272**:E877–E884.
24. Davoli, R., L. Fontanesi, M. Cagnazzo, E. Scotti, L. Buttazzoni, M. Yerle, and V. Russo 2003. Identification of SNPs, mapping and analysis of allele frequencies in two candidate genes for meat production traits: the *porcine* myosin heavy chain 2B (MYH4) and the skeletal muscle myosin regulatory light chain 2 (HUMMLC2B). *Anim. Genet.* **34**:221–225.
25. Davoli, R., P. Zambonelli, D. Bigi, L. Fontanesi, and V. Russo. 1998. Isolation and mapping of two *porcine* skeletal muscle myosin heavy chain isoforms. *Anim. Genet.* **29**:91–97.
26. De Gaetano, A., G. Mingrone, M. Castagneto, and M. Calvani. 1999. Carnitine increases glucose disposal in humans. *J. Am. Coll. Nutr.* **18**:289–295.
27. Dodson, M.V., D.C. McFarland, A.L. Grant, M.E. Doumit, and S.G. Velleman. 1996. Extrinsic regulation of domestic animal-derived satellite cells. *Domest. Anim. Endocrinol.* **13**:107–126.
28. Doi, J., Y. Shinbori, A. Ohtsuka, and K. Hayashi. 2002. Effects of the thyroid hormone on differentiation, growth, and proteolysis in cultured muscle cells during serum deprivation. *J. Nutr. Sci. Vitaminol.* **48**:265–269.
29. Dørup, I. 2004. The Impact of Minerals and Micronutrients on Growth Control, pp. 125–136. *In* M.F.W. Te Pas, M.E. Everts, and H.P. Haagsman, (eds), Muscle Development of Livestock Animals: Physiology, Genetics, and Meat Quality, CAB Int.,Wallingford, Oxon, UK.
30. Doumit, M.E., D.R. Cook, and R.A. Merkel. 1993. Fibroblast growth factor, epidermal growth factor, insulin-like growth factors, and platelet-derived growth factor-BB stimulate proliferation of clonally derived *porcine* myogenic satellite cells. *J. Cell Physiol.* **157**:326–332.
31. Doumit, M.E., D.R. Cook, and R.A. Merkel. 1996. Testosterone up-regulates androgen receptors and decreases differentiation of *porcine* myogenic satellite cells in vitro. *Endocrinology* **137**:1385–1394.
32. Doumit, M.E. and R.A. Merkel. 1991. Influence of insulin-like growth factor-I (IGF-I) and insulin on proliferation and differentiation of *porcine* satellite cells. *J. Anim. Sci.* **69** (Suppl.1):316.
33. Duchamp, C., K.A. Burton, P. Herpin, and M.J. Dauncey. 1996. Perinatal ontogeny of *porcine* growth hormone receptor gene expression is modulated by thyroid status. *Eur. J. Endocrinol.* **134**:524–531.
34. Ducsay, C.A., W.C. Buhi, F.W. Bazer, R.M. Roberts, and G.E. Combs. 1984. Role of uteroferrin in placental iron transport: effect of maternal iron treatment on fetal iron and uteroferrin content and neonatal hemoglobin. *J. Anim. Sci.* **59**:1303–1308.
35. Dwyer, C.M., J.M. Fletcher, and N.C. Stickland. 1993. Muscle cellularity and postnatal growth in the pig. *J. Anim. Sci.* **71**:3339–3343.
36. Dwyer, C.M. and N.C. Stickland. 1991. Sources of variation in myofiber number within and between litters of pigs. *Anim. Prod.* **52**:527–533.

37. Dwyer, C.M., N.C. Stickland, and J.M. Fletcher. 1994. The influence of maternal nutrition on muscle fibre number development in the *porcine* fetus and on subsequent postnatal growth. *J. Anim. Sci.* **72**:911–917.
38. Dziuk, P.J. 1992. Embryonic development and fetal growth. *Anim. Reprod. Sci.* **28**: 299–308.
39. Eder, K., A. Ramanau, and H. Kluge. 2001. Effect of L-carnitine supplementation on performance parameters in gilts and sows. *J. Anim. Physiol. Anim. Nutr.* **85**:73–80.
40. Ernst, C.W., V.D. Rilington, N.E. Raney, J. Yao, S.S. Sipkovsky, P.M. Saama, R.J. Tempelman, and P.M. Coussens. 2002. Use of cDNA microarrays to detect differentially expressed genes in developing pig skeletal muscle. Abstracts of the Plant, Animal & Microbe Genomes X Conference, January 12–16, San Diego, CA: 702.
41. Ernst, C.W., D.A. Vaske, R.G. Larson, M.E. White, and M.F. Rothschild. 1994. Rapid communication: MspI restriction fragment length polymorphism at the swine MYF6 locus. *J. Anim. Sci.* **72**:799.
42. Etienne, M., M. Bonneau, G. Kann, and F. Deletang. 1992. Effects of administration of growth hormone-releasing factor to sows during late gestation on growth hormone secretion, reproductive traits, and performance of progeny from birth to 100 kilograms live weight. *J. Anim. Sci.* **70**:2212–2220.
43. Fiedler, I., G. Dietl, C. Rehfeldt, J. Wegner, and K. Ender. 2004. Muscle fibre traits as additional selection criteria for muscle growth and meat quality in pigs – results of a simulated selection. *J. Anim. Breed. Genet.* **121**:331–344.
44. Fligger, J.M., P.V. Malven, M.E. Doumit, R.A. Merkel, and A.L. Grant. 1998. Increases in insulin-like growth factor binding protein-2 accompany decreases in proliferation and differentiation when *porcine* muscle satellite cells undergo multiple passages. *J. Anim. Sci.* **76**:2086–2093.
45. Florini, J.R. 1987. Hormonal control of muscle growth. *Muscle Nerve* **10**:577–598.
46. Florini, J.R., D.Z. Ewton, and S.A. Coolican. 1996. Growth hormone and the insulin-like growth factor system in myogenesis. *Endocr. Rev.* **17**:481–517.
47. Florini, J.R., D.Z. Ewton, and K.A. Magri. 1991. Hormones, growth factors, and myogenic differentiation. *Ann. Rev. Physiol.* **53**:201–216.
48. Florini, J.R. and S.B. Roberts 1979. A serum-free medium for the growth of muscle cells in culture. *In Vitro* **15**:983–992.
49. Freese, L.G., C. Rehfeldt, R. Fuerbass, G. Kuhn, C.S. Okamura, K. Ender, A.L. Grant, and D.E. Gerrard 2005. Exogenous somatotropin alters IGF axis in *porcine* endometrium and placenta. *Domest. Anim. Endocrinol.* **29**:457–475.
50. Gatford, K.L., J.M. Boyce, K. Blackmore, R.J. Smits, R.G. Campbell, and P.C. Owens. 2004. Long-term, but not short-term, treatment with somatotropin during pregnancy in underfed pigs increases the body size of progeny at birth. *J. Anim. Sci.* **82**:93–101.
51. Gatford, K.L., J.E. Ekert, K. Blackmore, M.J. De Blasio, J.M. Boyce, J.A. Owens, R.G. Campbell, and P.C. Owens. 2003. Variable maternal nutrition and growth hormone treatment in the second quarter of pregnancy in pigs alter semitendinosus muscle in adolescent progeny. *Brit. J. Nutr.* **90**:283–293.
52. Gatford, K.L., J.A. Owens, R.G. Campbell, J.M. Boyce, P.A. Grant, M.J. De Blasio, and P.C. Owens. 2000. Treatment of underfed pigs with GH throughout the second quarter of pregnancy increases fetal growth. *J. Endocrinol.* **166**:227–234.
53. Gerrard, D.E., C.S. Okamura, and A.L. Grant 1999. Expression and location of IGF binding proteins-2, -4, and -5 in developing fetal tissues. *J. Anim. Sci.* **77**: 1431–1441.
54. Giorgino, F. and R.J. Smith. 1995. Dexamethasone enhances insulin-like growth factor-I effects on skeletal muscle cell proliferation: role of specific intracellular signalling pathways. *J. Clin. Invest.* **96**:1473–1483.
55. Gluckman, P.D. 1986. The Regulation of Fetal Growth, pp. 85–104. *In* P.J. Buttery, D.B. Lindsay, and N.B. Haynes, (eds.) Control and Manipulation of Animal Growth, Butterworth, London.

56. Goldspink, G. 1996. Muscle growth and muscle function: a molecular biological perspective. *Res. Vet. Sci.* **60**:193–204.
57. Gondret, F., L. Lefaucheur, H. Juin, I. Louveau, and B. Lebret. 2006. Low birth weight is associated with enlarged muscle fiber area and impaired meat tenderness of the longissimus muscle in pigs. *J. Anim. Sci.* **84**:93–103.
58. Grobet, L., L.J. Martin, D. Poncelet, D. Pirottin, B. Brouwers, J. Riquet, A. Schoeberlein, S. Dunner, F. Menissier, J. Massabanda, R. Fries, R. Hanset, and M.A. Georges. 1997. Deletion in the bovine myostatin gene causes the double-muscled phenotype in cattle. *Nat. Genet.* **17**:71–74.
59. Grobet, L., D. Poncelet, L.J. Royo, B. Brouwers, D. Pirottin, C. Michaux, F. Menissier, M. Zanotti, S. Dunner, and M. Georges. 1998. Molecular definition of an allelic series of mutations disrupting the myostatin function and causing double-muscling in cattle. *Mamm. Genome* **9**:210–213.
60. Guise, H.J. and R.H. Penny. 1990. Influence of supplementary iron in late pregnancy on the performance of sows and litters. *Vet. Rec.* **127**:403–405.
61. Handel, S.E. and N.C. Stickland. 1987. Muscle cellularity and birthweight. *Anim. Prod.* **44**:311–317.
62. Handel, S.E. and N.C. Stickland. 1987. The growth and differentiation of *porcine* skeletal muscle fibre types and the influence of birthweight. *J. Anat.* **152**:107–119.
63. Harper, J.M.M. and P.J. Buttery. 1992. Muscle cell growth, pp. 27–58. *In* P.J. Buttery, K.N. Boorman, and D.B Lindsay, (eds), The Control of Fat and Lean Deposition, Butterworth-Heinemann, Oxford, UK.
64. Harper, J.M.M. and P.J. Buttery 1995. Muscle cell growth. *Meat Focus Int.* **4**:323–329.
65. Harper, J.M.M., J.B. Soar, and P.J. Buttery. 1987. Changes in protein metabolism of ovine primary muscle cultures on treatment with growth hormone, insulin, insulin-like growth factor I or epidermal growth factor. *J. Endocrinol.* **112**:87–96.
66. Hasty, P., A. Bradley, J.H. Morris, D.G. Edmondson, J.M. Venuti, E.E. Olson, and W.H. Klein 1993. Muscle deficiency and neonatal death in mice with a targeted mutation in the myogenin gene. *Nature* **364**:501–506.
67. Hathaway, M.R., J.R. Hembree, M.S. Pampusch, and W.R. Dayton. 1991. Effect of transforming growth factor beta-1 on ovine satellite cell proliferation and fusion. *J. Cell. Physiol.* **146**:435–441.
68. Hausman, D.B., G.J. Hausman, and R.J. Martin. 1994. Endocrine regulation of fetal adipose tissue metabolism in the pig: role of hydrocortisone. *Obes. Res.* **2**:314–320.
69. Hausman, D.B., G.J. Hausman, and R.J. Martin. 1999. Endocrine regulation of fetal adipose tissue metabolism in the pig: interaction of *porcine* growth hormone and thyroxine. *Obes. Res.* **7**:76–82.
70. Hausman, G.J. 1992. The influence of thyroxine on the differentiation of adipose tissue and skin during fetal development. *Pediatr. Res.* **32**:204–211.
71. Hausman, G.J. 1999. The interaction of hydrocortisone and thyroxine during fetal adipose tissue differentiation: CCAAT enhancing binding protein expression and capillary cytodifferentiation. *J. Anim. Sci.* **77**:2088–2097.
72. Hausman, G.J., D.R. Campion, and G.B. Thomas. 1982. Semitendinosus muscle development in fetally decapitated pigs. *J. Anim. Sci.* **55**:1330–1335.
73. Hausman, G.J., D.R. Campion, and G.B. Thomas. 1985. Enzyme histochemical studies in an ontogeny study of muscle development in Ossabaw and decapitated fetuses: cellular reactions. *J. Anim. Sci.* **60:**1562–1570.
74. Hausman, G.J., E.J. Hentges, and G.B. Thomas. 1987. Differentiation of adipose tissue and muscle in hypophysectomized pig fetuses. *J. Anim. Sci.* **64**:1255–1261.
75. Hausman, G.J. and S.P. Poulos. 2005. A method to establish co-cultures of myotubes and preadipocytes from collagenase digested neonatal pig semitendinosus muscles. *J. Anim. Sci.* **83**:1010–1016.

76. Hausman, G.J., R.L. Richardson, and F.A. Simmen. 2000. Expression of insulin-like growth factor binding proteins (IGFBPs) before and during the hormone sensitive period of adipose tissue development in the fetal pig. *Growth Dev. Aging* **64**:51–67.
77. Hausman, G.J. and G.B. Thomas. 1984. Histochemical and cellular aspects of adipose tissue development in decapitated pig fetuses: an ontogeny study. *J. Anim. Sci.* **58**:1540–1549.
78. Hausman, G.J. and R. Watson. 1994. Regulation of fetal muscle development by thyroxine. *Acta Anat.* **149**:21–30.
79. Hegarty, P.V. and C.E. Allen. 1978. Effect of pre-natal runting on the post-natal development of skeletal muscles in swine and rats. *J. Anim. Sci.* **46**:1634–1640.
80. Hembree, J.R., M.R. Hathaway, and W.R. Dayton. 1991. Isolation and culture of fetal *porcine* myogenic cells and the effect of insulin, IGF-I, and sera on protein turnover in *porcine* myotube cultures. *J. Anim. Sci.* **69**:3241–3250.
81. Hembree, J.R., M.S. Pampusch, F. Yang, J.L. Causey, M.R. Hathaway, and W.R. Dayton. 1996. Cultured *porcine* myogenic cells produce insulin-like growth factor binding protein-3 (IGFBP-3) and transforming growth factor beta-1 stimulates IGFBP-3 production. *J. Anim. Sci.* **74**:1530–1540.
82. Hu, Z.L., S. Dracheva, W. Jang, D. Maglott, J. Bastiaansen, M.F. Rothschild, and J.M. Reecy. 2005. A QTL resource and comparison tool for pigs: PigQTLDB. *Mamm. Genome* **16**: 792–800.
83. Hwa, V., J. Oh, and R.G. Rosenfeld. 1999. Insulin-like growth factor binding proteins: a proposed superfamily. *Acta Paediatr. Suppl.* **17**:37–45.
84. Jewell, D.E., G.J. Hausman, and D.R. Campion. 1989. Fetal hypophysectomy causes a decrease in preadipocyte growth and insulin like growth factor-1 in pigs. *Domest. Anim. Endocrinol.* **6**:243–252.
85. Ji, S., R.L. Losinski, S.G. Cornelius, G.R. Frank, G.M. Willis, D.E. Gerrard, F.F.S. Depreux, and M.E. Spurlock. 1998. Myostatin expression in *porcine* tissues: tissue specificity and developmental and postnatal regulation. *J. Am. Physiol. Regul. Integr. Comp. Physiol.* **275**:1265–1273.
86. Johnson, B.J., M.E. White, M.R. Hathaway, and W.R. Dayton. 1999. Decreased steady-state insulin-like growth factor binding protein-3 (IGFBP-3) mRNA level is associated with differentiation of cultured *porcine* myogenic cells. *J. Cell. Physiol.* **179**:237–243.
87. Johnson, B.J., M.E. White, M.R. Hathaway, and W.R. Dayton. 2003. Effect of differentiation on levels of insulin-like growth factor binding protein mRNAs in cultured *porcine* embryonic myogenic cells. *Domest. Anim. Endocrinol.* **24**:81–93.
88. Ka, H., L.A. Jaeger, G.A. Johnson, T.E. Spencer, and F.W. Bazer. 2001. Keratinocyte growth factor is up-regulated by estrogen in the *porcine* uterine endometrium and functions in trophectoderm cell proliferation and differentiation. *Endocrinology* **142**:2303–2310.
89. Ka, H., T.E. Spencer, G.A. Johnson, and F.W. Bazer. 2000. Keratinocyte growth factor: expression by endometrial epithelia of the *porcine* uterus. *Biol. Reprod.* **62**:1772–1778.
90. Kahlert, S., C. Grohe, R.H. Karas, K. Lobbert, L. Neyses, and H. Vetter. 1997. Effects of estrogen on skeletal myoblast growth. *Biochem. Biophys. Res. Commun.* **232**: 373–378.
91. Kalbe, C., M. Mau, K. Wollenhaupt, and C. Rehfeldt. 2007. Evidence for estrogen receptor alpha and beta expression in skeletal muscle of pigs. *Histochem. Cell Biol.* **127**:95–107.
92. Kamanga-Sollo, E., M.S. Pampusch, M.E. White, and W.R. Dayton 2003. Role of insulin-like growth factor binding protein (IGFBP)-3 in TGF-beta- and GDF-8 (myostatin)-induced suppression of proliferation in *porcine* embryonic myogenic cell cultures. *J. Cell Physiol.* **197**:225–231.
93. Kamanga-Sollo, E., M.S. Pampusch, G. Xi, M.E. White, M.R. Hathaway, and W.R. Dayton. 2004. IGF-I mRNA levels in bovine satellite cell cultures: effects of fusion and anabolic steroid treatment. *J. Cell Physiol.* **201**:181–189.
94. Kambadur, R., M. Sharma, T.P. Smith, and J.J. Bass. 1997. Mutations in myostatin (GDF8) in double-muscled Belgian Blue and Piedmontese cattle. *Genome Res.* **7**:910–916.

95. Kelley, R.L., S.B. Jungst, T.E. Spencer, W.F. Owsley, C.H. Rahe, and D.R. Mulvaney. 1995. Maternal treatment with somatotropin alters embryonic development and early postnatal growth of pigs. *Domest. Anim. Endocrinol.* **12**:83–94.
96. Kim, H.S., R.L. Richardson, and G.J. Hausman. 1998. The expression of insulin-like growth factor-1 during adipogenesis in vivo: effect of thyroxine. *Gen. Comp. Endocrinol.* **112**:38–45.
97. Kotts, C.E., F. Buonomo, M.E. White, C.E. Allen, and W.R. Dayton. 1987. Stimulation of in vitro muscle cell proliferation by sera from swine injected with *porcine* growth hormone. *J. Anim. Sci.* **64**:623–632.
98. Krempler, A. and B. Brenig. 1999. Zinc finger proteins: watchdogs in muscle development. *Mol. Gen. Genet.* **261**:209–15.
99. Kveragas, L., R.W. Seerley, R.J. Martin, and W.L. Vandergrift. 1986. Influence of exogenous growth hormone and gestational diet on sow blood and milk characteristics and on baby pig blood, body composition and performance. *J. Anim. Sci.* **63**:1877–1887.
100. Larzul, C., L. Lefaucheur, P. Ecolan, J. Gogue, A. Talmant, P. Sellier, P. Le Roy, and G. Monin 1997. Phenotypic and genetic parameters for longissimus muscle fiber characteristics in relation to growth, carcass, and meat quality traits in large white pigs. *J. Anim. Sci.* **75**:3126–3137.
101. Latimer, M., G.J. Hausman, R.H. McCusker, and F.C. Buonomo. 1993. The effects of thyroxine on serum and tissue concentrations of insulin-like growth factors (IGF-I and -II) and IGF-binding proteins in the fetal pig. *Endocrinology* **133**:1312–1319.
102. Lee, Y., C.S. Chung, and F.A. Simmen. 1993. Ontogeny of the *porcine* insulin-like growth factor system. *Mol. Cell. Endocrinol.* **93**:71–80.
103. Lefaucheur, L. 2006. Myofibre typing and its relationships to growth performance and meat quality. *Arch. Anim. Breed.* (Special Issue) **49**:4–17.
104. Lefaucheur, L., F. Edom, P. Ecolan, and G.S. Butlerbrowne. 1995. Pattern of muscle-fiber type formation in the pig. *Dev. Dyn.* **203**:27–41.
105. Lengerken, G.v., S. Maak, M. Wicke, I. Fiedler, and K. Ender. 1994. Suitability of structural and functional traits of skeletal muscle for the improvement of meat quality in pigs. *Arch. Anim. Breed.* **37**:133–143
106. Lengerken, G.v. and H. Pfeiffer. 1991. Reduction of stress susceptibility in pigs as prerequisite for quality and quantitatively high results. *Arch. Anim. Breed.* **34**:241–247.
107. Lengerken, G.v., M. Wicke, and S. Maak. 1997. Stress susceptibility and meat quality – Situation and prospects in animal breeding and research. *Arch. Anim. Breed.* **40,** Suppl.:163–171.
108. Le Roith, D., C. Bondy, S. Yakar, J.L. Liu, and A. Butler. 2001. The somatomedin hypothesis: 2001. *Endocr. Rev.* **22**:53–74.
109. Lin, C.S. and C.W. Hsu. 2005. Differentially transcribed genes in skeletal muscle of Duroc and Taoyuan pigs. *J. Anim. Sci.* **83**:2075–2086.
110. Lindemann, M.D. 1993. Supplemental folic acid: a requirement for optimizing swine reproduction. *J. Anim. Sci.* **71**:239–246.
111. Louveau, I., S. Combes, A. Cochard, and M. Bonneau. 1996. Developmental changes in insulin-like growth factor-I (IGF-I) receptor levels and plasma IGF-I concentrations in Large White and Meishan pigs. *Gen. Comp. Endocrinol.* **104**:29–36.
112. Lu, P.Y., M. Taylor, H.T. Jia, and J.H. Ni. 2004. Muscle LIM protein promotes expression of the acetylcholine receptor gamma-subunit gene cooperatively with the myogenin-E12 complex. *Cell. Mol. Life Sci.* **61**:2386–2392.
113. Mahan, C. and J.L. Vallet. 1997. Vitamin and mineral transfer during fetal development and the early postnatal period in pigs. *J. Anim. Sci.* **75**:2731–2738.
114. Malek, M., J.C. Dekkers, H.K. Lee, T.J. Baas, K. Prusa, E. Huff-Lonergan, and M.F. Rothschild. 2001. A molecular genome scan analysis to identify chromosomal regions influencing economic traits in the pig. II. Meat and muscle composition. *Mamm. Genome* **12**:637–645.

115. Maltin, A., M.I. Delday, K.D. Sinclair, J. Steven, and A.A. Sneddon. 2001. Impact of manipulations of myogenesis in utero on the performance of adult skeletal muscle. *Reproduction* **122**:359–374.
116. Mau, M., C. Kalbe, T. Viergutz, G. Nürnberg, and C. Rehfeldt. 2008. Effects of dietary isoflavones on proliferation and DNA integrity of myoblasts derived from newborn piglets. *Pediatr. Res.* **63**:39–45.
117. Mau, M., C. Kalbe, K. Wollenhaupt, and C. Rehfeldt. 2007. Isoflavones modify the growth response of *porcine* satellite cell cultures to IGF-I and EGF. *Arch. Anim. Breed.* (Special Issue) **50**:17–21.
118. McArdle, J., R. Danzeisen, C. Fosset, and L. Gambling. 2003. The role of the placenta in iron transfer from mother to fetus and the relationship between iron status and fetal outcome. *Biometals* **16**:161–167.
119. McCusker, R.H. and D.R. Clemmons. 1988. Insulin-like growth factor binding protein secretion by muscle cells: effect of cellular differentiation and proliferation. *J. Cell. Physiol.* **137**:505–512.
120. McGarry, J.D. and N.F. Brown. 1997. The mitochondrial carnitine palmitoyltransferase system. From concept to molecular analysis. *Eur. J. Biochem.* **244**:1–14.
121. McPherron, A.C. and S.J. Lee. 1997. Double muscling in cattle due to mutations in the myostatin gene. *Proc. Natl. Acad. Sci. USA* **94**:12457–12461.
122. McPherson, R.L., F. Ji, G. Wu, J.R. Blanton, Jr., and S.W. Kim. 2004. Growth and compositional changes of fetal tissues in pigs. *J. Anim. Sci.* **82**:2534–2540.
123. Merkel, R.A., M.E. Doumit, and D.R. Cook. 1994. Regulation of androgen receptors in cultured *porcine* satellite cells and satellite cell-derived myotubes. *J. Anim. Sci.* **72** (Suppl. 1):74.
124. Michael, K., B.S. Ward, and W.M.O. Moore. 1983. Relationship of fetal to placental size: the pig model. *Eur. J. Obstet. Gynecol. Reprod. Biol.* **16**:53–62.
125. Miller, B., E.A. Everitt, T.H. Smith, N.E. Block, and J.A. Dominov. 1993. Cellular and molecular diversity in skeletal muscle development: news from in vitro and in vivo. *Bioessays* **15**:191–196.
126. Milligan, N., D. Fraser, and D.L. Kramer. 2002. Within-litter birth weight variation in the domestic pig and its relation to pre-weaning survival, weight gain, and variation in weaning weights. *Livest. Prod. Sci.* **76**:181–191.
127. Milner, R.D.G. and D.J. Hill. 1984. Fetal growth control: the role of insulin and related peptides. *Clin. Endocrinol.* **21**:415–433.
128. Moss, F.P. and C.P. Leblond. 1971. Satellite cells as the source of nuclei in muscles of growing rats. *Anat. Rec.* **170**:421–435.
129. Mulvaney, D.R., D.N. Marple, and R.A. Merkel. 1988. Proliferation of skeletal muscle satellite cells after castration and administration of testosterone propionate. *Proc. Soc. Exp. Biol. Med.* **188**:40–45.
130. Murani, E., M. Muraniova, S. Ponsuksili, K. Schellander, and K. Wimmers. 2003. Expressed Sequence Tags derived from prenatal development of skeletal muscle in pigs: a source of functional candidate genes for pork quality. Abstracts of the 54th. Annual Meeting of the European Association for Animal Production (EAAP) **9**:247.
131. Murani, E., M. Muraniova, S. Ponsuksili, K. Schellander, and K. Wimmers. 2007. Identification of genes differentially expressed during prenatal development of skeletal muscle in two pig breeds differing in muscularity. *BMC Dev. Biol.* **7**:109.
132. Murani, E., M.F.W. Te Pas, K.C. Chang, R. Davoli, J.W.M. Merks, H. Henne, R. Wörner, H. Eping, S. Ponsuksili, K. Schellander, N. da Costa, D. Prins, B. Harlizius, Egbert Knol, M. Cagnazzo, S. Braglia, and K. Wimmers 2005. Analysis of effects of genes differentially expressed during myogenesis on pork quality. Abstracts of the 56th. Annual Meeting of the European Association for Animal Production (EAAP) **11**:370.
133. Musser, E., D.L. Davis, S.S. Dritz, M.D. Tokach, J.L. Nelssen, J.E. Minton, and R.D. Goodband. 2004. Conceptus and maternal responses to increased feed intake during early gestation in pigs. *J. Anim. Sci.* **82**:3154–3161.

134. Musser, E., R.D. Goodband, K.Q. Owen, D.L. Davis, M.D. Tokach, S.S. Dritz, and J.L. Nelssen. 2001. Determining the effect of increasing L-carnitine additions on sow performance and muscle fiber development of the offspring. *J. Anim. Sci.* (Suppl. 2) **79**:65.
135. Musser, E., R.D. Goodband, M.D. Tokach, K.Q. Owen, J.L. Nelssen, S.A. Blum, S.S. Dritz, and C.A. Civis. 1999. Effects of L-carnitine fed during gestation and lactation on sow and litter performance. *J. Anim. Sci.* **77**:3289–3295.
136. Nii, M., T. Hayashi, S. Mikawa, F. Tani, A. Niki, N. Mori, Y. Uchida, N. Fujishima-Kanaya, M. Komatsu, and T. Awata. 2005. Quantitative trait loci mapping for meat quality and muscle fiber traits in a Japanese wild boar x Large White intercross. *J. Anim. Sci.* **83**:308–315.
137. Nissen, P.M., P.F. Jorgensen, and N. Oksbjerg. 2004. Within-litter variation in muscle fiber characteristics, pig performance, and meat quality traits. *J. Anim. Sci.* **82**:414–421.
138. Nissen, P.M., I.L. Sorensen, M. Vestergaard, and N. Oksbjerg. 2005. Effects of sow nutrition on maternal and fetal serum growth factors and on fetal myogenesis. *Anim. Sci.* **80**:299–306.
139. Nutting, D.F. 1976. Ontogeny of sensitivity to growth hormone in rat diaphragm muscle. *Endocrinology* **98**:1273–1283.
140. O'Connor, L., M.F. Picciano, M.A. Roos, and R.A. Easter. 1989. Iron and folate utilization in reproducing swine and their progeny. *J. Nutr.* **119**:1984–1991.
141. Oksbjerg, N., F. Gondret, and M. Vestergaard. 2004. Basic principles of muscle development and growth in meat-producing mammals as affected by the insulin-like growth factor (IGF) system. *Domest. Anim. Endocrinol.* **27**:219–240.
142. Olson, E.N., T.J. Brennan, T. Chakraborty, T.C. Cheng, P. Cserjesi, D. Edmondson, G. James, and L. Li. 1991. Molecular control of myogenesis: antagonism between growth and differentiation. *Mol. Cell. Biochem.* **104**:7–13.
143. Owen, K.Q., H. Jit, C.V. Maxwell, J.L. Nelssen, R.D. Goodband, M.D. Tokach, G.C. Tremblay, and S.I. Koo. 2001. Dietary L-carnitine suppresses mitochondrial branched-chain keto acid dehydrogenase activity and enhances protein accretion and carcass characteristics of swine. *J. Anim. Sci.* **79**:3104–3112.
144. Pampusch, M.S., J.R. Hembree, M.R. Hathaway, and W.R. Dayton. 1990. Effect of transforming growth factor beta on proliferation of L6 and embryonic *porcine* myogenic cells. *J. Cell Physiol.* **143**:524–528.
145. Pampusch, M.S., E. Kamanga-Sollo, M.E. White, M.R. Hathaway, and W.R. Dayton. 2003. Effect of recombinant *porcine* IGF-binding protein-3 on proliferation of embryonic *porcine* myogenic cell cultures in the presence and absence of IGF-I. *J. Endocrinol.* **176**:227–235.
146. Pampusch, M.S., G. Xi, E. Kamanga-Sollo, K.J. Loseth, M.R. Hathaway, W.R. Dayton, and M.E. White. 2005. Production of recombinant *porcine* IGF-binding protein-5 and its effect on proliferation of *porcine* embryonic myoblast cultures in the presence and absence of IGF-I and Long-R3-IGF-I. *J. Endocrinol.* **185**:197–206.
147. Parker, M.H., P. Seale, and M.A. Rudnicki. 2003. Looking back to the embryo: defining transcriptional networks in adult myogenesis. *Nat. Rev. Genet.* **4**:497–507.
148. Peng, M., M.F. Palin, S. Veronneau, D. LeBel, and G. Pelletier. 1997. Ontogeny of epidermal growth factor (EGF), EGF receptor (EGFR) and basic fibroblast growth factor (bFGF) mRNA levels in pancreas, liver, kidney, and skeletal muscle of pig. *Domest. Anim. Endocrinol.* **14**:286–294.
149. Peng, M., G. Pelletier, M.F. Palin, S. Veronneau, D. LeBel, and T. Abribat. 1998. Ontogeny of IGFs and IGFBPs mRNA levels and tissue concentrations in liver, kidney and skeletal muscle of pig. *Growth Dev. Aging* **60**:171–187.
150. Père, M.C. 2003. Materno-fetal exchanges and utilisation of nutrients by the fetus: comparison between species. *Reprod. Nutr. Dev.* **43**:1–15.
151. Père, M.C., J.Y. Dourmad, and M. Etienne. 1997. Effect of number of pig embryos in the uterus on their survival and development and on maternal metabolism. *J. Anim. Sci.* **75**:1337–1342.
152. Père, M.C., M. Etienne, and J.Y. Dourmad. 2000. Adaptations of glucose metabolism in multiparous sows: effects of pregnancy and feeding level. *J. Anim. Sci.* **78**:2933–2941.

153. Picard, B., L. Lefaucheur, C. Berri, and M.J. Duclos. 2002. Muscle fibre ontogenesis in farm animal species. *Reprod. Nutr. Dev.* **42**:415–431.
154. Pond, W.G., R.R. Maurer, and J. Klindt. 1991. Fetal organ response to maternal protein deprivation during pregnancy in swine. *J. Nutr.* **121**:504–509.
155. Pond, W.G., J.T. Yen, and L.H. Yen. 1986. Response of nonpregnant versus pregnant gilts and their fetuses to severe feed restriction. *J. Anim. Sci.* **63**:472–483.
156. Pond, W.G., J.T. Yen, H.J. Mersmann, and R.R. Maurer. 1990. Reduced mature size in progeny of swine severely restricted in protein intake during pregnancy. *Growth Dev. Aging* **54**:77–84.
157. Powell, S.E. and E.D. Aberle. 1981. Skeletal muscle and adipose tissue cellularity in runt and normal birth weight swine. *J. Anim. Sci.* **52**:748–756.
158. Ramanau, A., H. Kluge, J. Spilke, and K. Eder. 2004. Supplementation of sows with L-carnitine during pregnancy and lactation improves growth of the piglets during the suckling period through increased milk production. *J. Nutr.* **134**:86–92.
159. Ramanau, A., R. Schmidt, H. Kluge, and K. Eder. 2006. Body composition, muscle fibre characteristics and postnatal growth capacity of pigs born from sows supplemented with L-carnitine. *Arch. Anim. Nutr.* **60**:110–118.
160. Randall, G.C.B. 1989. Effect of hypophysectomy on body and organ weights and subsequent development in the fetal pig. *Can. J. Anim. Sci.* **69**:655–661.
161. Rehfeldt, C., I. Fiedler, G. Dietl, and K. Ender. 2000. Myogenesis and postnatal skeletal muscle cell growth as influenced by selection. *Livest. Prod. Sci.* **66**:177–188.
162. Rehfeldt, C., I. Fiedler, and N.C. Stickland. 2004. Number and Size of Muscle Fibres in Relation to Meat Production. pp. 1–37. *In* M.F.W. Te Pas, M.E. Everts, and H.P. Haagsman, (eds), Muscle Development of Livestock Animals: Physiology, Genetics, and Meat Quality, CAB Int., Wallingford, Oxon, UK.
163. Rehfeldt, C., I. Fiedler, R. Weikard, E. Kanitz, and K. Ender. 1993. It is possible to increase skeletal muscle fibre number in utero. *Biosci. Rep.* **13**:213–220.
164. Rehfeldt, C. and G. Kuhn. 2006. Consequences of birth weight for postnatal growth performance and carcass quality in pigs as related to myogenesis. *J. Anim. Sci.* (E-Suppl). **84**:E113–E123.
165. Rehfeldt, C., G. Kuhn, G. Nürnberg, E. Kanitz, F. Schneider, M. Beyer, K. Nürnberg, and K. Ender. 2001. Effects of exogenous somatotropin during early gestation on maternal performance, fetal growth, and compositional traits in pigs. *J. Anim. Sci.* **79**:1789–1799.
166. Rehfeldt, C., G. Kuhn, J. Vanselow, R. Fürbass, I. Fiedler, G. Nürnberg, A.K. Clelland, N.C. Stickland, and K. Ender. 2001. Maternal treatment with somatotropin during early gestation affects basic events of myogenesis in pigs. *Cell Tiss. Res.* **306**:429–440.
167. Rehfeldt,C., P.M. Nissen, G. Kuhn, M. Vestergaard, K. Ender, and N. Oksbjerg. 2004. Effects of maternal nutrition and *porcine* growth hormone (pGH) treatment during gestation on endocrine and metabolic factors in sows, fetuses and pigs, skeletal muscle development, and postnatal growth. *Domest. Anim. Endocrinol.* **27**:267–285.
168. Rehfeldt, C., A. Tuchscherer, M. Hartung, and G. Kuhn. 2007. A second look at the influence of birth weight on carcass and meat quality in pigs. *Meat Sci.* **78**:170–175.
169. Robinson, D.W. 1969. The cellular response of *porcine* skeletal muscle to prenatal and neonatal nutritional stress. *Growth* **33**:231–240.
170. Robinson, J.J., K.D. Sinclair, and T.G. McEvoy. 1999. Nutritional effects on fetal growth. *Anim. Sci.* **68**:315–331.
171. Roe, J.A., A.S. Baba, J.M. Harper, and P.J. Buttery. 1995. Effects of growth factors and gut regulatory peptides on nutrient uptake in ovine muscle cell cultures. *Comp. Biochem. Physiol. A Physiol.* **110**:107–114.
172. Schiaffino, S. and C. Reggiani. 1996. Molecular diversity of myofibrillar proteins: gene regulation and functional significance. *Physiol Rev* **76**:371–423.
173. Schneider, F., E. Kanitz, D.E. Gerrard, G. Kuhn, K.P. Brüssow, K. Nürnberg, I. Fiedler, G. Nürnberg, K. Ender, and C. Rehfeldt. 2002. Administration of recombinant *porcine*

somatotropin (rpST) changes hormone and metabolic status during early pregnancy. *Domest. Anim. Endocrinol.* **23**:455–474.

174. Schoknecht, P.A., G.R. Newton, D.E. Weise, and W.G. Pond. 1994. Protein restriction in early-pregnancy alters fetal and placental growth and allantoic fluid proteins in swine. *Theriogenology* **42**:217–226.
175. Schoknecht, P.A., W.G. Pond, H.J. Mersmann, and R.R. Maurer. 1993. Protein restriction during pregnancy affects postnatal growth in swine progeny. *J. Nutr.* **123**:1818–1825.
176. Schultz, E. 1974. A quantitative study of the satellite cell population in postnatal mouse lumbrical muscle. *Anat. Rec.* **180**:589–595.
177. Simmen, R.C., F.A. Simmen, A. Hofig, S.J. Farmer, and F.W. Bazer. 1990. Hormonal regulation of insulin-like growth factor gene expression in pig uterus. *Endocrinology* **127**: 2166–2174.
178. Solanes, F.X., K. Grandinson, L. Rydhmer, S. Stern, K. Andersson, and N. Lundeheim. 2004. Direct and maternal influences on the early growth, fattening performance, and carcass traits of pigs. *Livest. Prod. Sci.* **88**:199–212.
179. Song, S., C.Y. Lee, M.L. Green, C.S. Chung, R.C. Simmen, and F.A. Simmen. 1996. The unique endometrial expression and genomic organization of the *porcine* IGFBP-2 gene. *Mol. Cell Endocrinol.* **120**:193–202.
180. Spencer, T.E. and F.W. Bazer. 2004. Uterine and placental factors regulating conceptus growth in domestic animals. *J. Anim. Sci.* **82** (E-Suppl):E4–E13.
181. Staun, H. 1972. The nutritional and genetic influence on number and size of muscle fibres and their response to carcass quality in pigs. *World Rev. Anim. Prod.* **3:**18–26.
182. Sterle, J.A., C. Boyd, J.T. Peacock, A.T. Koenigsfeld, W.R. Lamberson, D.E. Gerrard, and M.C. Lucy. 1998. Insulin-like growth factor (IGF)-I, IGF-II, IGF-binding protein- 2 and pregnancy-associated glycoprotein mRNA in pigs with somatotropin-enhanced fetal growth. *J. Endocrinol.* **159**:441–450.
183. Sterle, J.A., T.C. Cantley, W.R. Lamberson, M.C. Lucy, D.E. Gerrard, R.L. Matteri, and B.N. Day. 1995. Effects of recombinant *porcine* somatotropin on placental size, fetal growth, and IGF-I and IGF-II concentrations in pigs. *J. Anim. Sci.* **73**:2980–2985.
184. Stickland, N.C., S. Bayol, C. Ashton, and C. Rehfeldt. 2004. Manipulation of Muscle Fibre Number During Prenatal Development, pp. 69–82. *In* M.F.W. Te Pas, M.E. Everts, and H.P. Haagsman, (eds), Muscle Development of Livestock Animals: Physiology, Genetics, and Meat Quality, CAB Int., Wallingford, Oxon, UK.
185. Stickland, N.C. and S.E. Handel. 1986. The numbers and types of muscle fibres in large and small breeds of pigs. *J. Anat.* **147**:181–189.
186. Stratil, A. and S. Cepica. 1999. Three polymorphisms in the *porcine* myogenic factor 5 (MYF5) gene detected by PCR-RFLP. *Anim. Genet.***30**:79–80.
187. Straus, D.S. 1994. Nutritional regulation of hormones and growth factors that control mammalian growth. *FASEB J.* **8**:6–12.
188. Stryker, J.L. and P.J. Dziuk. 1975. Effects of fetal decapitation on fetal development, parturition and lactation in pigs. *J. Anim. Sci.* **40**:282–287.
189. Szabo, G., G. Dallmann, G. Muller, L. Patthy, M. Soller, and L. Varga. 1998. A deletion in the myostatin gene causes the compact (Cmpt) hypermuscular mutation in mice. *Mamm. Genome.* **9**:671–672.
190. Taylor-Roth, J.L., P.V. Malven, D.E. Gerrard, S.E. Mills, and A.L. Grant. 1998. Independent effects of food intake and insulin status on insulin-like growth factor-I in young pigs. *Comp. Biochem. Physiol. C. Pharmacol. Toxicol. Endocrinol.* **120**:357–363.
191. Te Pas, M.F.W.; M. Cagnazzo, A.A.C. de Wit, J. Priem, M. Pool, and R. Davoli. 2005. Muscle transcriptomes of Duroc and Pietrain pig breeds during prenatal formation of skeletal muscle tissue using microarray technology. *Arch. Anim. Breed.* (Special Issue) **48**:141–147.
192. Te Pas, M.F., F.L. Harders, A. Soumillion, L. Born, W. Buist, and T.H. Meuwissen 1999. Genetic variation at the *porcine* MYF-5 gene locus. Lack of association with meat production traits. *Mamm. Genome* **10**:123–127.

193. Te Pas, M.F.W.; M.H. Pool, I. Hulsegge, and L.L.G. Janss. 2006. Analysis of the differential transcriptome expression profiles during prenatal muscle tissue development in pigs *Arch. Anim. Breed.* **49** (Special Issue):110–115.
194. Te Pas, M.F.W. and A. Soumillion. 2001. Improvement of livestock breeding strategies using physiologic and functional genomic information of the muscle regulatory factors gene family for skeletal muscle development. *Curr. Genomics*. **2**:285–304.
195. Te Pas, M.F., A. Soumillion, F.L. Harders, F.J. Verburg, T.J. van den Bosch, P. Galesloot, and T.H. Meuwissen. 1999. Influences of myogenin genotypes on birth weight, growth rate, carcass weight, backfat thickness, and lean weight of pigs. *J. Anim. Sci.* **77**:2352–2356.
196. Te Pas, M.F., A.A.C. de Wit, J. Priem, M. Cagnazzo, R. Davoli, V. Russo, and M.H. Pool. 2005. Transcriptome expression profiles in prenatal pigs in relation to myogenesis. *J. Muscle. Res. Cell. Motil.* **26**:157–165.
197. Thissen, J.P., J.M. Ketelslegers, and L.E. Underwood. 1994. Nutritional regulation of the insulin-like growth factors. *Endocr. Rev.* **15**:80–101.
198. Town, S.C., J.L. Patterson, C.Z. Pereira, G. Gourley, and G.R. Foxcroft. 2005. Embryonic and fetal development in a commercial dam-line genotype. *Anim. Reprod. Sci.* **85**:301–316.
199. Urbanski, P., K. Flisikowski, R.R. Starzynski, J. Kuryl, and M. Kamyczek. 2006. A new SNP in the promoter region of the *porcine* MYF5 gene has no effect on its transcript level in m. longissimus dorsi. *J. Appl. Genet.* **47**:59–61.
200. Urbanski, P. and J. Kuryl. 2004. New SNPs in the coding and 5′ flanking regions of *porcine* MYOD1 (MYF3) and MYF5 genes. *J. Appl. Genet.* **45**:325–329.
201. Vyboh, P., D. Lamosova, M. Vanekova, and M. Jurani. 1994. Effects of thyroid hormones on chick embryo muscle cell culture. *Comp. Biochem. Physiol. C Pharmacol. Toxicol. Endocrinol.* **109**:269–276.
202. Wang, J., C.Y. Deng, Y.Z. Xiong, B. Zuo, L. Xing, F.E. Li, M.G. Lei, R. Zheng, and S.W. Jiang. 2005. cDNA cloning, sequence analysis of the *porcine* LIM and cysteine-rich domain gene. *Acta Biochim. Biophys. Sin. (Shanghai)* **37**:843–850.
203. Waylan, A.T., J.P. Kayser, D.P. Gnad, J.J. Higgins, J.D. Starkey, E.K. Sissom, J.C. Woodworth, and B.J. Johnson. 2005. Effects of L-carnitine on fetal growth and the IGF system in pigs. *J. Anim. Sci.* **83**:1824–1831.
204. Wesolowski, S.R., N.E. Raney, and C.W. Ernst. 2004. Developmental changes in the fetal pig transcriptome. *Physiol. Genomics* **16**:268–274.
205. Wicke, M., G. v. Lengerken, I. Fiedler, M. Altmann, and K. Ender. 1991. Einfluß der Selektion nach Merkmalen der Muskelstruktur des M. longissimus auf Belastungsempfindlichkeit und Fleischbeschaffenheit beim Schwein. *Fleischwirtsch.* **71**:437–442.
206. Wigmore, P.M. and D.J. Evans. 2002. Molecular and cellular mechanisms involved in the generation of fiber diversity during myogenesis. *Int. Rev. Cytol.* **216**:175–232.
207. Wigmore, P.M. and N.C. Stickland. 1983. Muscle development in large and small pig fetuses. *J. Anat.* **137**:235–245.
208. Wigmore, P.M. and N.C. Stickland. 1983. DNA, RNA and protein in skeletal muscle of large and small pig fetuses. *Growth* **47**:67–76.
209. Wimmers, K., I. Fiedler, T. Hardge, E. Murani, K. Schellander, and S. Ponsuksili. 2006. QTL for microstructural and biophysical muscle properties and body composition in pigs. *BMC Genet.* **7**:15.
210. Wimmers, K., E. Murani, K. Schellander, and S. Ponsuksili. 2005. Combining QTL- and expression-analysis: Identification of functional positional candidate genes for meat quality and carcass traits. *Arch. Anim. Breed.* **48** (Special Issue):23–31.
211. Wimmers, K., E. Murani, M.F.W. Te Pas, K.C. Chang, R. Davoli, J.W.M. Merks, H. Henne, M. Muraniova, N. Da Costa, B. Harlizius, K. Schellander, I. Böll, S. Braglia, A.A.C. de Wit, M. Cagnazzo, L. Fontanesi, D. Prins, and S. Ponsuksili. 2007. Associations of functional candidate genes derived from gene expression profiles of prenatal *porcine* muscle tissue with meat quality and carcass traits. *Anim. Genet.* **38**:474–484.

212. Wimmers, K., N.T. Ngu, D.G.J. Jennen, D. Tesfaye, E. Murani, K. Schellander, and S. Ponsuksili 2008. Relationship between myosin heavy chain isoform expression and muscling in several diverse pig breeds. *J. Anim. Sci.* **86**:795–803.
213. Wimmers, K., N.T. Ngu, E. Murani, K. Schellander, and S. Ponsuksili 2006. Linkage and expression analysis to elucidate the genetic background of muscle structure and meat quality in the pig. *Arch. Anim. Breed.* **49** (Special Issue):116–125.
214. Wimmers, K., M.F.W. Te Pas, K.C. Chang, R. Davoli, J.W.M. Merks, R. Wörner, H. Eping, E. Murani, S. Ponsuksili, K. Schellander, J.M. Priem, M. Cagnazzo, L. Fontanesi, B. Lama, B. Harlizius, and H. Henne. 2002. A European initiative towards identification of genes controlling pork quality. Proc. 7th WCGALP:11–13.
215. Wootton, R., I.R. McFayden, and J.E. Cooper. 1977. Measurement of placental blood flow int the pig and its relation to placental and fetal weight. *Biol. Neonate* **31**:333–339.
216. Wu, G.Y., W.G. Pond, S.P. Flynn, T.L. Ott, and F.W. Bazer. 1998. Maternal dietary protein deficiency decreases nitric oxide synthase and ornithine decarboxylase activities in placenta and endometrium of pigs during early gestation. *J. Nutr.* **128**:2395–2402.
217. Wu, G.Y., W.G. Pond, T.L. Ott, and F.W. Bazer. 1998. Maternal dietary protein deficiency decreases amino acid concentrations in fetal plasma and allantoic fluid of pigs. *J. Nutr.* **128**: 894–902.
218. Wyszynska-Koko, J. and J. Kuryl. 2004. Porcine MYF6 gene: sequence, homology analysis, and variation in the promoter region. *Anim. Biotechnol.* **15**:159–73.
219. Wyszynska-Koko, J., M. Pierzchala, K. Flisikowski, M. Kamyczek, M. Rozycki, and J. Kuryl. 2006. Polymorphisms in coding and regulatory regions of the *porcine* MYF6 and MYOG genes and expression of the MYF6 gene in m. longissimus dorsi versus productive traits in pigs. *J. Appl. Genet.* **47**:131–138.
220. Yao, J., P.M. Coussens, P. Saama, S. Suchyta, and C.W. Ernst. 2002. Generation of expressed sequence tags from a normalized *porcine* skeletal muscle cDNA library. *Anim. Biotechnol.* **13**:211–222.
221. Yi, Z., M.R. Hathaway, W.R. Dayton, and M.E. White. 2001. Effects of growth factors on insulin-like growth factor binding protein (IGFBP) secretion by primary *porcine* satellite cell cultures. *J. Anim. Sci.* **79**:2820–2826.
222. Young, L.G., G.J. King, J.S. Walton, I. McMillan, M. Klevorick, and J. Shaw. 1990. Gestation energy and reproduction in sows over 4 parities. *Can. J. Anim. Sci.* **70**:493–506.
223. Zhao, S.H., D. Nettleton, W. Liu, C. Fitzsimmons, C.W. Ernst, N.E. Raney, and C.K. Tuggle. 2003. Complementary DNA macroarray analyses of differential gene expression in *porcine* fetal and postnatal muscle. *J. Anim. Sci.* **81**:2179–88.

Part III
Regulators of Fetal and Neonatal Nutrient Supply

Chapter 8
Placental Vascularity: A Story of Survival

Stephen P. Ford

Introduction

In this review, I have been asked to provide insight into the structure and related function of the chorioallantoic placentae of different livestock species, with emphasis on their role in mediating conceptus growth and survival. As there have been many excellent books and reviews written on comparative placentation, I have decided to focus this paper on a comparison of two livestock species with which I have a long research history, namely the pig and the sheep, but will present comparisons with other species where appropriate and warranted. Further, I will stress the importance of placental vascularity and blood flow in dictating normal growth, development and survival of the fetus. Throughout my research career, I have conducted studies aimed at gaining a better understanding of conceptus-uterine interactions, in an attempt to understand how intrauterine and extrauterine events impact the conceptus. Significant data have accumulated suggesting that negative impacts on the fetus, occurring during critical periods of gestation, result in developmental adaptations that can permanently change the growth, physiology and metabolism of offspring [3, 28, 41, 49].

Regardless of species, the placenta is a highly specialised organ whose role is to provide for physiological exchange between mother and fetus in the support of normal fetal growth and development [50, 72]. The importance of the utero-placental vasculature in supporting normal pregnancy is emphasised by the close relationships among fetal weight, placental weight, as well as uterine and placental blood flows across many mammalian species [1, 73, 74]. Further, uterine and umbilical flows increase exponentially throughout gestation, essentially keeping pace with fetal growth and development [45, 51, 72]. On the basis of a number of studies, it appears that increased blood flow is a primary determinant of transplacental nutrient

S.P. Ford (✉)
Center for the Study of Fetal Programming and Department of Animal Science, University of Wyoming, Laramie, WY, USA
e-mail: spford@uwyo.edu

P.L. Greenwood et al. (eds.), *Managing the Prenatal Environment to Enhance Livestock Productivity*, DOI 10.1007/978-90-481-3135-8_8,

and waste product exchange throughout gestation in livestock species [18, 20, 50, 51, 69, 72–74]. With these concepts in mind, I will now discuss placental growth and development, first in the pig and then in the sheep, with emphasis on the impact of changes in placental vascular development and blood flow on fetal growth, development and survival.

8.1 Pig Placental Growth and Development

In the pig, the placenta is termed epitheliochorial and adeciduate, is minimally invasive, and thus the uterine luminal epithelium remains intact with no marked alteration in the morphology of the maternal tissue throughout gestation [4, 30, 44, 58, 59]. In this species the chorionic epithelium, the outermost tissue layer of the fetal chorioallantoic membrane, is in direct contact with the uterine luminal epithelium [52, 64], and nutrient transfer occurs over the entire surface of the chorionic sac. The placenta is defined as diffuse, and the allantoic capillaries responsible for the absorption of nutrients and oxygen and the expulsion of waste products including carbon dioxide ramify throughout the entire surface of the placenta excluding the necrotic tips [7, 44, 59]. The epitheliochorial placenta of the pig is comprised of six tissue layers that separate the maternal and fetal blood streams: (1) endometrial capillary endothelium, (2) connective tissues, (3) uterine luminal epithelium, (4) chorionic epithelium, (5) connective tissues and (6) placental capillary endothelium [29]. To facilitate maternal-fetal nutrient exchange, the surface area of the chorioallantoic membrane increases rapidly from day 35 to day 70 of gestation in the pig [37]. From day 70 to 100, there is little change in placental surface area [37], however, sometime after day 100 there is a further marked and variable increase in the weight and surface area of the pig placenta [61, 5, 91]. Additionally, beginning on day 35–40, there is folding of the chorionic membrane into the adjacent permanent folds of the endometrium through macroscopic folds called plicae, of different orders, and by microscopic folds or ridges called rugae, also of different orders [7, 12, 26, 43]. This degree of interdigitation of the chorion and associated uterine luminal surface increases progressively with the advancement of gestation, markedly increasing the area of fetal-maternal exchange. By days 60–70 of gestation, the chorionic and uterine luminal epithelial cells are provided with interdigitating microvilli that further increase the area of exchange by a factor of 10 [7]. Finally, by the end of gestation the effective placental barrier separating the fetal and maternal bloodstreams has decreased from approximately 20 μm to 2 μm or less, which further facilitates nutrient flux [26, 27].

In addition to the progressively increasing surface area of contact between the placenta and the uterine wall, the number, density, orientation and distance separating chorioallantoic and adjacent uterine endometrial capillaries, is critical in determining the quantity of nutrients conveyed to the fetal-maternal interface and thus to the rapidly growing fetus [5, 26, 43, 72]. By midgestation in the pig, placental vascularity is relatively constant, averaging between 3 and 4% of the total volume of the chorioallantoic membrane, but by the end of gestation, placental size and vascularity differ markedly between individual conceptuses both within and between litters of U.S. commercial pig breeds and are highly negatively associated [5, 94].

8.1.1 Periods of Conceptus Loss in the Pig

There are three main waves of conceptus mortality that occur during gestation in U.S. commercial pig breeds [22]. The first wave of conceptus loss is thought to be a result of littermate conceptus asynchrony, and occurs between days 11 and 18 of gestation when a relatively constant 30% loss occurs [10, 14, 15]. Further, it appears that diversity in development within a litter is the culprit, as the least developed littermate conceptuses are the ones lost [62, 63, 92]. Data suggest that conceptus losses after day 18 are significant, and that variation in placental size and vascularity may contribute to these losses [94–96]. Conceptus losses are particularly critical in U.S. commercial crossbred sows, in which ovulation rates exceeding 26 oocytes have been reported [82]. Consistent with these observations, the numbers of conceptuses surviving to day 25 far exceeds the number of piglets farrowed [25, 84, 96].

The second wave of conceptus mortality occurs between day 30 and day 40 of gestation when the surface area of the chorioallantoic membrane of each conceptus begins to expand rapidly, forcing the conceptuses to compete for limited uterine space [14, 17, 33]. This limitation of uterine space for implantation is referred to as uterine capacity, and is generally considered to be the major limitation to increasing litter size in the pig. During this interval, an additional 10–15% of the remaining conceptuses are lost. The final wave of conceptus loss occurs between day 90 and farrowing, when there is potential for an additional 5–10% conceptus loss [10, 35]. This loss results from the markedly increasing nutrient requirements of the exponentially growing pig fetus, and is associated with as much as a doubling of placental surface area, leading to an additional period of littermate competition [5, 91].

8.1.2 Use of the Chinese Meishan Pig in Understanding the Limitations to Litter Size

To gain a better understanding of the controls of litter size, our group compared the prolific Chinese Meishan pig with the American Yorkshire pig. In a series of studies we determined that these two breeds had the same uterine size and ovulation rate, but the Meishans farrowed 3–4 more piglets/litter than Yorkshire females [19, 99]. These studies also demonstrated that the Meishan conceptus exhibited reduced growth rates and less littermate asynchrony during early gestation when compared to Yorkshire conceptuses [21, 75, 93, 99, 100], and possessed markedly smaller and more vascular placenta in late gestation [5, 21, 94]. In addition, the increased placental vascularity of the Meishan versus that of the Yorkshire breed was associated with an elevated placental expression of the potent angiogenic factor vascular endothelial growth factor and its receptors [86]. To estimate placental efficiency (**PE**) across these two breeds, we divided piglet weight by the weight of its placenta at farrowing, which provides an estimate of the grams of piglet that could be supported per gram of placenta. Interestingly, the PE of the Meishan breed was over twice as great as that of the Yorkshire breed, averaging 8.7 ± 0.4 versus 3.4 ± 0.8, respectively [21].

8.1.3 Increasing Litter Size in an American Pig Breed

It was felt that because the Chinese had been selecting the Meishan breed for their prolificacy for thousands of years [48, 60], this breed might have been selected indirectly for smaller, more vascular placentae. To test this hypothesis, Wilson et al. [95] determined within and between litter PE for a group of Yorkshire females. The PE across all females in this study averaged 4.2, which was similar to that reported previously for Yorkshire females [21]. More importantly, Wilson et al. [95] reported that the variation in PE for individual conceptuses across all Yorkshire litters ranged from 2.7 to 7.4, and the PE within a single litter on that study ranged from 3.8 to 7.4. Vonnahme et al. [84] used a large group ($n = 190$ litters) of commercial sows (Camborough line 6–02, PIC) to study ovulation rate and PE in relation to litter size. These researchers reported that ovulation rate was not limiting to litter size, and that the largest litters present on days 25, 36 and 44 of gestation in this study were composed of fetuses of similar size to those of smaller litters, but with significantly smaller placentae. It was determined in this study [84] and in other studies [6, 95, 96] that PE was not determined by changes in fetal weight (prenatally) or piglet weight (postnatally), but was dependent on changes in placental weight. In fact, Vonnahme et al. [84] determined that the negative relationship between PE and placental weight occurred as early as day 25 of gestation, a time point prior to when the limitations of uterine capacity are exerted [14, 17, 33]. In support of this concept, Leenhouwers [42] reported that an increase in estimated breeding values for piglet survival of 507 piglets from 46 litters (crossbred line D12, Netherlands) was significantly associated with a decrease in average placental weight and within-litter variation in placental weight, but not piglet weight, with an increase in average PE.

It was speculated that selection of Yorkshire boars and gilts for differences in PE might alter litter size due to associated impacts in placental growth and vascularity of littermate conceptuses. With this hypothesis in mind, Wilson et al. [95] selected Yorkshire boars and gilts with above average and below average PE for study. It is important to note that these researchers held piglet weight constant between the above average and below average PE groups. In contrast, the placentae were approximately 30% lighter for piglets selected in the high PE group than for the low PE group and much more vascular. When boars and gilts from the high PE group were bred and allowed to farrow, they produced litters of greater than three more live piglets than those resulting from matings of low PE boars and gilts (12.8 ± 0.7 versus 9.5 ± 0.5 piglets, respectively). Further, after four generations of selection on PE, the litter size differences remained, as did differences in PE and vascularity [83]. Perhaps more importantly, while piglet weight was only reduced about 20% in the high PE group when compared to the low PE group, placental weight was reduced by approximately 40%. Further, the heritability of PE has been estimated at between 0.29 and 0.37% [80, 95] which is the highest heritability of any reproductive trait yet determined in the pig. These data suggest that selection of individual piglets in a litter on the basis of smaller and more efficient placentae may provide a useful method for rapidly increasing litter size in the pig.

8.2 Sheep Placental Growth and Development

Except for differences in time and minor structural features, the development of the fetal membranes of domesticated ruminants is similar. In ruminants, including the cow, goat and sheep, the placenta is often called epitheliochorial, but can be further categorised as a subset of this placental type called syndesmochorial to indicate that the chorionic binucleate cells are fused with the uterine luminal epithelium to form a feto-maternal syncytium at sites called placentomes [97]. The placentomes are the sites of fetal-maternal nutrient and waste product exchange in ruminant species. The placenta is defined as cotyledonary and is characterised by separate tufts of chorionic villi called cotyledons, which are scattered widely over the surface of the chorion. These localised dense areas of villi develop only on those parts of the chorion that are adjacent to preexisting aglandular proliferations of connective tissue on the uterine luminal mucosa called caruncles. There are between 75 and 120 caruncles in the cow and between 80 and 100 in the sheep [13, 53]. These caruncles are arranged in two dorsal and two ventral rows throughout the length of the uterine horns. When contacted by fetal membranes, they enlarge to form swellings with a convex surface in cattle, and a concave surface in sheep. Each cotyledon develops highly branched villi which interdigitate with corresponding crypts that develop in the proliferating caruncle. As previously stated, ruminant species exhibit a syndesmochorial placental type at the caruncular-cotyledonary interface, with only five layers separating the fetal and maternal blood streams.

In the ewe major placentomal growth is complete by mid-gestation, before the major period of fetal growth begins [2, 16, 73, 74, 88]. However, individual placentomes may continue to undergo morphologic and functional transformations as fetal demand for nutrients increases in the second half of gestation [32, 57]. A morphological classification system was developed for placentomal differentiation in the sheep by Vatnick et al. [81], and reflects distinct differences in placentome appearance as follows: (1) caruncular tissue completely surrounding the cotyledonary tissue (Type A), (2) cotyledonary tissue beginning to grow over the surrounding caruncular tissue (Type B), (3) flat placentomes with caruncular tissue on one surface and cotyledonary tissue on the other (Type C), and (4) everted placentomes resembling bovine placentomes (Type D). With the advancement of gestation, individual placentomes may or may not progress from Type A to Type B, C or D, as the nutrient requirements of the fetus progressively increase.

In the ewe, vascular growth of placentomal tissues has been reported to increase throughout gestation. Stegeman [78] reported that vascular density of caruncular tissues increases substantially from day 40 through mid-gestation, then more slowly thereafter. In contrast, the vascular density of the cotyledonary bed remains relatively constant until mid-gestation, then increases exponentially thereafter [2, 79]. These data are consistent with the observation that umbilical blood flow increases more rapidly than uterine blood flow during the last half of gestation in the ewe [70, 76, 77]. Both vascular growth and the relaxation and dilation of existing vessels are considered necessary to ensure adequate placental blood flow to support fetal growth [20, 71].

Of the overall increase in utero-placental blood flow to the gravid ovine uterus throughout gestation, an increasing proportion of the total uterine and umbilical blood flows is directed toward the caruncular and cotyledonary vascular beds with advancing gestation [46, 47, 50, 76]. This may help to explain the fact that while placental growth slows at mid-gestation, placental nutrient transport capacity keeps pace with fetal growth throughout gestation in a variety of species including the sheep [31, 34, 90]. As previously mentioned, it seems that increased blood flow, rather than increased extraction, is the primary mechanism of increased transplacental exchange throughout gestation in ruminant species [18, 50, 51, 69].

The idea that the primary role of blood flow is determining fetal nutrient delivery in the sheep is supported by the fact that uterine and/or umbilical blood flows have been shown to decrease in virtually every case where fetal growth has been compromised [9, 39, 68, 89]. Further, the reduction in uterine blood flow and decreased fetal growth is associated with an initial decrease in the expression of placental angiogenic and vasodilatory factors, including vascular endothelial growth factor and endothelial nitric oxide synthase [66, 67], and a later alteration of placental vascular architecture [65, 67]. Thus, angiogenic and vasodilatory factors that stimulate increased caruncular and/or cotyledonary vascularity and blood flow undoubtedly have a tremendous impact on fetal growth and development as well as neonatal growth and survival.

8.2.1 Divergent Ewe Selection, Placentomal Differentiation and Fetal Growth

Results from several laboratories demonstrate an increased conversion of Type A placentomes to Types B, C, or D in undernourished ewes when compared to their well-fed counterparts, however, this conversion has been thought to occur only in late gestation in association with an exponentially growing fetus [32, 56, 57]. Osgerby et al. [57] reported that the conversion of Type A placentomes to more advanced placentomal types was associated with a flattening of the placentome and an increased ratio in the area of interdigitated maternal and fetal villi to unattached fetal allantochorion. Further, these researchers speculated that this increase in surface area of association may enhance nutrient delivery to the fetus. Hoet and Hansen [32] suggested that the advancement in placentomal morphologic type may also be associated with a compensatory increase in placentomal vascularisation in response to nutrient restriction. We have indeed confirmed that as ovine placentomes progress from Type A through Types B, C, or D, they increase in size and vascularity [23], as well as blood flow per gram of tissue [24].

Data in rodents suggest that subtle genetic or environmental differences within or between breed (or strain) may exist in the susceptability of the conceptus to a variety of maternal stressors [8, 36, 38]. Several laboratories have evaluated the impact of similar levels of nutrient restriction from early to mid-gestation on fetal growth and development in the sheep [11, 55, 86] and have reported markedly different results. These divergent results may relate to differences in ewe breed, age, or parity, but may also relate to differences in nutritional or environmental exposures of experimental animals either before or after birth. Our laboratory compared two divergent flocks of multiparous Rambouillet-Columbia cross ewes, to determine if the pre-pregnancy environment under which ewes are selected altered their ability to protect their fetus against a bout of gestational nutrient restriction. The first flock located near Baggs, Wyoming (Baggs ewes) was adapted for approximately 30 years to a nomadic existence, with very limited nutritional inputs. The second flock was also selected for approximately 30 yr by the University of Wyoming, and in contrast to the Baggs ewes had a relatively sedentary lifestyle and consumed a diet from birth that met or exceeded National Research Council [54] recommendations (UW ewes). From day 28 to day 78 of gestation, both Baggs and UW ewes were divided into control (C, fed 100% of NRC requirements), and nutrient restricted (NR, fed 50% of NRC requirements) groups, and necropsied on day 78 of gestation. Glucose and essential amino acid concentrations on day 78 were significantly decreased in both maternal and fetal blood of NR UW ewes [40]. While there was a significant decrease in the glucose and essential amino acid concentrations in the blood of NR Baggs ewes on day 78, there was no corresponding decrease in the blood of their fetuses [98]. Thus the fetuses of NR Baggs ewes were not subjected to a reduction in glucose or essential amino acids during maternal undernutrition, even though their mothers were. On day 78 fetal weight in the NR UW ewes was about 30% lower than in the C UW ewes, and these fetuses exhibited enlarged hearts [85], and reduced numbers of secondary skeletal muscle fibers when compared to fetuses from C UW ewes [101]. Since muscle fiber numbers are set prenatally, this suggests the possibility that offspring of NR UW ewes could exhibit a decreased muscle mass in adulthood. Interestingly, in the Baggs ewes, there was absolutely no effect of maternal undernutrition on fetal weight or composition [87]. It would appear that the Baggs fetuses have been spared the negative impact of maternal undernutrition on fetal growth and development. In evaluating PE, Vonnahme et al. [87] reported that PE was significantly decreased in the NR UW ewes compared to their adequately fed control group. In contrast, there was no change whatsoever in the PE of undernourished Baggs ewes when they were compared to their control-fed group. These data suggested that the placentomes of the NR Baggs ewes were able to carry on normal nutrient delivery to the fetus even in the face of a reduced glucose and amino acid concentrations in maternal blood. One possibility for this maintenance of fetal nutrient delivery is an increased blood flow through the cotyledonary and/or caruncular components of the placentome. Vonnahme et al. [87] found no difference in the types of placentomes (virtually all type A) in the uterus on day 78 between C and NR UW ewes. However, there was a significant decrease in the numbers of Type A placentomes and a significant increase in the numbers of B, C, and D placentomal types in the NR Baggs ewes when compared to C Baggs ewes on day 78. These data suggest that there was a difference in the ability of these females or their conceptuses to respond to a bout of early gestational nutrient restriction by stimulating an early conversion of Type A placentomes to more efficient type B, C and D placentomal types. Similar to the results obtained with nutrient restricted UW ewes in our study, Osgerby et al. [56] reported that a less severe 30% global nutrient restriction from day 26 of gestation in a group of multiparous Welsh mountain ewes, which failed to alter placentomal growth or type by day 90, led to alterations in fetal organ growth. Interestingly, when these researchers allowed a group of these undernourished ewes to gestate their fetuses through to day 135 of gestation, they exhibited greater numbers of Type C and D placentomes than adequately fed control ewes, but fetal body weight and fetal organ weights remained growth retarded. These data are consistent with the concept that placentomal conversion must occur before the period of exponential fetal growth to alleviate the impact of maternal nutrient restriction on the fetus.

8.3 Summary

Both the pig and the sheep have adeciduate, minimally invasive placental types which are often referred to as epitheliochorial. As previously discussed, placental vascularity and blood flow are thought to be major determinants of maternal-fetal nutrient exchange in these species. The pig has a diffuse placentation where the chorionic epithelium is in direct contact with the uterine luminal epithelium, and nutrient exchange takes place over the entire surface of the chorionic sac. In this species, both the area of placental-endometrial attachment and placental vascularity are important for nutrient and waste product exchange between a conceptus and its mother. As early as day 25 of gestation, each conceptus appears to establish its own fetal/placental weight ratio, previously referred to as PE, which is negatively correlated with placental size, and positively correlated with placental vascularity. Placental efficiency is an individual conceptus trait, where placental vascularity dictates how large the placenta will have to grow to harvest adequate nutrients to support the normal growth and development of a fetus. Thus, within a litter of pigs, there can be a large variation in littermate PE, which results in some piglets being attached to relatively large, less vascular placentae and some piglets being attached to proportionally smaller and more vascular placentae. The prolific Chinese Meishan pig, which has been selected for thousands of years for increased litter size, experiences little conceptus mortality, in association with small highly vascular placentae, which differ very little in size, resulting in little variation in littermate placental efficiencies. In contrast, the American Yorkshire breed experiences high conceptus mortality, with a marked variation in littermate placental sizes, vascularities, and efficiencies. The fact that PE is an individual conceptus trait and is highly variable within litters in commercially relevant U. S. pig breeds, has prevented us from significantly increasing litter size by simply selecting breeding stock from large litters. Perhaps it is time for the swine industry to consider selecting for PE, a trait with is highly associated with the capacity for producing larger litters. We are currently working on establishing genetic markers for PE, which should make selection for this trait considerably easier.

The sheep has a cotyledonary placentation where nutrient and waste product exchange occurs at discrete sites called placentomes. While placentomal growth is completed by mid-gestation, individual placentomes continue to exhibit morphological changes which result in increased size, vascularity and efficiency. Recent research in our laboratory has indicated that the environment and/or nutrition under which an individual is selected can alter their physiological responses to a bout of maternal nutrient restriction. More specifically, ewes which were selected under adverse environmental conditions and limited nutrition developed the ability to initiate placentomal conversion to more efficient forms, when subjected to a bout of early gestational nutrient restriction. This conversion of Type A placentomes to more efficient forms (Types B, C, and D) was associated with increased placentomal vascularity, and the maintainence of normal amino acid and glucose delivery to the fetus, and a normal PE, in the face of reduced concentrations of these nutrients in maternal blood. In contrast, ewes which were selected to a relatively sedentary lifestyle, and more than adequate nutrition, failed to convert placentomes to more efficient forms in response to a similar bout of nutrient restriction, resulting in reduced amino acid and glucose concentration in fetal blood, as well as reduced PE and fetal weights. An understanding of how these two flocks of ewes evolved in approximately 30 years to be either "nutrient responsive" or "nutrient resistant" could be critically relevant to our understanding of factors dictating nutrient requirements during gestation.

Thus, the pig conceptus appears to utilise and increase placental vascularity to offset the need for continual placental growth throughout gestation, thereby reducing the constraints of uterine capacity, with a resultant increase in litter size. In contrast, the sheep, which has established permanent sites of fetal-maternal nutrient exchange by midgestation called placentomes, utilises morphologic conversion and vascularisation of these sites to maintain nutrient delivery to the fetus at levels which support normal growth and development. The variable ability of a conceptus to initiate the proliferation of capillaries at the fetal-maternal interface in response to a nutrient deficit appears to be the common thread between these two species, and dictates the ability of the fetus to harvest adequate nutrients for normal growth and development. An understanding of the factors which stimulate optimal placental vascularity and thus efficiency in the face of maternal nutrient deficits is the focus of ongoing investigations in our laboratory.

References

1. Alexander, G. 1964. Studies on the placenta of sheep. Placental size. *J. Reprod. Fertil.* **7**:289.
2. Barcroft, J. and D.H. Barron. 1946. Observations on the form and relations of the maternal and fetal vessels in the placenta of the sheep. *Anat. Rec.* **94**:569–595.
3. Barker, D.J. and P.M. Clark. 1997. Fetal undernutrition and disease in later life. *Rev. Reprod.* **2**:105–112.
4. Benirschke, K. 1983. Placentation. *J. Exp. Zool.* **228**:385–389.
5. Biensen, N.J., M.E. Wilson, and S.P. Ford. 1998. The impact of either a Meishan or Yorkshire uterus on Meishan or Yorkshire fetal and placental development to days 70, 90, and 110 of gestation. *J. Anim. Sci.* **76**:2169–2176.
6. Biensen, N.J., M.E. Wilson, and S. P. Ford. 1999. The impacts of uterine environment and fetal genotype on conceptus size and placental vascularity during late gestation in pigs. *J. Anim. Sci.* **77**:954–959.
7. Bjorkman, N. and V. Dantzer. 1987. Placentation, pp. 340–36. *In* H.-D. Dellmann and E. M. Brown, (eds.), Textbook of Veterinary Histology, Lea & Febiger, Philadelphia.
8. Buhimschi, I.A., S.Q. Shi, G.R. Saade, and R.E. Garfield. 2001. Marked variation in responses to long-term nitric oxide inhibition during pregnancy in outbred rats from two different colonies. *Am. J. Obstet. Gynecol.* **184**:686–693.
9. Chandler, K.D., B.J. Leury, A.R. Bird, and A.W. Bell. 1985. Effects of undernutrition and exercise during late pregnancy on uterine, fetal and uteroplacental metabolism in the ewe. *Br. J. Nutr.* **53**:625–635.
10. Christenson, R.K., K.A. Leymaster, and L.D. Young. 1987. Justification of unilateral hysterectomy-ovariectomy as a model to evaluate uterine capacity in swine. *J. Anim. Sci.* **65**:738–744.
11. Clarke, L., L. Heasman, D.T. Juniper, and M.E. Symonds. 1998. Maternal nutrition in early-mid gestation and placental size in sheep. *Br. J. Nutr.* **79**:359–364.
12. Dantzer, V. and R. Leiser. 1994. Initial vascularisation in the pig placenta: I. Demonstration of nonglandular areas by histology and corrosion casts. *Anat. Rec.* **238**:177–190.
13. Dellmann, H.D. and E.M. Brown. 1987. Textbook of Veterinary Histology (3rd ed.). p. 468. Lea & Febiger, Philadelphia.
14. Dziuk, P.J. 1968. Effect of number of embryos and uterine space on embryo survival in the pig. *J. Anim. Sci.* **27:**673–676.
15. Dziuk, P.J. 1987. Embyonic loss in the pig: An enigma. *Paper presented at the inaugural Australasian Pig Science Association conference*, November 23–25, Albury, NSW, Australia.
16. Ehrhardt, R.A. and A.W. Bell. 1995. Growth and metabolism of the ovine placenta during mid-gestation. *Placenta* **16**:727–741.
17. Fenton, F.R., F.L. Schwartz, F.W. Bazer, O.W. Robison, and L.C. Ulberg. 1972. Stage of gestation when uterine capacity limits embryo survival in gilts. *J. Anim. Sci.* **35**:383–388.
18. Ferrell, C.L. 1989. Placentation of fetal growth, pp. 1–19. *In* Animal growth regulations, D.R. Campion, G.J. Hausman, R.J. Martin, (eds.), Plenum, New York.
19. Ford, S.P. and C.R. Youngs. 1993. Early embryonic development in prolific Meishan pigs. *J. Reprod. Fertil.* (Suppl.) **48**:271–278.
20. Ford, S.P. 1995. Control of blood flow to the gravid uterus of domestic livestock species. *J. Anim. Sci.* **73**:1852–1860.
21. Ford, S.P. 1997. Embryonic and fetal development in different genotypes in pigs. *J. Reprod. Fertil.*(Suppl.) **52**:165–176.
22. Ford, S.P., K.A. Vonnahme, and M.E. Wilson. 2002. Uterine capacity in the pig reflects a combination of uterine environment and conceptus genotype effects. *J. Anim. Sci.* 80: E66–E73.
23. Ford, S.P., Vonnahme, K.A. Vonnahme, M.C. Drumhiller, L.P. Reynolds, M.J. Nijland, and P.W. Nathanielz. 2004. Arteriolar density and capillary volume increase as placentomes

advance from type A through type D developmental stages. *Biol. Reprod.* (Suppl.) **71**:509 (Abstract).

24. Ford, S.P., M.J. Nijland, M.M. Miller, B.W. Hess, and P.W. Nathaniesz. 2006. Maternal undernutrition advance placentomal type, in association with increase placentomal size and cotyledonary (cot) blood flow. *J. Soc. Gynecol. Invest.* (Suppl.) **12**:212A (Abstract).
25. Foxcroft, G.R. 1997. Mechanisms mediating nutritional effects on embryonic survival in pigs. *J. Reprod. Fertil.* (Suppl). **52**:47–61.
26. Friess, A.E., F. Sinowatz, R. Skolek-Winnisch, and W. Traautner. 1980. The placenta of the pig. I. Fine structural changes of the placental barrier during pregnancy. Anat. Embryol. (Berl) **158**:179–191.
27. Friess, A.E., F. Sinowatz, R. Skolek-Winnisch, and W. Trautner. 1982. Structure of the epitheliochorial porcine placenta. *Bibl. Anat.* **42**:40–143.
28. Gilbert, J.S., A.L. Lang, A.R. Grant, and M.J. Nijland. 2005. Maternal nutrient restriction in sheep: Hypertension and decreased nephron number in offspring at 9 months of age. *J. Physiol.* **565**:137–147.
29. Grosser, O. 1927. *Fruhentwick, eihautbidung and placentation des menshen und der santier.* J.F. Bergmann, Munchen.
30. Grosser, O. 1933. Human and comparative placentation: including the early stages of human development. *Lancet* **111**:999–1001.
31. Hammond, J. 1935. The changes in the reproductive organs of the rabbit during pregnancy. *Institut. Zhivotnovodsva Trudy Po Dinamike Razvitiya (Trans Dynmaic Develop, Moscow)* **10**:93.
32. Hoet, J.J. and M.A. Hanson. 1999. Intrauterine nutrition: Its importance during critical periods for cardiovascular and endocrine development. *J. Physiol.* (Pt 3) **514**:617–627.
33. Huang, Y.T., R.K. Johnson, and G.R. Eckardt. 1987. Effect of unilateral hysterectomy and ovariectomy on puberty, uterine size and embryo development in swine. *J. Anim. Sci.* **65**:1298–1305.
34. Ibsen, H.L. 1928. Prenatal growth in guinea-pigs with special reference to environmental factors affecting weight at birth. *J. Exp. Zool.* **51**:51.
35. Johnson, R.K., M.K. Nielsen, and D.S. Casey. 1999. Responses in ovulation rate, embryonal survival, and litter traits in swine to 14 generations of selection to increase litter size. *J. Anim. Sci.* **77**:541–557.
36. Kawakami, T., R. Ishimura, K. Nohara, K. Takeda, C. Tohyama, and S. Ohsako. 2006. Differential susceptibilities of holtzman and sprague-dawley rats to fetal death and placental dysfunction induced by 2,3,7,8-teterachlorodibenzo-p-dioxin (tcdd) despite the identical primary structure of the aryl hydrocarbon receptor. *Toxicol. Appl. Pharmacol.* **212**:224–236.
37. Knight, J.W., F.W. Bazer, W.W. Thatcher, D.E. Franke, and H.D. Wallace. 1977. Conceptus development in intact and unilaterally hysterectomized-ovariectomized gilts: Interrelations among hormonal status, placental development, fetal fluids and fetal growth. *J. Anim. Sci.* **44**:620–637.
38. Knight, B., C. Pennell, and S. Lye. 2005. Strain difference in the impact of maternal dietary restriction on fetal growth, pregnancy and postnatal development in mice. *Pediatric Res.* **58**:1031.
39. Krebs, C., L.D. Longo, and R. Leiser. 1997. Term ovine placental vasculature: Comparison of sea level and high altitude conditions by corrosion cast and histomorphometry. *Placenta* **18**:43–51.
40. Kwon, H., S.P. Ford, F.W. Bazer, T.E. Spencer, P.W. Nathanielsz, M.J. Nijland, B.W. Hess, and G. Wu. 2004. Maternal nutrient restriction reduces concentrations of amino acids and polyamines in ovine maternal and fetal plasma and fetal fluids. *Biol. Reprod.* **71**: 901–908.
41. Langley-Evans, S.C. 2004. Fetal nutrition and adult disease. CABI Publishing, Cambridge.
42. Leenhouwers, J.I. 2001. Biological aspects of genetic differences in piglet survival. Doctoral Thesis. Wagenigen Institute of Animal sciences, The Netherlands. ISBN: 90-5808-517-1.

43. Leiser, R. and V. Dantzer. 1988. Structural and functional aspects of porcine placental microvasculature. *Anat. Embryol. (Berl)* **177**:409–419.
44. Macdonald, A.A. and A.A. Bosma. 1985. Notes on placentation in the suina. *Placenta* **6**: 83–91.
45. Magness, R.R. 1998. Maternal cardiovascular and other physiological responses to the endocrinology of pregnancy. *In*: The Endocrinology of Pregnancy. F.W. Bazer (ed.). pp. 507–539. Humana Press, Totowa, NJ.
46. Makowski, E.L., G. Meschia, W. Droegemueller, and F.C. Battaglia. 1968. Distribution of uterine blood flow in the pregnant sheep. *Am. J. Obstet. Gynecol.* **101**:409–412.
47. Makowski, E.L., G. Meschia, W. Droegemueller, and F.C. Battaglia. 1968. Measurement of umbilical arterial blood flow to the sheep placenta and fetus in utero. Distribution to cotyledons and the intercotyledonary chorion. *Circ. Res.* **23**: 623–631.
48. Mao, J.D. 1995. Mechanisms for the high prolificacy of Taihu pigs. *Pig. News. Info.* **16**:55 N–59N.
49. McMillen, I.C., M.B. Adams, J.T. Ross, C.L. Coulter, G. Simonetta, J.A. Owens, J.S. Robinson, and L.J. Edwards. 2001. Fetal growth restriction: Adaptations and consequences. *Reproduction* **122**:195–204.
50. Meschia, G. 1983. Circulation to female reproductive organs. *In*: Am. Physiol. Soc. Sec. 2. Sheperd, J.T., Abboud, F. M. (eds.). *Vol. III, Part 1*. pp. 241–269.
51. Metcalfe, J., M.K. Stock and D.H. Barron. 1988. Maternal physiology during gestation, pp. 2145–2176. *In* The physiology of reproduction, E. Knobil, J. Neill, L.L. Ewing, G.S. Greenwald, C.L. Markert, E.W. Pfaff, (eds.), Raven Press, New York.
52. Mossman, H.W. 1987. Vertebrate fetal membranes: Comparative ontogeny and morphology; evolution; phylogenetic significance; basic functions; research opportunities. Macmillan Press, Ltd., London.
53. Noden D.M. and A. de Lahunta. 1985. Extraembryonic membranes and placentation, pp. 47–69. In: The Embryology of Domestic Animals, Williams and Wilkins, Baltimore.
54. NRC. 1985. Nutrient Requirements of Sheep (6th ed.). National Academy Press, Washington, DC.
55. Osgerby, J.C., T.S. Gadd, and D.C. Wathes. 1999. Expression of insulin-like growth factor binding protein-1 (igfbp-1) mrna in the ovine uterus throughout the oestrous cycle and early pregnancy. *J. Endocrinol.* **162**:279–287.
56. Osgerby, J.C., D. C. Wathes, D. Howard, and T.S. Gadd. 2002. The effect of maternal undernutrition on ovine fetal growth. *J. Endocrinol.* **173**:131–141.
57. Osgerby, J.C., D.C. Wathes, D. Howard, and T.S. Gadd. 2004. The effect of maternal undernutrition on the placental growth trajectory and the uterine insulin-like growth factor axis in the pregnant ewe. *J. Endocrinol.* **182**:89–103.
58. Patten, B.M. 1948. Embryology of the Pig. McGraw-Hill Book Company, New York.
59. Patten, B.R. 1964. Foundations of Embryology (2nd ed.). McGraw-Hill Book Company, New York.
60. Peilieu, C. 1985. Pig breeds, p. 159–194. *In* Livestock Breeds of China, FAO and China Academic Publishers, Rome.
61. Pomeroy, R.W. 1960. Infertility and Neonatal Mortality in the Sow. *J. Agric.* **54**:31–56.
62. Pope, W.F., R.R. Maurer, and F. Stormshak. 1982. Survival of porcine embryos after asynchronous transfer (41495). *Proc. Soc. Exp. Biol. Med.* **171**:179–183.
63. Pope, W.F., M.H. Wilde, and S. Xie. 1988. Effect of electrocautery of nonovulated day 1 follicles on subsequent morphological variation among day 11 porcine embryos. *Biol. Reprod.* **39**:882–887.
64. Ramsey, E. 1982. The placenta, human and animal. Praeger, New York.
65. Redmer, D.A., R.P. Aitken, J.S. Milne, L.P. Reynolds, J.M. Wallace. 2004. Influence of maternal nutrition on placental vascularity during late pregnancy in adolescent ewes. *Biol. Reprod.* (Suppl 1)**70**:150.

66. Redmer, D.A., R.P. Aitken, J.S. Milne, L.P. Reynolds, and J.M. Wallace. 2005. Influence of maternal nutrition on messenger RNA expression of placental angiogenic factors and their receptors at midgestation in adolescent sheep. *Biol. Reprod.* **72**:1004–1009.
67. Regnault, T.R., H.L. Galan, T.A. Parker, and R.V. Anthony. 2002. Placental development in normal and compromised pregnancies- a review. *Placenta* **23** Suppl A:S119–129.
68. Regnault, T.R., B. de Vrijer, H.L. Galan, M.L. Davidson, K.A. Trembler, F.C. Battaglia, R.B. Wilkening, and R.V. Anthony. 2003. The relationship between transplacental o2 diffusion and placental expression of plgf, vegf and their receptors in a placental insufficiency model of fetal growth restriction. *J. Physiol.* **550**:641–656.
69. Reynolds, L.P. 1986. Utero-ovarian interactions during early pregnancy: Role of conceptus-induced vasodilation. *J. Anim. Sci. 62 Suppl.* **2**:47–61.
70. Reynolds, L.P. and C.L. Ferrell. 1987. Transplacental clearance and blood flows of bovine gravid uterus at several stages of gestation. *Am. J. Physiol.* **253**:R735–739.
71. Reynolds, L.P., S.D. Killilea, and D.A. Redmer. 1992. Angiogenesis in the female reproductive system. *Faseb. J.* **6**:886–892.
72. Reynolds, L.P. and D.A. Redmer. 1995. Utero-placental vascular development and placental function. *J. Anim. Sci.* **73**:1839–1851.
73. Reynolds, L.P., P.P. Borowicz, K.A. Vonnahme, M.L. Johnson, A.T. Grazul-Bilska, D.A. Redmer, J.S. Caton. 2005. Placental angiogenesis in sheep models of compromised pregnancy. *J. Physiol.* **565**:43–58.
74. Reynolds, L.P., P.P. Borowicz, K.A. Vonnahme, M.L. Johnson, A.T. Grazul-Bilska, J.M. Wallace, J.S. Caton, and D.A. Redmer. 2005. Animal models of placental angiogenesis. *Placenta* **26**:689–708.
75. Rivera, R.M., C.R. Youngs, and S.P. Ford. 1996. A comparison of the number of inner cell mass and trophectoderm cells of preimplantation Meishan and Yorkshire pig embryos at similar developmental stages. *J. Reprod. Fertil.* **106**:111–116.
76. Rosenfeld, C.R., F.H. Morriss, Jr., E.L. Makowski, G. Meschia, and F.C. Battaglia. 1974. Circulatory changes in the reproductive tissues of ewes during pregnancy. *Gynecol. Invest.* **5**:252–268.
77. Rudolph, A.M. and M.A. Heymann. 1970. Circulatory changes during growth in the fetal lamb. *Circ. Res.* **26**:289–299.
78. Stegeman, J.H.J. 1974. Placental development in the sheep and its relation to fetal development. *Bijdragen Tot De Dierkunde (Contrib. Zool.)* **44**:3.
79. Teasdale, F. 1976. Numerical density of nuclei in the sheep placenta. *Anat. Rec.* **185**: 181–196.
80. Vallet, J.L., K.A. Leymaster, J.P. Cassady, and R.K. Christenson. 2001 Are hematocrit and placental selection tools for uterine capacity in swine? *J. Anim. Sci. Suppl.* **79**:64 (Abstract).
81. Vatnick, I., P.A. Schoknecht, R. Darrigrand, and A.W. Bell. 1991. Growth and metabolism of the placenta after unilateral fetectomy in twin pregnant ewes. *J. Dev. Physiol.* **15**:351–356.
82. Vonnahme, K.A., M.E. Wilson, and S.P. Ford. 2000. The role of vascular endothelial growth factor (VEGF) in increasing placental vascularity and efficiency during gestation in the pig, Abstract 299. *In* Proceedings of the Society for the study of Reproduction Annual Meeting, Madison, WI.
83. Vonnahme, K.A. and S.P. Ford. 2001. Selection for increased placental efficiency (PE) results in increased placental expression of vascular endothelial growth factor (VEGF) in the pig. *J. Anim. Sci.* **79**:64 (Abstract).
84. Vonnahme, K.A., M.E. Wilson, G.R. Foxcroft, and S.P. Ford. 2002. Impacts on conceptus survival in a commercial swine herd. *J. Anim. Sci.* **80**:553–559.
85. Vonnahme, K.A., B.W. Hess, T.R. Hanson, R.J. McCormick, D.C. Rule, G.E. Moss, W.J. Murdoch, M.J. Nijland, D.C. Skinner, P.W. Nathanielsz, and S.P. Ford. 2003. Maternal undernutrition from early- to mid-gestation leads to growth retardation, cardiac ventricular hypertrophy, and increased liver weight in the fetal sheep. *Biol. Reprod.* **69**:133–140.

86. Vonnahme, K.A. and S.P. Ford. 2004. Differential expression of the vascular endothelial growth factor-receptor system in the gravid uterus of Yorkshire and Meishan pigs. *Biol. Reprod.* **71**:163–169.
87. Vonnahme, K.A., B.W. Hess, M.J. Nijland, P.W. Nathanielsz, and S.P. Ford. 2006. Placentomal differentiation may compensate for maternal nutrient restriction in ewes adapted to harsh range conditions. *J. Anim. Sci.* **84**:3451–3459.
88. Wallace, L.R. 1948. The growth of lambs before and after birth in relation to the level of nutrition. Part III. *Biol. Reprod.* **38**:367.
89. Wallace, J.M., D.A. Bourke, R.P. Aitken, N. Leitch, and W.W. Hay, Jr. 2002. Blood flows and nutrient uptakes in growth-restricted pregnancies induced by overnourishing adolescent sheep. *Am. J. Physiol. Regul. Integr. Comp. Physiol.* **282**:R1027–1036.
90. Warwick, B.L. 1928. Prenatal growth of swine. *J. Morphol. Physiol.* **46**:59.
91. Wigmore, P.M. and N.C. Stickland. 1985. Placental growth in the pig. *Anat. Embryol. (Berl)* **173**:263–268.
92. Wilde, M.H., S. Xie, M.L. Day, and W.F. Pope. 1988. Survival of small and large littermate blastocysts in swine after synchronous and asynchronous transfer procedures. *Theriogenology.* **30**:1069–1074.
93. Wilson, M.E. and S.P. Ford. 1997. Differences in trophectoderm mitotic rate and p450 17alpha-hydroxylase expression between late preimplantation meishan and yorkshire conceptuses. *Biol. Reprod.* **56**:380–385.
94. Wilson, M.E., N.J. Biensen, C.R. Youngs, and S.P. Ford. 1998. Development of meishan and yorkshire littermate conceptuses in either a Meishan or Yorkshire uterine environment to day 90 of gestation and to term. *Biol. Reprod.* **58**:905–910.
95. Wilson, M.E., N.J. Biensen, and S.P. Ford. 1999. Novel insight into the control of litter size in pigs, using placental efficiency as a selection tool. *J. Anim. Sci.* **77**:1654–1658.
96. Wilson, M.E. and S.P. Ford. 2000. Effect of estradiol-17beta administration during the time of conceptus elongation on placental size at term in Meishan pigs. *J. Anim. Sci.* **78**: 1047–1052.
97. Wooding F.B.P. and A.P.F. Flint. 1994. Placentation, pp. 233–460. *In* G.E. Lamming, (ed.), Marshall's Physiology of Reproduction (4th ed.), Vol. 3, Part 1, Chapman & Hall, London.
98. Wu, G., T. Spencer, B. Hess, P. Nathanielsz, and S. Ford. 2005. Production system under which ewes are selected alters nutrient availability to the fetus in response to early pregnancy undernutrition. *J. Anim. Sci.* **83** Suppl.1:297 (Abstract).
99. Youngs, C.R., S.P. Ford, L.K. McGinnis, and L.H. Anderson. 1993. Investigations into the control of litter size in swine: I. Comparative studies on in vitro development of Meishan and Yorkshire preimplantation embryos. *J. Anim. Sci.* **71**:1561–1565.
100. Youngs, C.R., L.K. Christenson, and S.P. Ford. 1994. Investigations into the control of litter size in swine: Iii. A reciprocal embryo transfer study of early conceptus development. *J. Anim. Sci.* **72**:725–731.
101. Zhu, M.J., S.P. Ford, P.W. Nathanielsz, and M. Du. 2004. Effect of maternal nutrient restriction in sheep on the development of fetal skeletal muscle. *Biol. Reprod.* **71**:1968–1973.

Chapter 9
Management and Environmental Influences on Mammary Gland Development and Milk Production

Anthony V. Capuco and R. Michael Akers

Introduction

The vast majority of mammary growth occurs postnatally. For example, in dairy cows mammary parenchymal mass increases approximately 10,000-fold, from less than 0.5 g per gland at birth to approximately 5.5 kg during lactation, and the mammary epithelium differentiates from a primitive branching ductal network to a fully differentiated lobuloalveolar network that is capable of synthesising copious quantities of milk. These growth and developmental processes are hormonally regulated, but also involve extensive interactions among cell types (e.g. stromal, epithelial, myoepithelial, stem and progenitor cells) and local regulation by paracrine/autocrine factors. Appropriate nutrient intake and balance is important for mammary growth and development, and these processes may be influenced by under- or over-nutrition. Because of interactions between the environment and the endocrine system, mammary development and differentiation can be influenced by additional factors such as bioactive substances in feed and by photoperiod. Ultimately, milk yield is determined by the number of fully differentiated mammary epithelial cells and by homeorhetic mechanisms that regulate functions of other organs to support milk production.

Because mammary gland biology of the dairy cow has been more intensely studied than that of other livestock species, the primary focus of this chapter will be on the dairy cow. This information will be supplemented with information derived from rodent models, and pertinent information for other species will be provided as feasible. This review addresses the nature of prenatal and postnatal mammary growth and development, mammary epithelial stem cells, the impact of nutrition, bioactive factors in the environment (including phytoestrogens, mycotoxins and endocrine disruptors) and photoperiod.

A.V. Capuco (✉)
Bovine Functional Genomics Laboratory, USDA-ARS, Beltsville, MD 20705, USA
e-mail: tony.capuco@ars.usda.gov

P.L. Greenwood et al. (eds.), *Managing the Prenatal Environment to Enhance Livestock Productivity*, DOI 10.1007/978-90-481-3135-8_9,

9.1 Fetal Mammary Development

Despite the lack of soft tissue fossils, a comparison of mammary development among various mammals supports the idea that the ectoderm-derived mammary gland arose from sweat or sebaceous glands. Reproductive secretions associated with incubation and care of eggs were likely the precursors of nutritive milk-like secretions produced by brood patches on the abdominal or inguinal surface of primitive mammals. Such secretions, similar to mammary colostrum, may also contain antibacterial substances such as lactoferrin, immunoglobulins, and other bactericidal components [98, 99].

Hypotheses for the adaptive value of these proto-lacteal secretions include: thermoregulatory, antibiotic, behavioral, or nutritive functions. A reasonable conjecture is that lactation arose in endothermic, oviparous ancestors that exhibited a degree of maternal care. Early anatomists tried to define the origin of the mammary gland by classifying the secretion mechanism for the alveolar cells. Sebaceous glands exhibit a holocrine mode of secretion in which sloughed cells become a part of the secretion. Sweat glands follow an apocrine mode of secretion in which a portion of the cell is budded off with the secretion. Other common glands follow a merocrine mode of secretion in which products are secreted but the secretory cells remain intact. Mammary secretory cells utilize both apocrine and merocrine modes of secretion. For example, as lipid droplets form in the cytoplasm of the cells they progressively enlarge, migrate to the apical end of the cell, and protrude into the alveolar lumen until the membrane bound droplets pinch off to become the butterfat of milk. Because the membrane surrounding the lipid droplet comes from the plasma membrane, secretion of milk lipid is apocrine in nature. Specific milk proteins and milk sugar (lactose) are packaged into secretory vesicles in the Golgi apparatus. These vesicles fuse with the apical plasma membrane and release their contents via exocytosis. Because only the contents of the secretory vesicle are lost from the cell, this is a merocrine mode of secretion. These details for secretion patterns of mammary cells were not settled until the early 1960s, when mammary tissue from lactating mammals was studied with transmission electron microscopy.

Evolution of a skin-gland product to provide support for a newly hatched offspring in addition to protecting the egg with antibacterial secretions prior to hatching seems logical and is intuitively appealing. For example, the primitive echidna has two mammary glands on either side of the abdomen with each gland containing clusters of secretory lobules, each of which has a separate duct that opens to the surface of the skin in a small depression. But there are no teats or nipples. Thus it seems likely that the primitive mammary glands arose from a hybrid combination of both sebaceous and sweat glands [15].

The first signs of mammary development in the fetus appear as a slight thickening of the ventro-lateral ectoderm in the embryo at about the time limb buds begin to lengthen. These thickening areas are referred to as the mammary band, streak, or line (Fig. 9.1, top panel). In mammals with mammary glands along the entire ventral surface, e.g. rodents, a mammary line becomes apparent from the forelimb to hind limbs. For mammals with pectoral glands, the mammary lines appear only in the

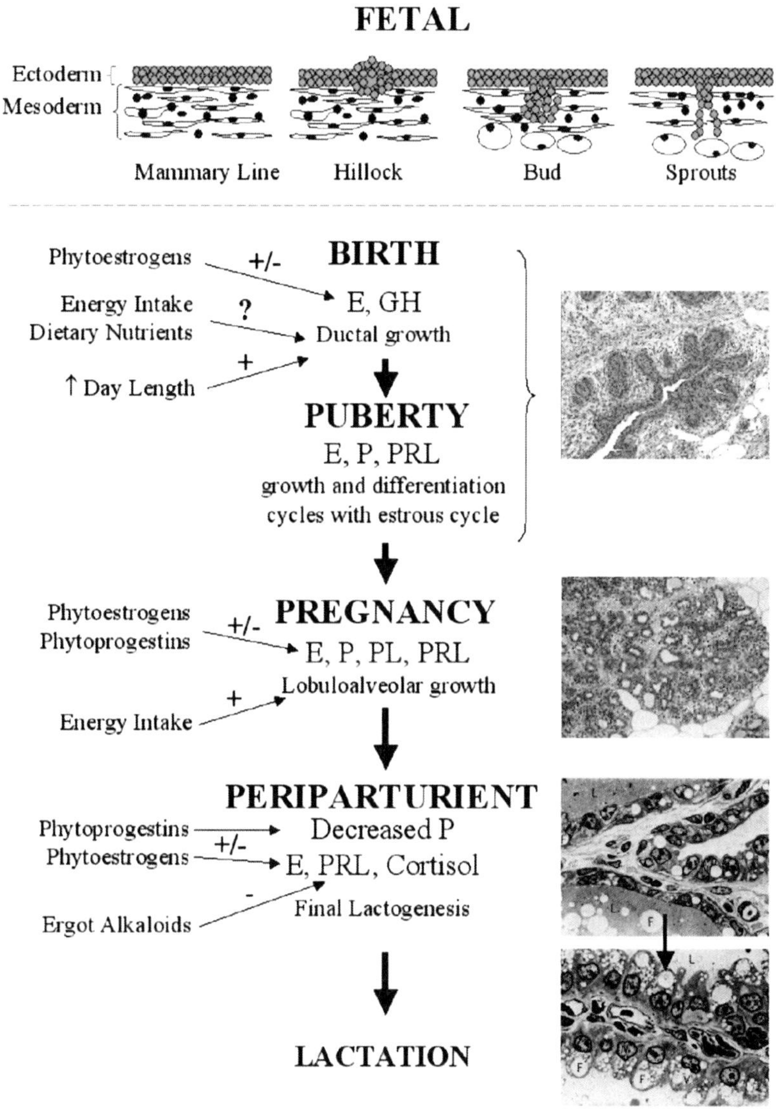

Fig. 9.1 Fetal and postnatal mammary gland development. Key stages of mammary development are depicted, with the primary hormones and potential management parameters that influence growth or differentiation. Micrographs shown are images of bovine mammary tissue. In the flow diagram: E = estrogen, GH = growth hormone, PL = placental lactogen, P = progesterone, PRL = prolactin. In the micrographs: F = fat droplet, L = alveolar lumen, N = nucleus, V = secretory vesicle

thoracic area. The cells of the mammary line begin to condense and progress through a series of somewhat arbitrary stages typified by structures called the mammary crest and mammary hillock. This culminates with the appearance of mammary buds, each of which is the precursor for an individual mammary gland. In the mouse, for example, each of five mammary buds on either side of the ventral midline is positioned at the location of each of the future nipples.

The mammary epithelium arises from the germinal ectoderm and the primitive mammary buds. The bovine mammary bud appears at about day 40 of gestation. By day 80 of gestation the teat and primary sprout have formed. The primary sprout subsequently produces the teat cistern (Fig. 9.1). Secondary sprouts occur by day 90 and by day 100 the primary and secondary spouts are believed to be canalized by a combination of apoptosis and cell migration. At the time of birth, the teat, teat cistern, and gland cistern have formed. In most species, mammary structure at birth is similarly rudimentary. Regardless, the mammary cells that are present are capable of proliferation and a degree of mammary specific cytological differentiation. Indeed, the endocrine changes at parturition are believed to be responsible for precocious development and fluid secretion ("witches milk") sometimes observed in human infants.

For mammals without mammary glands arranged into an udder, a teat or gland cistern is absent but there is a nipple and a cluster of primary and secondary sprouts for each individual mammary gland. Further growth of secondary sprouts yields the major ducts that drain groups of alveoli (lobules) in the mature gland. In those species studied, development of the alveoli is usually restricted to pregnancy. Mammary glands are compound alveolar glands. However, clusters of alveoli and the ducts that drain them are arranged into units called lobules. The ducts that drain individual alveoli lead to progressive larger ducts, which connect with the nipple or teat openings. Within the alveolus, lining epithelial cells are a single cell layer thick (simple epithelium) but the non-secretory ducts are stratified with two or more layers of cells. Creation of the lobuloalveolar structure during gestation does not automatically lead to the onset of milk secretion by the alveolar cells. These cells must undergo progressive biochemical and structural differentiation to prepare the cells for onset of copious milk secretion at parturition [3].

A critical question is how early mammary development – fetal, neonatal, peripubertal – can impact future lactational performance. Of course this question is important to the dairy industry. In brief, can management decisions and choices, feeding rate, ration composition, environment (photoperiod, housing, social setting, etc.), exogenous treatments (GH, early induced lactation), impact either mammary development or the functional capacity of the mature mammary gland? The short answer is yes, but there are qualifiers.

For example, as outlined in several reviews [4, 8, 112, 125], changes in feeding rate and/or diet composition, hormone treatments or endocrine alterations can alter mammary growth and development in peripubertal heifers and some of these effects involve alterations in GH- and IGF-I-related hormones, receptors or binding proteins [9]. However, as discussed subsequently, many of these studies were confounded by comparing mammary growth at similar body weight, but differing ages.

Additionally, it is true that many studies that have demonstrated impacts on mammary development have not necessarily shown effects on subsequent lactational performance. This reflects, in part, the fact that there are relatively few such longitudinal studies and also that mammary development and functionality of the mature gland may exhibit more plasticity than previously appreciated.

Unfortunately there are markedly less data in cattle compared with rodents to evaluate impacts of fetal treatments or manipulation on subsequent mammary development and function. Nonetheless, there is compelling evidence to indicate that alterations or insults during fetal development can impact the subsequent development of many organs and tissues. Indeed the human literature is replete with epidemiological studies linking exposure to various chemicals and subsequent malformation. In a 1999 review, Hilakivi-Clarke et al. [61] proposed that high fat diets in human females may produce circumstances that raise maternal estrogen production and thereby expose female fetuses to conditions that lead to an increased risk of breast cancer in their daughters. Furthermore, they reviewed both epidemiological and rat experimental data to support this hypothesis. Of course the focus of this work was disease risk in humans but if inappropriate development can be produced, why not mammary impacts that would be beneficial in farm animals?

Mellor [89] reviewed impacts of nutrition on fetal development and mammary growth during pregnancy but the emphasis was mammary development in the pregnant animal. For example, in sheep, prenatal mammary growth in the dam and fetal growth are linked because the mammary to lamb weight ratio is consistent across a range of litter sizes and feeding levels. Especially for sheep, this effect is apparently dependent on the placenta and is believed to be associated with secretion of ovine placental lactogen (oPL). In sheep, oPL concentration in maternal serum increases markedly during gestation with maximal sustained concentrations corresponding to periods of greatest mammary development. Concentration at each stage of gestation is also positively correlated with placental weight. Placental lactogens (PLs) have been identified in a number of mammals, including cows, goats, and sheep, and they have been implicated in mammogenesis, fetal growth, and metabolic alterations to support the fetus. However, as reviewed several years ago [2], the impact of PLs especially related to mammary development (fetal or maternal) likely varies markedly in the bovine compared with other ruminants.

Although it is reasonable to predict that nutrition (and other factors) during pregnancy impact fetal mammary development, data, especially for farm animals, is lacking. On the other hand, fetal mammary tissue certainly can respond to known mammogenic and/or lactogenic hormones. Ceriani [27] showed that fetal rat mammary fragments, taken 17 days after conception and maintained in organ culture, could be stimulated to produce mammary ducts and initiate mammary secretion. Specifically, insulin induced growth of the primitive epithelium into the surrounding mesenchymal tissue and the effect was increased in the presence of prolactin. Aldosterone enhanced duct formation and branching as well as appearance of secretions and the effect was enhanced with further addition of prolactin. It is probable that effects produced by the addition of insulin are actually dependent on IGF-I in vivo [4]. However, progesterone added with insulin, prolactin, and aldosterone further increased growth and induced the appearance of profuse granular secretion. Interestingly, growth hormone failed to replace prolactin in its growth promoting capacity. Companion studies [28] demonstrated that secretions produced in response to hormone additions were casein-like.

We are unaware of similar work with ruminant fetal mammary tissue, but it is clear that this tissue expresses receptors to important mammogenic hormones. Knabel et al. [74] provide convincing evidence using RT-PCR, in situ hybridisation, and immunohistochemistry for expression of growth hormone receptor (GHR) in the fetal bovine gland. Specifically, GHR was detected beginning in the third month of gestation through to the ninth month. Moreover, the pattern of distribution for GHR at the mRNA and protein level was similar. As early as three months of gestation, GHR mRNA and protein were found in the ductal epithelium, stromal cells, endothelial cells of the vascular system and the epidermis. Forsyth et al. [49] used in situ hybridisation to characterize temporal and spatial expression of mRNA for IGF-I, II and IGF-I receptor in female ovine fetal mammary tissue. At all stages tested (between 10 and 20 weeks of intrauterine life), IGF-I and II mRNA were expressed in stromal cells of the mammary parenchymal region but appeared especially abundant in stromal cells adjacent to the epithelium. In contrast, expression of IGF-I receptor was abundant in the epithelial cells. Expression of IGF-I and II significantly increased with gestational age and there was a tendency for greater IGF-I receptor expression during late gestation than early gestation. These examples suggest that various mammogenic hormones and growth factors normally influence fetal mammary development. By extension, elements that modulate the endocrine system have the potential to influence fetal mammary development.

Since the late 1980s, there has been increasing evidence that mechanisms other than mutation and natural selection can impact gene expression and ultimately tissue and cell development. As reviewed [115] some of these additional mechanisms include: endosymbiosis, genetic and genomic duplication, polyploidy, hybridisation, horizontal gene transfer, metabolic imprinting and epigenetics. These seemingly neo-Lamarckian ideas are collectively referred to as genomic creativity. In part this reflects knowledge that the one-gene/one-protein model is overly simplistic. Individual genes contain multiple exons disrupted by the noncoding introns. An extreme example, cited by Ryan [115] is for the Drosophila Dscam gene, which controls axon elongation. This gene has 24 exons, which theoretically can produce 38,000 protein splice variants. Utilisation of exons from one or more genes can also produce protein isotypes due to alternative splicing. As Ryan concludes, "Evolution derives from genomic (not solely genetic) novelty, under the influence of natural selection ..." As Crews and McLachlan summarize in a review focused on epigenetics and endocrine disruptors [35], "a gene is not expressed in isolation but rather in the context of other genes, and their products, cells, and tissues in a temporal/spatial dimension." In this case, epigenetic imprinting is a term used to describe how estrogen (or estrogen-like endocrine disruptors) during development can cause persistent, long-lived alterations in gene expression and thereby reprogram cell fate as well as gene silencing. These authors provide compelling arguments and examples to support the notion that epigenetic mechanisms explain the transgenerational effects of hormonally active chemicals linked to the steroid/retinoid/thyroid superfamily of receptors, e.g. the impact of exposure to diethylstilbestrol during pregnancy on the incidence of cancer in female offspring. That altered exposure to steroid hormones during fetal development can impact subsequent development and function should not be surprising to animal scientists or dairy producers. Problems with malformation of the reproductive system in females born twin to a male (freemartin) are well known.

Multiple cellular modifications are known to impact the capacity of DNA (genes) to be transcribed without altering the nucleotide sequence of the gene. These alterations can include both changes in the DNA and/or changes in DNA-binding proteins. Thus epigenetics is the study of heritable changes in gene expression that occur without a change in DNA sequence. Examples of such alterations include: methylation of DNA, active demethylation of DNA, biotinylation of histones, and ADP-ribosylation of histones and other DNA-binding proteins [100, 116, 142]. It is logical to suggest that changes in histones that modify chromatin structure or methylation of the promoter region of a gene would likely dramatically impact gene expression. The key to our discussion is the question of how the "landscape" surrounding the genes in the developing mammary gland might be modified to impact future development and lactational performance, as epigenetic alterations in gene expression are critical events in organogenesis and tissue/cell differentiation.

Many human diseases are apparently influenced by events during fetal development. In Western societies, low birth weight is associated with increased risk of cardiovascular disease, hypertension, type-two diabetes, and obesity. The developing concept is that the unfavorable prenatal conditions that produce low birth weight trigger epigenetic changes that improve the odds of postnatal survival but have the unfortunate consequence of increasing some disease susceptibility. Rats treated during pregnancy with diethylstilbestrol exhibit multi-generational susceptibility to uterine cancer, persistent uterine expression of the proto-oncogene c-fos, lactoferrin, and permanent repression of the Hoxa-10 and 11 genes. Specifically, CpG sites in the promoter and regulatory regions of the c-fos and lactoferrin genes are hypomethylated [62]. Gluckman et al. [55] solidified the concept that early life history directly impacts later disease risk in what they call a DOHaD ("developmental origins of health and disease") paradigm. The central idea is that DOHaD reflects a subset of broader mechanism(s) whereby individual genotypic variation is preserved during varying environmental (internal, external, social, etc.) changes. Furthermore, signals to drive this plasticity are proposed to be especially prominent in early development and can affect individual organs or systems, and, at least indirectly, the entire organism.

Given the unique nature of mammary development, i.e. major postnatal developmental milestones (peripubertal, pregnancy, parturition), it is logical that the mammary gland (like other reproductive organs) is liable to have multiple windows of opportunity to be impacted by "environmental" stimuli. Much of the evidence to link epigenetics with mammary development or mammary function is focused, not on variation in normal development and function, but on disease. Meta-analysis of multiple large studies has confirmed that human birthweight is positively correlated with premenopausal incidence of breast cancer [93]. The underlying mechanism(s) to explain the influence of birthweight on future breast cancer are unknown but is believed to involve interactions between the number of stem cells in the mammary gland, steroid hormones and growth factors, and epigenetic events. As a growth factor example, IGF-II is an imprinted gene, i.e. normally expressed only by the paternal allele and also expressed predominantly prenatally. Loss of imprinting, i.e. both alleles active, potentially doubles IGF-II expression. In mice IGF-II impacts the number of embryonic cells and over expression of IGF-II (transgenic mouse models) produces enhanced fetal growth and therefore placental development. Because placental size and birthweight are correlated factors this likely increases placental steroid production. Thus the fetus is likely to be exposed to more estrogens, which provides a possible functional link between fetal IGF-II production and birth weight. Interestingly, biallelic expression of IGF-II is detected in most mammary tumors (but not adjacent normal mammary tissue) [88].

Epigenetic reprogramming occurs normally during early development as genome-wide demethylation occurs when germ cells and other specific cell lineages are established [116]. Methylation markers on imprinted genes of germ cells are heritable in the usual sense, but clearly subject to failure.

Similar epigenetic markers on somatic cells are acquired and are likely tissue- or niche-specific. Success of studies to identify and isolate mammary stem cells have sparked even greater interest in how changes in the number and activity of these cells might impact incidence or susceptibility to breast cancer [127, 133, 134, 139]. Logically, many proposed epigenetic mechanisms related to mammary cancer focus on stem cells because it is hypothesized that many tumors have their origins in stem cells that are supposed to be especially susceptible to transformation, or that simply changing the number of available stem cells alters susceptibility by increasing the number of susceptible targets. Work on bovine mammary stem cells has lagged, but Capuco and Ellis [24] provided initial putative evidence for morphologically identifiable stem cells in the bovine mammary gland. A recent report [20], as well as other unpublished data (Daniels et al.), confirm the existence of BrdU (bromodeoxyuridine, a thymidine analog) label retaining cells in the mammary glands of prepubertal heifers. While it was initially supposed that mammary stem cells were quiescent, i.e. in G_0, it now seems that the long-term label retaining cells remain in the cell cycle but that the cells divide asymmetrically so that template DNA stands are retained by the stem cell. A similar phenomenon has been noted in the mouse mammary gland [134].

From a dairy or animal science perspective is there evidence for epigenetic involvement to explain differences in mammary development or function? Clearly, environment of the animal in its broad sense e.g. management, feeding, treatments, etc. can have an impact on mammary development and future performance but what are the mechanisms? Park [101] provides an intriguing hypothesis that a long-term effect of compensatory mammary growth produced by the stair-step nutritional scheme that he developed (to improve lactational performance in dairy heifers) is just such an example.

9.2 Prepubertal Mammary Growth

9.2.1 Nature of Prepubertal Mammary Growth

The secretory and ductular tissues present in the mammary gland are epithelial tissues collectively referred to as parenchyma. The connective tissues that support the parenchyma are referred to as stroma, which contains cellular and noncellular (e.g., collagen, elastin) elements. Cells of the stroma are primarily fibroblasts and, depending upon stage of development, may contain a considerable number of adipocytes; however, stroma also contains immune cells that are resident or transitory, as well as cells of blood vessels and lymphatics. During mammary gland development, the epithelial ducts grow into the surrounding stroma, which, from birth until late in first pregnancy, contains many adipocytes and is referred to as the mammary fat pad.

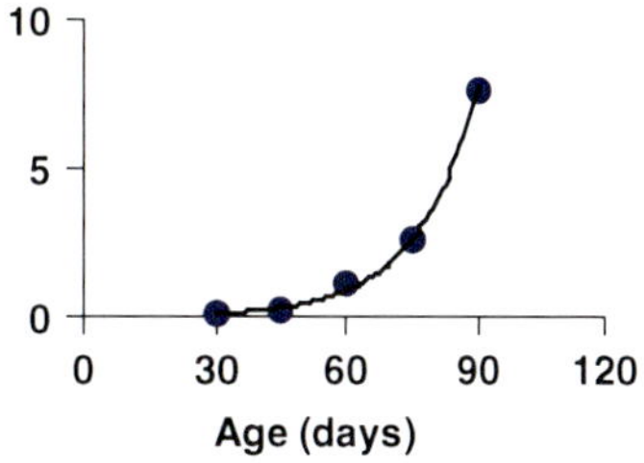

Fig. 9.2 Parenchymal tissue development in Holstein calves (Akers and Capuco, unpublished)

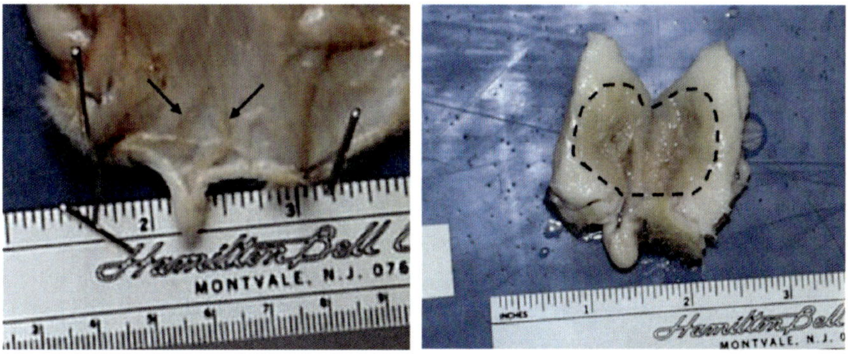

Fig. 9.3 Gross bovine mammary gland development at day 45 (*left panel*) and day 75 (*right*). Note the very thin strand of parenchymal tissue adjacent to the teat at day 45 (*arrows*). One month later by day 75 there is a dense mass of very evident parenchymal tissue (*dashed line*). Note that most of the surrounding stromal tissue has been removed (Akers, unpublished)

At birth, bovine mammary parenchyma consists of a rudimentary duct network connected to a small cisternal cavity that connects to the teat cistern and ultimately communicates with the teat duct. At 2–3 months of age, parenchymal growth in heifers has been shown to be positively allometric [25, 132], but rapid mammary growth likely occurs at earlier postnatal stages [45, 91, 91]. The development of bovine mammary parenchymal tissue between 30 and 90 days of age (n =2 per stage) is shown in Fig. 9.2. At 30 days of age, it is difficult to detect parenchymal tissue in the developing udder. The degree of tissue development at this time appears similar to that observed in bulls [47]. In the region adjacent to each teat, mammary ducts radiate toward the body wall in a thin arrangement that corresponds with the future gland cistern. This small amount of parenchymal tissue averages about 150 mg per gland. By approximately 75 days of age, a rounded, walnut-like mass of mammary parenchymal tissue becomes very evident on palpation (Fig. 9.3). By 90 days of age, this mass of tissue has grown to approximately 10 g (approximately 60-fold). In relative terms this change does not simply reflect a change of body weight, which only doubles during this time period.

After puberty, the number of mammary epithelial cells appears to remain relatively stable in nonpregnant heifers, with cyclical changes in cell number and a degree of lobuloalveolar development during the estrous cycle. Some cumulative net isometric growth occurs during the first few cycles [132]. In contrast to the long, infrequently branching ducts and terminal end buds (TEB) of the prepubertal murine mammary gland, ruminant mammary ducts develop as compact, highly arborescent structures within loose connective tissue. Ductal elongation is accomplished through the coordinated growth, branching and extension of terminal ductal units (TDU) and growth of the loose connective tissue that surrounds the TDU as it invades the mammary fat pad.

Bovine mammary TDU consist initially of solid cords of epithelial cells that penetrate into the mammary stroma. As the primary cord of epithelial cells within the TDU (and surrounding loose connective tissue) extends into the mammary fat pad, the TDU contains 5–10 separate ductule outgrowths arranged around the central epithelial cord. Each epithelial cord contains approximately 4–8 layers of epithelial cells. The basal cell layer is populated by undifferentiated cells and presumptive myoepithelial precursors [25]. These basal cells express cytokeratin-19, but little if any α-smooth muscle actin. They form a nearly continuous stratum upon which 1–3 additional layers of epithelial cells are typically positioned. However, some of the cells in the basal layer span the entire distance from the basement membrane to the mid-line of the ductule [45]. Although the TDU consist primarily of solid epithelial cords, the subtending ducts contain lumena that are formed by undefined mechanisms. Whereas apoptosis in the trailing region of the terminal end bud seemingly accounts for lumen formation in murine ducts [67], apoptosis does not account for duct canalisation in the bovine mammary gland [25]. New TDU are formed as outgrowths from these small newly developed ducts while lumenal spaces enlarge to form a mature ductal structure.

Mammary gland histology is similar in other ruminants. However, differences are evident in the histological appearance of the prepubertal porcine mammary gland. Although the mammary epithelium develops within a sheath of loose connective tissue, as in the bovine, the developing ducts appear less branched than in ruminant mammary gland. Similar to ruminants, allometric growth of the mammary parenchyma occurs during the prepubertal period and during gestation. In contrast, the first postnatal phase of allometric growth of the murine mammary gland occurs peripubertally with onset of puberty at approximately 4 weeks of age and ends at 6–8 weeks of age [19]. Thus, as is the case with other aspects of mammary development, rodents are likely to be of limited value as models for mammary development in farm animals.

9.2.2 Hormonal Regulation of Prepubertal Mammary Growth

Regulation of prepubertal mammary growth is dependent upon a number of hormones (Fig. 9.1) and growth factors, so that regulation is both systemic and paracrine in nature. Early research indicated that mammary growth during the prepubertal period is primarily influenced by estrogens and growth hormone (GH) or somatotropin [34, 85, 153].

Estrogen secretion by the ovary is necessary for prepubertal mammary growth in heifers. Ovariectomy of calves prevents mammary growth and normal mammary growth can be restored by estrogen administration [153]. This estrogen effect has been reinforced by studies in rodents, which extended the observation and demonstrated that estrogens are required for growth and morphogenesis of mammary ducts. Classical endocrine ablation/replacement studies in rats [85] and more recent estrogen receptor knockout studies in mice [17] indicated that estrogens are essential for the normal mammary ductal growth that occurs from birth to sexual maturity. Two major forms of estrogen receptor (ER) exist, and the receptor that mediates ductal growth and morphogenesis appears to be the ERα isoform [17] rather than the ERβ isoform [33]. In prepubertal heifers, the predominant isoform in mammary gland is ERα [25, 31]. ERα is expressed by a portion of epithelial cells within the parenchymal region of the mammary gland. A small percentage of cells in the stroma between lobules (i.e., not immediately adjacent to epithelial cells) may express ERα, but the stromal cells surrounding the epithelium do not [25]. Within the mammary fat pad, roughly one-third of fibroblasts and adipocytes express ERα, accounting for the ability of the fat pad to synthesise IGF-I in response to estrogen stimulation [90]. After estrogen stimulation of mammary growth, proliferating epithelial cells were almost exclusively (>99%) ERα-negative [25]. Data suggest that epithelial proliferation in response to estrogen is initiated within ERα-positive epithelial cells of the developing mammary gland and that the signal was propagated in paracrine fashion to stromal cells and ERα-negative epithelial cells. A paracrine mediation of estrogen-induced mammary ductal growth is also indicated in mouse mammary gland, where ERα is expressed in a portion of epithelial and stromal cells. Recent tissue transplantation studies of epithelium and stromal elements between ERα knockout mice and wild-type mice support the conclusion that ERα-positive cells in the epithelium stimulate growth of mammary epithelial cells through paracrine mechanisms [86].

Growth hormone is an essential hormone for normal mammary ductal growth in rats and mice; the minimal hormone requirement for ductal growth being a combination of GH, estrogen and adrenal glucocorticoid [72, 85]. In ruminants, GH appears to be similarly important because administration of exogenous GH stimulates prepubertal mammary growth in heifers and ewes [111, 122]. Many effects of GH on the mammary gland appear to be mediated by the insulin-like growth factors (IGF), most prominently IGF-I. Insulin-like growth factor-I has mitogenic effects on mammary epithelial cells of numerous species [9]. Rather than systemic IGF-I, locally produced IGF-I may be of primary importance for paracrine regulation of mammary gland function. Additionally, IGF-regulated functions are modulated by local production of IGF-binding proteins. Six high-affinity IGF binding proteins (IGFBP-1 to 6) have been identified and nine low-affinity IGFBPs, also known as IGFBP-related proteins (IGFBP-rp1 to 9), have been identified. Most IGFs are sequestered into ternary complexes containing IGF, IGFBP-3 or IGFBP-5, and an acid-labile subunit (ALS) that also serves to regulate IGF availability [18]. Depending upon the specific IGFBP, the binding proteins may reduce IGF activity by competing with IGF-receptors for ligand, increase IGF-activity by serving as delivery vehicles to the target cell, or serve as a reservoir for IGFs, causing their slow release and reducing IGF turnover. Furthermore, IGFBPs may have activities that are independent of their interaction with IGFs and are subject to enzymatic modifications that may alter their various activities.

In addition to actions of estrogen and GH/IGF-I on prepubertal mammary growth, a number of growth factors are likely to be modulating mammary growth. Members of the epidermal growth factor (EGF) family are known to be key regulators of rodent mammogenesis. Stromal cells produce EGF and it serves as an important paracrine regulator of ductal elongation in mice, as indicated by the fact that ductal elongation does not occur in EGF-receptor knockout mice [157]. Amphiregulin, an EGF-receptor ligand, is highly responsive to estrogen and has been proposed as a paracrine mediator of ERα signaling [86, 138]. Other EGF family members, such as TGFα and TGFβ appear to serve in various capacities to stimulate or inhibit growth, ductal branching, and differentiation in rodents [42] and heifers [106, 107]. Other growth factors, such as hepatocyte growth factor/scatter factor (HGF), keratinocyte growth factor (KGF), may also regulate of mammogenesis in cattle [112]. The roles of these growth factors, their regulation and interactions with other regulators of mammogenesis continue to be an area of intense research, as it is clear that potentially important interactions occur. For example, there is evidence for interactions between the GH/IGF-I and estrogen axes. Growth hormone fails to stimulate ductal growth in ovariectomised heifers [111], estrogen enhances GH-induced production of IGF-I by mammary stroma [72] and GH induces ERα expression [46, 73] in rodents. Because of the importance of both GH and estrogens in mammogenesis, greater understanding of these interactions is critical to a more thorough understanding of mammary growth.

In a recent experiment, we identified estrogen-responsive genes in the prepubertal bovine mammary gland [80]. After 54 hours of estrogen treatment, gene expression was analyzed in the parenchymal tissue (but also containing stroma) and fat pad (stroma only) using a high-density oligonucleotide microarray representing approximately 45,000 bovine unique sequences/genes. A total of 124 genes were noted to be estrogen responsive, many of which were not previously categorised as estrogen-responsive genes. Of particular note is the finding that the genes regulated in the fat pad and parenchymal portions of the mammary gland were mostly distinct. Of the 124 responsive genes only 16 were concomitantly regulated in both the fat pad and the parenchyma. Data are consistent with the observed increase in epithelial proliferation. The pattern of gene expression was consistent with a general turnover of extracellular matrix (ECM) in the parenchymal region and deposition of ECM in the fat pad. This would facilitate the penetration of mammary epithelium into the fat pad, which occurs within a sheath of connective tissue. Data also demonstrate that estrogen treatment induces expression of IGF-I transcripts in both the mammary parenchyma and fat pad, with the greatest response in the fat pad [90]. This supports the concept that IGF-I is intimately related to estrogen stimulation of cell proliferation within the mammary gland [9, 66] and that the fat pad modulates proliferation of the neighboring epithelium [80, 90].

9.2.3 Mammary Stem Cells

It is generally accepted that mammary stem cells are responsible for mammary growth and development, as well as the turnover and regeneration of mammary epithelial cells for the life of the organism. Evidence for the existence of mammary stem cells is largely derived from mouse studies. Transplantation experiments have shown that isolated segments from any portion of the developing or lactating mammary gland are capable of regenerating a complete mammary ductal and alveolar network [40, 64, 136]. Experiments by Kordon and Smith [75] showed that an entire mammary gland could be regenerated with the progeny of a single cell. Recently, multiparameter cell sorting techniques have been used to isolate murine mammary stem cells, capable of regenerating the mammary gland from a single cell [127, 139]. In human breast, the existence of mammary stem cells is supported by the observations that entire mammary lobules are often comprised of cells showing identical X-inactivation patterns [144] and from cancer studies where mammary tumors comprised of a variety of cell types are frequently found to be of clonal origin [44, 135, 143].

Epithelial stem cells have been shown to retain labeled DNA strands for extended periods of time [14, 97, 108]. This retention of label was suggested to be due to segregation and selective retention of template DNA strands by stem cells undergoing asymmetric division [108, 109, 128, 134]. Division of a stem cell to produce a stem cell and a more differentiated cell progeny is termed asymmetric division, and division to form two stem cells is termed symmetric division. For example, selective retention of template DNA strands occurs during asymmetric division of adult murine intestinal crypt cells.

Studies of stem cells and progenitor cells in bovine mammary gland have been limited. We previously demonstrated that proliferative cells in prepubertal bovine mammary glands displayed light cytoplasmic staining and were ERα-negative [25, 45]. We suggested that the most undifferentiated population of cells likely contains mammary stem cells or primitive progenitor cells. However, because this population accounts for approximately 10% of mammary epithelial cells prepubertally, it undoubtedly contains more than epithelial stem cells. Recently, we adapted a procedure to identify and characterise putative bovine mammary stem cells based upon their ability to retain labeled DNA for extended periods. These label-retaining cells were evident in the prepubertal bovine mammary gland. We showed that these cells were cycling, both ERα-positive and ERα-negative and were of epithelial lineage [20]. Data are consistent with identification of putative mammary ERα-negative stem cells and ERα-positive progenitor cells. The methodology provides a mechanism to further characterise these putative stem cells and to evaluate factors and physiological events that might alter stem cell number. However, additional research is necessary to characterise the epithelial lineage in bovine and other livestock species. In addition to stem cells, progenitor cells with more limited division and differentiation capacity are evident [24]. Increased knowledge about the characteristics of stem cells and progenitor cells during mammary development can lead to management schemes that enhance livestock productivity and profitability. The ability to regulate mammary gland development has clear application to rearing heifers or other livestock species in a manner that supports maximal lactational performance, and ample renewal of mammary progenitor cells may be key to eliminating or shortening the dry period, and to achieving a more persistent lactation by promoting cell turnover and maintenance of mammary epithelial number during lactation. [22, 24].

9.2.4 Effect of Prepubertal Level of Nutrition on Future Milk Production

Rapid rearing of replacement dairy heifers has the potential to increase dairy profitability by bringing heifers to puberty and milk production at an early age. This reduces the time period during which the heifer produces no revenue. However, rapid rearing during the prepubertal period can result in decreased milk production [125] and calving difficulty [63]. Data indicate that overfeeding during the prepubertal period causes deposition of mammary fat and reduced mammary epithelium at puberty [124]. This has led to the hypothesis that impairment of prepubertal mammary growth causes a lifetime reduction in the number of mammary secretory cells and hence lactation potential. However, this hypothesis remains controversial and recent evidence indicates that epithelial growth is not reduced by a high plane of nutrition [91, 92]. It has also been suggested that prepubertal overfeeding may cause a permanent adjustment in nutrient partitioning within the animal, resulting in a propensity toward fat deposition rather than lactational performance [26, 51]. Finally, it has been suggested that lactational effects of prepubertal nutrition can be largely attributed to effects on skeletal growth and body weight [30, 63, 87]. These hypotheses are not mutually exclusive and effects of prepubertal nutrition on lifetime production may be a consequence of more than one of these factors, which in turn may be influenced by other management factors.

Because nearly one-third of dairy cows are replaced yearly at a cost that exceeds four billion dollars, this area of research is of considerable interest to the dairy cattle industry. Rapid rearing of other species used for milk and meat production is a common management practice and implications of these rearing practices to milk production and growth of suckling offspring are of similar interest. However, because this research in domestic animals is costly, the time required lengthy and the problem complex, a clear understanding of the relationship between prepubertal rate of gain, mammary gland development and resulting milk production remains elusive.

Compared with rates of gain and related body condition in many neonates, dairy calves are reared on milk replacers and feeding rates that produce lean animals with minimal body fat reserves. Some dairy nutritionists have questioned the wisdom of this approach from an animal health and performance viewpoint. Although a high rate of gain in the period closer to puberty may or may not have negative impacts, there is increasing evidence that higher rates of gain and body condition during the neonatal period prior to weaning may have a positive impact on mammary development and subsequent lactation performance.

9.2.4.1 Prepubertal Mammary Growth and Future Milk Production

Milk production of heifers calving at an early age is low [50, 82, 141] and may be related to high energy intake to permit early breeding [82]. Other studies indicated that high energy consumption prepubertally, but not postpubertally, reduces growth of mammary parenchyma [119, 124]. This prepubertal period of sensitivity appeared to coincide with the phase of allometric mammary growth, suggesting a time-frame of critical mammary development. Examination of a number of hormones that are known to affect mammary growth and development led to the finding that serum concentrations of growth hormone (GH) are reduced in prepubertal heifers on a high plane of nutrition [124] and concentrations of plasma leptin increased [16]. GH is required for ductal growth in rodents and prepubertal administration of GH increased the amount of mammary parenchyma in the bovine [122]. Intramammary leptin infusions have blocked epithelial cell proliferation in the prepubertal bovine mammary gland [130]. These observations have led to the hypotheses that elevated nutrient intake reduces prepubertal mammary development by impairing epithelial cell proliferation because of reduced circulating growth hormone [123, 126] or elevated circulating levels of leptin [131].

It has been hypothesised that decreased parenchymal tissue at puberty equates to reduced lifetime milk production. However, this relationship has not been tested rigorously. Although Harrison et al. [59] demonstrated a permanent reduction in mass and morphology of mammary parenchyma as a consequence of rapid prepubertal weight gain; only one study [26] has quantitatively assessed the influence of diet on both prepubertal mammary development and on subsequent milk production. Of particular importance is the recent demonstration that that proliferation of prepubertal bovine mammary epithelial cells and parenchymal DNA accretion rate are not negatively impacted by elevated nutrient intake [91]. In fact, a trend towards increased epithelial cell proliferation in response to elevated nutrient intake was observed in first few months of life (i.e. before 3 months). The lack of effect of elevated nutrient intake on these parameters of prepubertal mammary development occurred despite increased plasma leptin and altered nutrient deposition [92]. These data demonstrated that the developing mammary parenchyma is largely refractory to subtle changes in homeorhetic signals involved in coordinating changes in nutrient partitioning that are brought about by elevated nutrient intake. Finally, because age at slaughter described most of the variation in parenchyma DNA observed in this experiment, age at slaughter appears to be the single greatest determinant of prepubertal mammary development and therefore appears to have been a confounding factor in studies that evaluated the impact of nutrient intake at a constant body weight (i.e. age differential across treatments).

Most studies dealing with effects of overfeeding dairy heifers prepubertally have utilised high concentrate diets. Potential effects of diet composition during the prepubertal period of mammary development have received limited attention [26, 113, 121, 151]. In addition to feed and energy intake, diet composition may alter energy deposition, influence secretion of key hormones or tissue responsiveness, and produce permanent effects on development of mammary and adipose tissues.

Waldo and colleagues evaluated the impact of two diets and rates of gain on mammary growth and first lactation milk yields [26, 151, 152]. The diets compared were alfalfa or corn silage based diets, which differed in protein and energy content. The alfalfa silage diet contained 22% crude protein and 3.1 Mcal of digestible energy/kg of DM, whereas the corn silage diet provided 16% crude protein and 3.4 Mcal of digestible energy/kg DM. Differences in fat deposition were clearly evident in both carcass and mammary gland composition. More fattening occurred in corn silage- than alfalfa silage-fed heifers and puberty was accelerated in the corn silage group. Greatest fattening occurred in heifers fed the corn silage diet for accelerated growth, and mammary parenchymal growth was reduced in this group. In the study, slaughter was at a constant physiological age (second estrous cycle) rather than chronological age. Decreased mammary parenchymal growth was correlated with reduced concentrations of GH in the circulation. Concentrations of IGF-I in the circulation were greatest in the group with reduced GH and parenchymal growth (accelerated growth rate, corn silage-fed), arguing against IGF-I mediated regulation of mammary ductal growth.

Although accelerated rearing of heifers on a corn silage-based diet caused excessive body fattening and had a negative impact on mammary parenchymal growth, there was no decline in first lactation milk yield. Assuming that impaired mammary growth prepubertally accounts for the reduced milk production accompanying accelerated rearing on some experiments, one might hypothesise that there was compensatory mammary growth in the Waldo study. This is possible, for the mammary gland exhibits a degree of plasticity, as evidenced by the compensatory growth and development of glands adjacent to a blind quarter (i.e., a quarter without a teat that connects to the lactiferous sinus, resulting in atrophy of that gland). In fact, data indicate that a degree of compensation occurs even in those experiments where a permanent effect on milk production has been observed. The inhibition of mammary growth and cell number during the prepubertal period has been of greater magnitude than the negative impact on milk production observed in later life. High energy intake during the prepubertal period has been reported to reduce mammary cell number at the end of the allometric phase of mammary growth by 30–45% [26, 121, 124], whereas the milk production deficit reported for non-contemporaneous heifers reared with similar accelerated rates are in the order of 7–14% for recent investigations using modern U.S. Holsteins [78, 113]. Although data are consistent with the concept that accelerated rearing of dairy heifers can induce excessive fattening and reduced mammary development at puberty, data relating to the impact of these prepubertal events on mammary cell number and function during ensuing lactations are needed.

It has been suggested that crude protein is a limiting factor for developing accelerated heifer rearing programs [147, 149]. By supplying a diet of high protein and energy from 4 months of age until luteal phase of the fifth estrous cycle, Radcliff et al. increased growth rate to 1200 g/d (controls at 800 g/d) without negatively impacting mammary development, and reduced age at puberty without impact on BW or skeletal size at puberty [114]. Administration of GH to heifers on either high-gain or control diets increased BW, skeletal size and mammary growth (47%). In a subsequent experiment [113], heifers were reared on analogous diets for BW gains of 800 vs. 1200 g/d. A third group of heifers reared on the high-gain diet were injected daily with GH (25 μg/d). Heifers were bred after body weight exceeded 363 kg and treatments (dietary and GH) were continued until pregnancy was confirmed. Heifers in both high-gain groups were 90 days younger than control heifers at first breeding and parturition. Postpartum body weight, body condition score and skeletal size did not differ among treatments. Milk production of heifers reared for high rate of gain produced 14% less milk than heifers reared at the standard rate of gain, even though the diet was formulated for high protein content. However, GH treatment prepubertally prevented the decline in milk production observed in the high-gain group. These results were contrary to expectations. In light of results from their first experiment, it was hypothesised that the high-gain group would not produce less milk than heifers in the low gain group and that GH injection would increase milk production beyond that of heifers on the standard diet.

Van Amburgh et al. [146] evaluated the impact of body weight gain and different sources of protein (fed at 18% crude protein) during the prepubertal period. Three growth rates were employed from 90 to 320 kg of body weight: 600, 800 and 1000 g/d. Among diets, energy was balanced to achieve the target rates of gain, and protein was formulated to meet ruminal nitrogen requirements and to exceed the tissue requirements. Ages at first calving were 24.5, 22.0 and 21.3 months for heifers reared at 600, 800 and 1000 g/d, respectively. First insemination occurred when heifers weighed 340 kg. Without regression analysis, heifers reared at 1000 g/d produced 9% less milk than those reared at 600 g/d. However, when calving weight was taken into account, there was no difference in milk production due to rate of gain or protein source. Post-treatment factors, such as postcalving weight, accounted for more variation in milk yield than prepubertal body weight gain. This of course emphasises the importance of postpubertal heifer management. However, the impact of accelerated postpubertal body weight gain has provided conflicting results. Finally, no effects of prepubertal dietary treatments were detected for milk yield of cows in second lactation.

Another recent study evaluated the effects of accelerated gain (700 vs. 1000 g/d) and estrogen (estradiol) treatment prepubertally (4.5–9.5 months of age) on mammary development and milk production [78, 79]. The experimental rationale was to determine if estrogen treatment could be used to increase prepubertal mammary growth and overcome potential negative effects of accelerated body growth on milk production. Only indirect, non-invasive measurements of mammary development were employed. Although data suggested that estrogen treatment enhanced mammary development during the treatment period, the effects were subsequently lost. Consistent with its apparent increase in mammary gland growth, estrogen treatment increased circulating concentrations of GH and IGF-I during the treatment period. However, both accelerated rate of gain and estrogen treatment prepubertally decreased first lactation milk yield (7.1 and 5.2%, respectively), with heifers being of similar body weight at calving regardless of treatment.

Knowledge of the interactions between rate of gain, the GH/IGF axis and estrogen axis are essential to our understanding of prepubertal development. We and others have hypothesised that a negative impact of accelerated heifer rearing on mammary development may be partly mediated by estrogen. As summarised earlier, GH stimulates ERα expression. Consequently, decreased GH concentrations in high-gain heifers may cause decreased expression of ERα in epithelial cells and a reduction in epithelial sensitivity to the mitogenic effects of estrogen. A recent study evaluating the impact of GH treatment on ERα expression, found no change in the percentage of cells expressing ERα by immunohistochemistry, but protein and transcript levels were not evaluated [13]. In another recent study, ERα transcript abundance and immunohistochemical staining were consistent with the drop in mammary epithelial cell proliferation and parenchymal accretion observed over development, but did not support a negative effect of nutrition on parenchymal growth [91, 92].

If underlying changes by which alterations in mammary growth prepubertally exert effects on subsequent lactation are subtle, perhaps greater insight can be gained by knowledge of the underlying proliferative population in the prepubertal bovine mammary gland. We recently characterised the proliferative epithelial populations on the basis of BrdU labeling and differential staining characteristics [45]. The proliferative populations in the prepubertal bovine gland consist of two classes of lightly staining cells in histological sections that may reflect adult stem cells and their progeny of progenitor cells. Total number or lineage of these cells may be altered by nutritional "overfeeding" so that their ability to fully populate or maintain a cohort of fully differentiated secretory cells in the mammary gland throughout lactation is impaired. Further characterisation of these progenitor cells and their regulatory mechanisms may provide opportunities to alter proliferation and cell renewal in prepubertal and mature cattle [24].

9.2.4.2 Body Growth and Milk Production

A goal of rearing replacement dairy heifers is to decrease the interval from birth to first calving, but to do so while promoting skeletal growth and minimising negative impacts on lactation potential. It is well established that there is a positive relationship between body weight at calving and milk production in first lactation dairy cows [30, 58, 63]. First lactation milk yield appears to be maximal for Holstein heifers weighing between 590 and 635 kg at calving [70]. Furthermore, adequate skeleton size is needed to minimise dystocia during the first parturition and is more positively related to first lactation milk yield than body weight [87, 129]. The majority of skeletal growth occurs during the prepubertal period [60] when rates of withers and hip height growth are as much as 3-fold greater than after puberty [12]. Relative rates of skeletal growth as measured by changes in withers or hip height decrease gradually over time from as much as 5 cm/month at 2 months of age to around 1 cm/month during the post-pubertal period [60]. This suggests that the greatest opportunity for enhancing skeletal growth is during the prepubertal period. Increases in rates of skeletal and body weight in prepubertal dairy heifers can be achieved by increasing the energy density of diets. However increasing body weight gain above 1 kg/d reduces mammary parenchymal tissue and increases mammary fat deposition [26, 124], and both factors are associated with lower milk production during the first lactation. Thus methods to increase skeletal growth rates without increasing fat deposition might provide appropriate strategies to accelerate growth without potential implications of accelerated growth on mammary development.

Somatotropin, particularly when used in the presence of increased intestinal protein such as abomasal infusion of casein [65], has been shown to enhance N retention in Holstein steers, suggesting that lean tissue and possibly skeletal deposition may be enhanced by combined treatment with recombinant bovine somatotropin (bST) and dietary rumen undegradeable protein (RUP). Previous studies demonstrated that prepubertal treatment with bST enhances mammary growth [81, 120] and increases withers height at puberty [56, 114]. Thus, bST in combination with added RUP might provide a practical means to optimise skeletal growth rates during the prepubertal period without having negative impact on mammary development.

This provided the objective of a study to test effects of dietary rumen-undegradable protein (RUP) and administration of recombinant bovine somatotropin (bST) during the prepubertal period on mammary growth, skeletal growth and milk yield of dairy heifers [23, 95, 96]. Seventy-two Holstein heifers were used in the experiment. At 90 days of age, 8 heifers were slaughtered before initiation of treatment. Remaining heifers were assigned randomly to 1 of 4 treatments. Treatments consisted of a control diet (5.9% RUP, 14.9% CP, DM basis) or RUP-supplemented diet (control diet plus 2% added RUP) with or without 0.1 mg of bST/kg of BW per day applied in a 2 × 2 factorial design. A total of 6 heifers per treatment were slaughtered at specific chronological ages (3 each at 5 and 10 months) for mammary tissue and body composition analysis. Remaining heifers were bred to evaluate impact of treatment on subsequent milk yield and composition. Mammary parenchymal growth (mass or epithelial cell proliferation index) was not affected by RUP or bST treatment, nor was mammary gland composition

[23]. Neither deleterious effects of increased rates of gain nor positive effects of bST were evident in prepubertal mammary growth. Subsequent milk production and mammary composition did not differ among treatments [23]. Growth curves showed that effects of RUP on rates of body weight and skeletal growth (height at withers and hip) were greatest from 90 to 150 days of age and diminished thereafter, suggesting that protein was limiting during this time period [95]. Conversely, bST effects tended to be greater as the heifers approached puberty, but only in the presence of added RUP. Age at puberty was not affected by treatment, averaging 314 days of age across treatments. From 314 to 644 days of age, rates of BW, and skeletal growth were similar among treatment groups. However, treatment differences present at 314 days of age persisted through 644 days of age, more than 10 months after treatments ceased. The results from this experiment suggest that protein during the early postweaning period and bST during the 200–300 days of age period just prior to puberty could be used to accelerate simultaneous increases in both body weight and skeletal growth rates in dairy heifers without reducing age at puberty [95]. Additionally, there were no treatment effects on rates of body fat, protein, and energy deposition [96]. This suggests that nutritional and endocrine manipulations of prepubertal heifers could increase growth rates of skeletal tissues without increasing fat deposition or altering mammary growth and subsequent milk production.

9.2.4.3 Nutritional Imprinting and Milk Production

Prepubertal dietary regimen may influence secretion of key hormones or tissue responsiveness, thus producing permanent effects on development of mammary and adipose tissues or endocrine controls of lactogenesis and lactation. Certainly nutritional deficiencies can influence the lifetime function of an organ, such as the brain, by failure to supply necessary nutrients for complete development of the organ in question. However, it has been suggested that nutrients might serve as critical signals for more subtle developmental effects, such as by affecting stem cell proliferation and thereby permanently altering the cellular composition of a tissue or organ [83]. Lifetime metabolic patterns may be altered. For example, rats fed low protein diets during gestation gave birth to pups with greater gluconeogenic capacity [41], an early postnatal shift in energy supply to suckling rats from a fat-rich milk to carbohydrate rich caused metabolic programming that resulted in permanent hyperinsulinemia, adult onset diabetes and obesity [137], and protein restriction of suckling dams caused a permanent alteration in lipid metabolism of the neonate, resulting in permanent reductions in cholesterol and triacylglycerol concentrations [84]. The critical windows for these effects are restricted to the period of gestation or early postnatal life. However, it is possible that nutrition during pregnancy may impose imprinting effects on mammary gland function of the dam, for this is a critical period of allometric mammary growth and tissue differentiation. This latter aspect will be discussed in the subsequent section dealing with pregnancy.

Whether metabolic programming occurs in livestock species has received scant attention. An experiment was designed to evaluate the effects of rate of body weight gain and type of silage before puberty on the partitioning of excess dietary energy between synthesis of milk and body weight gain [51]. Cows that had been fed at two rates of gain (725 vs. 950 g/d) from 175 to 325 kg of body weight on an alfalfa or corn silage based diet were used. Lactating cows were switched from a control to a high-energy diet according to a double-reversal experimental design with 6-week periods. Neither body weight gain nor diet influenced the magnitude of change in DMI, milk yield, milk composition or circulating hormone concentrations. These data argue against a metabolic programming induced by accelerated heifer rearing. However, because there was no effect of the prepubertal treatment on milk yield, the results do not fully address whether metabolic programming can explain potential reductions in milk yield from heifers subjected to accelerated prepubertal growth rates.

Park and colleagues have developed and utilised a nutritional regimen for rearing heifers, which capitalised on the nature of compensatory growth [48, 102, 103] and suggest that the lasting benefits on milk production are due to epigenetic effects of the heifer rearing program [101]. The regimen alternated periods of energy restriction with periods of realimentation, to achieve high rates of body and mammary growth. During dietary energy restriction, energy costs were reduced due to reduction of metabolic processes that are not essential for growth. Realimentation then induced compensatory growth, i.e. growth with reduced maintenance requirements and altered endocrine status resulting in deposition of leaner tissues. The proposed program included three phases of compensatory growth, referred to as stair-steps. The first stair-step occurred prepubertally and consisted of energy restriction [17% crude protein and 2.35 Mcal/kg of ME] for 3 months followed by realimentation (12% crude protein and 3.05 Mcal/kg of ME) for 2 months. The second step was puberty to breeding and consisted of energy restriction for 4 months followed by realimentation for 3 months. The third and final step occurred during pregnancy, with energy restriction for 4 months and realimentation for the final 2 months of gestation. Animals reared on the stair-step program produced more milk than controls during ensuing lactations. The increase was approximately 6% for beef heifers [103] during their first lactation and 21% or 15% for dairy cows [48] during first and second lactation, respectively. The persistence of increased milk production across lactation cycles suggests that maintenance of altered gene expression in mammary cell progeny is due to epigenetic factors, such as gene methylation or histone modification. Indeed, reduced DNA-methylation was observed in Holstein heifers during the gestational phase of compensatory growth [29].

9.2.5 Effect of Biologically Active Agents in Feed

9.2.5.1 Phytoestrogens

Phytoestrogens are nonsteroidal plant molecules that exhibit estrogenic or anti-estrogenic activity. There are three primary classes of phytoestrogens: isoflavanoids, lignans, and coumestans. The most widely studied dietary phytoestrogens are the soy isoflavones and the flaxseed lignans, which have primarily been evaluated with regard to human health and breast cancer risk [145]. These compounds typically bind to both ER isoforms with affinities for ERα that are far less than those for the primary endogenous ligand, estradiol-17β, but with an affinity for ERβ that may be nearly equal to that for estradiol-17β [71, 76]. The relative affinity for ER isoforms depends upon the specific phytoestrogen, which may therefore act as a selective estrogen receptor modulator (SERM). In addition to the ERα and ERβ isoforms of the estrogen receptor, there is a related orphan receptor, known as estrogen-related receptor (ERR), which exists in three isoforms, α, β, γ [52]. Flavone and isoflavone phytoestrogens are agonists of ERRα [140]. The nature of the response to ligand binding by nuclear receptors is determined by the structure of the ligand, the receptor subtype involved (ERα, ERβ, ERRα, ERRβ, ERRγ), the nature of the hormone-responsive gene promoter, and the balance of co-activators and co-repressors that modulate these cellular responses in the tissue [69]. Additionally, whether the response to a phytoestrogen is viewed as estrogenic or anti-estrogenic depends upon the concentration of the phytoestrogen and of endogenous estrogens (Fig. 9.1).

Expression of estrogen receptor isoforms in mammary tissues of domestic animals has not been fully evaluated. However, in bovine mammary gland, ERα is the predominantly expressed isoform throughout mammary development, with little or no expression of ERβ mRNA [31, 117] or protein [31]. Because ERRα is also present in bovine mammary gland throughout all developmental stages evaluated [31], the bovine mammary gland appears capable of responding to phytoestrogens through ERα- or ERRα-mediated pathways. ERRα is probably expressed in mammary tissues of other ruminants and other domestic species (as it is in rodent and human mammary tissue); however, the relative importance of ERβ as an estrogenic transcription factor in these various species remains to be determined.

Because circulating concentrations of phytoestrogens may exceed those of endogenous estrogen, they may have a significant impact on mammary gland growth and differentiation. For example, concentrations of isoflavones in plasma of Japanese women, consuming a typical Japanese diet that is high in isoflavones, may contain five-times the concentration of isoflavones as estradiol-17β (700 vs. 150 nM) and plasma of infants consuming soy milk may contain 7000 nM phytoestrogens and undetectable levels of estradiol-17β [11]. Considering the quantity of phytoestrogens consumed by cattle and other livestock species, mammary developmental processes are likely to be influenced or perhaps even depend upon dietary phytoestrogens (Fig. 9.1). The low concentration of endogenous estrogens in ruminants also suggests that estrogenic effects of phytoestrogens will predominate over antiestrogenic effects in these species. Indeed, a well-documented effect of isoflavone consumption is reproductive disorders reported for sheep consuming clover [1]. Reproductive disorders in cows and sheep resulting from consumption of forages that are high in phytoestrogens are typically due to impaired ovarian function, which is reversed 4–6 weeks after consumption of the estrogenic forage is discontinued. Mammary hypertrophy and mammary secretions are often evident along with the ovarian dysfunction. Prolonged consumption of forages that are high in phytoestrogens can lead to permanent infertility due to impaired function of the cervix [1].

Rodent studies have been employed to evaluate the impact of early consumption of dietary isoflavones on mammary carcinogenesis. Prenatal administration of genistein (the primary and potent isoflavone in soybean) to the dam or neonatal administration to the pups reduced the induction of mammary tumors in the pups subsequent to DMBA treatment at 50 days of age, an effect that has been attributed to increased differentiation of the mammary gland prior to DMBA administration [77]. Similarly, neonatal diets of soy protein isolate (containing isoflavones) rendered rats less susceptible to mammary carcinogenesis, even when challenged with DMBA (28 days of age) prior to any apparent effects on tissue growth and differentiation [11]. Of relevance to animal agriculture, is not only the effects on mammary growth and differentiation, but the implication that isoflavones produce lasting effects on mammary gland progenitor populations and hence life-long effects on regulation of mammary growth [11, 77].

9.2.5.2 Mycotoxins

Molds can interfere with animal performance by reducing the nutritional value and palatability of feed, but more importantly by producing secondary metabolic byproducts known as mycotoxins. The mycotoxins cause noninfectious diseases or mycotoxicoses, but also can have endocrine disruptive effects on animals that consume tainted feed. Major mycotoxins that impair growth and reproductive efficiency of livestock are aflatoxins, trichothecenes (includes T-2 toxin, deoxynivalenol, diacetyscerpenol) fumonisins, ochratoxins, zearalenone and ergot alkaloids. Mold infection of crops may occur in the field or in storage. In addition to the impact on animal productivity, risk to human health from mycotoxin residue in milk or meat is a potential concern.

The severity of toxicity resulting from consumption of mycotoxins is dependent upon the mycotoxin, its concentration, mode of action, duration of exposure, as well as the species, age and physiological state of the animal ingesting the contaminated feed. Ruminants are generally more resistant to mycotoxins than are monogastic mammals [158], seemingly due to the ability of rumen microbes to metabolise the toxins. Among monogastrics, swine appear to be particularly sensitive (for review see [68]). Clinical symptoms of mycotoxicoses vary according to the toxin and species of animal. For example in swine, Fumonsisin B_1 can cause hepatotoxicity, pulmonary edema, heart and respiratory dysfunction, and death [43]. In cattle symptoms of mycotoxicoses are typically nonspecific, including reduced feed intake and milk production, poor body condition, rough hair coat, immunosupression and reproductive problems [68]. Symptoms of mycotoxicoses typically disappear shortly after ingestion of the mycotoxins is reduced. However, effects on mammary development may produce lasting effects on lactational performance. Because immunosupression may result from ingestion of mycotoxins, and this often involves reduced cellular immunity, macrophage migration and phagocytosis [68, 94, 148], it is possible that mycotoxicoses during phases of mammary development can impair future productivity. This hypothesis is based on the importance of mammary gland macrophages to mammary development [118]. Effects on mammary gland development and differentiation are clearly evident for mycotoxins that exhibit estrogenic or endocrine disruptor activities, such as zearalenone and ergot alkaloids.

Zearalenone is an estrogenic mycotoxin produced primarily by fungi of the *Fusarium* genus, which contaminate cereal crops throughout the world [159]. Zearalenone is rapidly biotransformed and excreted by animals (for reviews see [68, 159]). Key biotranstransformations include the formation of reduced metabolites, α-zearalenone and β-zearalenone, which have increased estrogenic activity, and the conjugation of zearalenone and its metabolites with glucuronic acid, which facilitates excretion. Additional metabolites, including α- and β-zearalenol, are formed; with the profile of zearalenone metabolism depending upon the animal species. Although rapidly excreted, concentrations of zearalenone and its metabolites are typically undetectable in milk (< 0.5 ppb; [110]).

Biological effects of zearalenone and its metabolites appear to occur primarily via interaction with estrogen receptors. In this regard, affinity by the rat uterine estrogen receptor-α for these mycotoxins is in the order: α-zearalenol > α-zearalenone > β-zearalenol > zearalenone > β-zearalenone [159]. The α-zearalenol metabolite has been used as an anabolic feed additive (Ralgro) for sheep and cattle. Investigations have demonstrated estrogenic properties of zearalenone in the feed. Although concentrations are typically low, severe contamination can occur, with ensuing adverse effects on livestock. Pigs are most sensitive, and effects can be detected at 1.5 – 3 mg/kg diet [37]. In gilts, zearalenone can cause vulvovaginal swelling, anestrus or lengthened postpartum anestrus. During pregnancy, zearalenone reduces embryonic survival and can decrease fetal weight; however, effects on mammary development and lactational performance have not been reported. Because cattle are less sensitive than pigs to zearalenone, more highly contaminated feeds are necessary to produce adverse effects [154, 155]. Zearalenone contamination of corn (1.5 mg zearaleonone and 1.0 mg deoxynivalenol/kg of feed) increased incidence of behavioral estrus, vaginitis, and a reported increase in prepubertal mammary gland development when fed to dairy cattle [32]. Zearaleonone has also been implicated as an agent producing infertility in sheep and cattle grazing on pasture in New Zealand [37]. The impact of zearaleonone on mammary gland development and milk production has received little attention. However, the potential for such effects is evident when contamination is sufficient, or perhaps when augmented by interactions with other mycotoxins or phytoestrogens in the diet.

Ergot alkaloids are produced by ubiquitous ergot fungi (Claviceps species) that are parasitic to cereals and by grass endophytes (Epichloe, Neotyphodium and Balansia species) that are found in pastures grasses. The ergot alkaloids produce varied physiological effects resulting from their interactions with neuroreceptors and endocrine disruption. The variety of effects of ergot alkaloids result from their actions as partial agonists or antagonists at adrenergic, dopaminergic, and tryptaminergic receptors, with the effect depending upon the specific ergot alkaloid and its concentration as well as the animal species, tissue and physiological state. Generally, these compounds promote vasoconstriction and uterine contraction

[39]. Acting as a dopamine agonist, ergot alkaloids may also inhibit prolactin secretion. This endocrine disruptor activity is of relevance to mammary development. In particular, the lobuloalveolar development that occurs during pregnancy is highly dependent on prolactin, as is the final differentiation (Fig. 9.1; see section about gestation effects). Prolactin is considerably less important during prepubertal mammary development in ruminant livestock species [3]

9.2.6 Effects of Photoperiod

Photoperiod serves as an environmental cue for reproductive activity in many species. The first reported effect of photoperiod on mammary gland function was the demonstration that exposing lactating dairy cows to long days (18 hours of light) stimulated milk yield above that of cows exposed to short days [104]. This effect of day length on lactating cows has been confirmed in numerous experiments and commercial applications. More recently the utility of managing photoperiod has been extended by findings that short days during the dry period (the nonlactating interval between successive lactations) increased milk production in the ensuing lactation, and seemingly had beneficial health effects [38].

In addition to documented effects in the mature cow, photoperiod manipulation has been shown to influence mammary development. Relative to short days, parenchymal cell number was greater in heifers exposed to long days during the prepubertal period [105] (Fig. 9.1). Because long days hasten puberty (for review see [57]), a portion of that growth may be associated with the acceleration in gonadal steroid secretion associated with long day treatment. However, long days also increase parenchymal cell number after puberty, suggesting that other factors in addition to gonadal steroids play a role [105]. Certainly the effects of photoperiod on PRL and IGF-I would be consistent with greater mammary growth, though definitive experiments have not been reported.

9.3 Mammary Growth During Gestation

9.3.1 Gestational Development and Lactogenesis

Extensive mammary development occurs during gestation, at which time mammary growth is exponential and driven by hormones of pregnancy, most importantly estrogen, progesterone and prolactin [3, 10]. Epithelial development during pregnancy gives rise to true alveoli that emanate from the distal termini of ducts. The resulting structures have been likened to clusters of grapes, wherein the grapes represent alveoli and the stems represent ducts that drain these secretory units. Alveoli consist of a single layer of epithelial cells overlain and engulfed by a few myoepithelial cells and their processes. During pregnancy, mammary epithelial cells undergo extensive cytological and biochemical differentiation necessary for transition to an organ that is capable of producing copious quantities of milk during lactation [7]. The process of cellular differentiation to a secretory state is termed lactogenesis. The timing of lactogenic events differs among species, but generally some synthesis of milk protein and fat is initiated during early lactogenesis (last trimester of pregnancy), whereas synthesis of α-lactalbumin is more tightly coupled to parturition. Because α-lactalbumin is a cofactor for lactose synthetase, its synthesis is coupled to lactose synthesis. Being the primary osmotically active molecule in milk, lactose synthesis draws water into milk and accounts for the onset of copious milk secretion. These processes are hormonally regulated and timed to meet the nutritional needs of the neonate through interaction of the dam's endocrine system and fetus-placenta during pregnancy and parturition.

Endocrine disruptors can impair the balance of growth and differentiation that occurs during gestation. As indicated above, ovarian steroids and prolactin are key to promoting growth and differentiation of mammary epithelial cells. Impairment of the effects of estrogen or progesterone by phytoestrogens or by less studied phytoprogestins, can potentially reduce mammary development and subsequent lactational performance. Dysregulation of prolactin secretion by ergot alkaloids also has the potential to impair mammary development and lactogenesis (Fig. 9.1). Indeed, prohibiting this prepartum increase in plasma prolactin concentration by administration of ergot alkaloids prevented the final stage of final stage of differentiation by mammary secretory cells, thus inhibiting lactogenesis [5, 6, 156]. Although the impact was severe in dairy cows, milk production slowly increased, though it remained well below control levels for the duration of the study [5, 6]. In non-dairy species, a failure to rapidly initiate secretion of ample quantities of milk can severely compromise the ability of suckling offspring to thrive. Although endocrine disruptors may have a significant effect throughout gestation, ingestion of feed or forage that is contaminated with ergot alkaloids during the periparturient period can have dire consequences.

9.3.2 Mammary Gland Progenitor Cells

Significant changes in the stem cell/progenitor cell population of the mammary gland occur during pregnancy. The impetus for many of these studies has been the observation that early pregnancy reduces the risk of breast cancer, i.e. susceptibility is greater in nulliparous than multiparous women. In mice, pregnancy has been shown to cause persistent changes in the transcriptome of the parous mammary gland [36, 53, 54] and it was hypothesised that epigenetic factors, such as DNA methylation and histone modification may be responsible for these persistent changes in gene expression. Using Cre-lox technology, Wagner and colleagues demonstrated that parity induces a population of mammary epithelial cells that appear to serve as alveolar progenitor cells during successive lactation cycles [150]. However, in the appropriate environment these parity-induced mammary epithelial cells (PI-MEC) are pluripotent. On the other hand, expression of TGF-β1 inhibits proliferation and apoptosis of PI-MEC. The resulting lack of cell turnover induces senescence of mammary epithelial cells and renders the gland unable to support lactation. This is consistent with our hypothesis that turnover of progenitor cells during the interval between successive lactations is critical for lactational success [21], as these cells are responsible for replacement and maintenance of the population of mammary secretory cells [24]. Evaluation of the dynamics of such progenitor cells should permit development of methods to decrease the necessity for a dry period in dairy cattle, and generally to promote lactational success and to improve persistency of lactation.

9.4 Summary

Fetal and postnatal growth of the mammary gland is regulated by endocrine and paracrine signals. These may be influenced by external stimuli such as level of nutrition, biologically active components in the diet, and photoperiod. Thus management factors can alter mammary growth and differentiation. These effects may be immediate and short-lived, or long-term, depending upon the time when the stimulus is applied. Endocrine modulating factors, including nutritional influences, may produce long term effects when applied during critical periods of mammary gland development. This may induce epigenetic effects on the proliferation and differentiation of mammary stem cells and progenitor cells. This in turn may enhance milk yield and persistency of lactation. Short-term effects such as stimulation or inhibition of lactogenesis may have a strong influence on survival and growth of neonates that are dependent upon suckling for nourishment. Recent development of tools for studying mammary growth, differentiation and gene expression hold the promise for continued advances in this important area for promoting livestock productivity and efficiency through improved management of growing animals.

References

1. Adams, N.R. 1995. Detection of the effects of phytoestrogens on sheep and cattle. *J. Anim. Sci.* **73**:1509–1515.
2. Akers, R.M. 1985. Lactogenic hormones: binding sites, mammary growth, secretory cell differentiation, and milk biosynthesis in ruminants. *J. Dairy Sci.* **68**:501–519.
3. Akers, R. M. 2002. Lactation and the Mammary Gland. Iowa State Press, Ames, IA.
4. Akers, R.M. 2006. Major advances associated with hormone and growth factor regulation of mammary growth and lactation in dairy cows. *J. Dairy Sci.* **89**:1222–1234.
5. Akers, R.M., D.E. Bauman, A.V. Capuco, G.T. Goodman, and H.A. Tucker. 1981. Prolactin regulation of milk secretion and biochemical differentiation of mammary epithelial cells in periparturient cows. *Endocrinology* **109**:23–30.
6. Akers, R.M., D.E. Bauman, G.T. Goodman, A.V. Capuco, and H.A. Tucker. 1981. Prolactin regulation of cytological differentiation of mammary epithelial cells in periparturient cows. *Endocrinology* **109**:31–40.
7. Akers, R. M. and A. V. Capuco 2002. Lactogenesis, p. 1442–1446. *In* H. Roginski, J. W. Fuquay, and P. F. Fox, (eds.), Encyclopedia of Dairy Sciences, Academic Press, London.

8. Akers, R.M., S.E. Ellis, and S.D. Berry. 2005. Ovarian and IGF-I axis control of mammary development in prepubertal heifers. *Domest. Anim. Endocrinol.* **29**:259–267.
9. Akers, R.M., T.B. McFadden, S. Purup, M. Vestergaard, K. Sejrsen, and A.V. Capuco. 2000. Local IGF-I axis in peripubertal ruminant mammary development. *J. Mammary Gland Biol. Neoplasia* **5**:43–51.
10. Anderson, R. R. 1985. Mammary Gland. p. 3–38. *In* B. L. Larson, (ed), Lactation, The Iowa State University Press, Ames, IA.
11. Badger, T.M., M.J. Ronis, R. Hakkak, J.C. Rowlands, and S. Korourian. 2002. The health consequences of early soy consumption. *J. Nutr.* **132**:559S–565S.
12. Barash, H., Y. Barmeir, and I. Bruckental. 1994. Effects of low-energy diet followed by a compensatory diet on growth, puberty and milk production in dairy heifers. *Livest. Prod. Sci.* **39**:263–268.
13. Berry, S.D.K., P.M. Jobst, S.E. Ellis, R.D. Howard, A.V. Capuco, and R.M. Akers. 2003. Mammary epithelial proliferation and estrogen receptor alpha expression in prepubertal heifers: effects of ovariectomy and growth hormone. *J. Dairy Sci.* **86**:2098–2105.
14. Bickenbach, J.R. 1981. Identification and behavior of label-retaining cells in oral mucosa and skin. *J. Dent. Res.* Sp.Iss.C **60**:1611–1620.
15. Blackburn, D.G., V. Hayssen, and C.J. Murphy. 1989. The origins of lactation and evolution of milk: A review with new hypothesis. *Mamm. Rev.* **19**:1–26.
16. Block, S.S., J.M. Smith, R.A. Ehrhardt, M.C. Diaz, R.P. Rhoads, M.E. Van Amburgh, and Y.R. Boisclair. 2003. Nutritional and developmental regulation of plasma leptin in dairy cattle. *J. Dairy Sci.* **86**:3206–3214.
17. Bocchinfuso, W.P. and K.S. Korach. 1997. Mammary gland development and tumorigenesis in estrogen receptor knockout mice. *J. Mammary Gland Biol. Neoplasia* **2**:323–334.
18. Boisclair, Y.R., R.P. Rhoads, I. Ueki, J. Wang, and G.T. Ooi. 2001. The acid-labile subunit (ALS) of the 150 kDa IGF-binding protein complex: an important but forgotten component of the circulating IGF system. *J. Endocrinol.* **170**:63–70.
19. Borellini, F. and T. Oka. 1989. Growth control and differentiation in mammary epithelial cells. *Environ. Health Perspect.* **80**:85–99.
20. Capuco, A.V. 2007. Identification of putative bovine mammary epithelial stem cells by their retention of labeled DNA strands. *Exp. Biol. Med. (Maywood)* **232**:1381–1390.
21. Capuco, A.V. and R.M. Akers. 1999. Mammary involution in dairy animals. *J. Mammary Gland Biol. Neoplasia* **4**:137–144.
22. Capuco, A. V., E. Annen, A. C. Fitzgerald, S. E. Ellis, and R. J. Collier 2006. Mammary cell turnover: relevance to lactation persistency and dry period management. p. 363–388. *In* Ruminant physiology: Digestion, metabolism and impact of nutrition on gene expression, immunology and stress, Wageningen Academic Publishers, Wageningen, The Netherlands..
23. Capuco, A.V., G.E. Dahl, D.L. Wood, U. Moallem, and R.E. Erdman. 2004. Effect of bovine somatotropin and rumen-undegradable protein on mammary growth of prepubertal dairy heifers and subsequent milk production. *J. Dairy Sci.* **87**:3762–3769.
24. Capuco, A.V. and S. Ellis. 2005. Bovine mammary progenitor cells: current concepts and future directions. *J. Mammary Gland Biol. Neoplasia* **10**:5–15.
25. Capuco, A.V., S. Ellis, D.L. Wood, R.M. Akers, and W. Garrett. 2002. Postnatal mammary ductal growth: three-dimensional imaging of cell proliferation, effects of estrogen treatment, and expression of steroid receptors in prepubertal calves. *Tissue Cell* **34**:143–154.
26. Capuco, A.V., J.J. Smith, D.R. Waldo, and C.E.,Jr. Rexroad. 1995. Influence of prepubertal dietary regimen on mammary growth of Holstein heifers. *J. Dairy Sci.* **78**:2709–2725.
27. Ceriani, R.L. 1970. Fetal mammary gland differentiation in vitro in response to hormones. I. Morphological findings. *Dev. Biol.* **21**:506–529.
28. Ceriani, R.L. 1970. Fetal mammary gland differentiation in vitro in response to hormones. II. Biochemical findings. *Dev. Biol.*. **21**:530–546.
29. Choi, Y.J., K. Jang, D.S. Yim, M.G. Baik, D.H. Myung, Y.S. Kim, H.J. Lee, and J.S. Kim. 1998. Effects of compensatory growth on the expression of milk protein gene and biochemical changes of the mammary gland in Holstein cows. *J. Nutr. Biochem.* **9**:380–387.

30. Clark, R.D. and R.W. Touchberry. 1962. Effect of body weight and age at calving on milk production in Holstein cattle. *J. Dairy Sci.* **45**:1500–1510.
31. Connor, E.E., D.L. Wood, T.S. Sonstegard, A.F. da Mota, G.L. Bennett, J.L. Williams, and A.V. Capuco. 2005. Chromosomal mapping and quantitative analysis of estrogen-related receptor alpha-1, estrogen receptors alpha and beta and progesterone receptor in the bovine mammary gland. *J. Endocrinol.* **185**:593–603.
32. Coppock, R.W., M.S. Mostrom, C.G. Sparling, B. Jacobsen, and S.C. Ross. 1990. Apparent zearalenone intoxication in a dairy herd from feeding spoiled acid-treated corn. *Vet. Hum. Toxicol.* **32**:246–248.
33. Couse, J.F. and K.S. Korach. 1999. Estrogen receptor null mice: what have we learned and where will they lead us? *Endocr. Rev.* **20**:358–417.
34. Cowie, A.T., J.S. Tindal, and A. Yokoyama. 1966. The induction of mammary growth in the hypophysectomised goat. *J. Endocrinol.* **34**:185–195.
35. Crews, D. and J.A. McLachlan. 2006. Epigenetics, evolution, endocrine disruption, health, and disease. *Endocrinology* **147**:S4–10.
36. D'Cruz, C.M., S.E. Moody, S.R. Master, J.L. Hartman, E.A. Keiper, M.B. Imielinski, J.D. Cox, J.Y. Wang, S.I. Ha, B.A. Keister, and L.A. Chodosh. 2002. Persistent parity-induced changes in growth factors, TGF-beta3, and differentiation in the rodent mammary gland. *Mol. Endocrinol.* **16**:2034–2051.
37. D'Mello, J.P.F., C.M. Placinta, and A.M.C. Macdonald. 1999. Fusarium mycotoxins: a review of global implications for animal health, welfare and productivity. *Anim. Feed Sci. Techn.* **80**:183–205.
38. Dahl, G.E. and D. Petitclerc. 2003. Management of photoperiod in the dairy herd for improved production and health. *J. Anim. Sci.* (Suppl. 3) **81**:11–17.
39. de Groot, A.N., P.W. van Dongen, T.B. Vree, Y.A. Hekster, and J. van Roosmalen. 1998. Ergot alkaloids. Current status and review of clinical pharmacology and therapeutic use compared with other oxytocics in obstetrics and gynaecology. *Drugs* **56**:523–535.
40. DeOme, K.B., L.J.Jr. Faulkin, H.A. Bern, and P.B. Blair. 1959. Development of mammary tumors from hyperplastic alveolar nodules transplanted into gland-free mammary fat pads of female C3H mice. *Cancer Res.* **19**:515–520.
41. Desai, M., N.J. Crowther, S.E. Ozanne, A. Lucas, and C.N. Hales. 1995. Adult glucose and lipid metabolism may be programmed during fetal life. *Biochem. Soc. Trans.* **23**:331–335.
42. DiAugustine, R.P., R.G. Richards, and J. Sebastian. 1997. EGF-related peptides and their receptors in mammary gland development. *J. Mammary Gland Biol. Neoplasia* **2**:109–117.
43. Diaz, G.J. and H.J. Boermans. 1994. Fumonisin toxicosis in domestic animals: a review. *Vet. Hum. Toxicol.* **36**:548–555.
44. Dontu, G., M. Al-Hajj, W.M. Abdallah, M.F. Clarke, and M.S. Wicha. 2003. Stem cells in normal breast development and breast cancer. *Cell. Prolif.* (Suppl. 1) **36**:59–72.
45. Ellis, S. and A.V. Capuco. 2002. Cell proliferation in bovine mammary epithelium: identification of the primary proliferative cell population. *Tissue Cell* **34**:155–163.
46. Feldman, M., W. Ruan, I. Tappin, R. Wieczorek, and D.L. Kleinberg. 1999. The effect of GH on estrogen receptor expression in the rat mammary gland. *J. Endocrinol.* **163**:515–522.
47. Filep, R. and R.M. Akers. 2000. Casein secretion and cytological differentiation in mammary tissue from bulls of high or low genetic merit. *J. Dairy Sci.* **83**:2261–2268.
48. Ford, J.A. Jr and C.S. Park. 2001. Nutritionally directed compensatory growth enhances heifer development and lactation potential. *J. Dairy Sci.* **84**:1669–1678.
49. Forsyth, I.A., G. Gabai, and G. Morgan. 1999. Spatial and temporal expression of insulin-like growth factor-I, insulin-like growth factor-II and the insulin-like growth factor-I receptor in the sheep fetal mammary gland. *J. Dairy Res* **66**:35–44.
50. Gardner, R.W., J.D. Schuh, and L.G. Vargas. 1977. Accelerated growth and early breeding of Holstein heifers. *J. Dairy Sci.* **60**:1941–1948.
51. Gaynor, P.J., D.R. Waldo, A.V. Capuco, R.A. Erdman, and L.W. Douglass. 1995. Effects of prepubertal growth rate and diet on lipid metabolism in lactating Holstein cows. *J. Dairy Sci.* **78**:1534–1543.

52. Giguere, V. 2002. To ERR in the estrogen pathway. *Trends Endocrinol. Metab.* **13**:220–225.
53. Ginger, M.R., M.F. Gonzalez-Rimbau, J.P. Gay, and J.M. Rosen. 2001. Persistent changes in gene expression induced by estrogen and progesterone in the rat mammary gland. *Mol. Endocrinol.* **15**:1993–2009.
54. Ginger, M.R. and J.M. Rosen. 2003. Pregnancy-induced changes in cell-fate in the mammary gland. *Breast Cancer Res.* **5**:192–197.
55. Gluckman, P.D., M.A. Hanson, and A.S. Beedle. 2007. Early life events and their consequences for later disease: A life history and evolutionary perspective. *Am. J. Hum. Biol.* **19**:1–19.
56. Grings, E.E., D.M. deAvila, R.G. Eggert, and J.J. Reeves. 1990. Conception rate, growth, and lactation of dairy heifers treated with recombinant somatotropin. *J. Dairy Sci.* **73**:73–77.
57. Hansen, P.J. 1985. Seasonal modulation of puberty and the postpartum anestrus in cattle: a review. *Livest. Prod. Sci.* **12**:309–328.
58. Hardville, D.A. and C.R. Henderson. 1966. Interrelationships among age, body weight, and production traits during first lactation of dairy cattle. *J. Dairy Sci.* **49**:1254–1261.
59. Harrison, R.D., I.P. Reynolds, and W. Little. 1983. A quantitative analysis of mammary glands of dairy heifers reared at different rates of live weight gain. *J. Dairy Res.* **50**: 405–412.
60. Heinrichs, A.J. and G.L. Hargrove. 1987. Standards of weight and height for Holstein heifers. *J. Dairy Sci.* **70**:653–660.
61. Hilakivi-Clarke, L., R. Clarke, and M. Lippman. 1999. The influence of maternal diet on breast cancer risk among female offspring. *Nutrition* **15**:392–401.
62. Hilakivi-Clarke, L., A. Shajahan, B. Yu, and S. de Assis. 2006. Differentiation of mammary gland as a mechanism to reduce breast cancer risk. *J. Nutr.* **136**:2697S–2699S.
63. Hoffman, P.C. 1997. Optimum body size of Holstein replacement heifers. *J. Anim. Sci.* **75**:836–845.
64. Hogg, N.A., C.J. Harrison, and C. Tickle. 1983. Lumen formation in the developing mouse mammary gland. *J. Embryol. Exp. Morphol.* **73**:39–57.
65. Houseknecht, K.L., D.E. Bauman, D.G. Fox, and D.F. Smith. 1992. Abomasal infusion of casein enhances nitrogen retention in somatotropin- treated steers. *J. Nutr.* **122**:1717–1725.
66. Hovey, R.C., J.F. Trott, and B.K. Vonderhaar. 2002. Establishing a framework for the functional mammary gland: from endocrinology to morphology. *J. Mammary Gland Biol. Neoplasia* **7**:17–38.
67. Humphreys, R.C., M. Krajewska, S. Krnacik, R. Jaeger, H. Weiher, S. Krajewski, J.C. Reed, and J.M. Rosen. 1996. Apoptosis in the terminal endbud of the murine mammary gland: A mechanism of ductal morphogenesis. *Development* **122**:4013–4022.
68. Hussein, H.S. and J.M. Brasel. 2001. Toxicity, metabolism, and impact of mycotoxins on humans and animals. *Toxicology* **167**:101–134.
69. Katzenellenbogen, B.S., M.M. Montano, T.R. Ediger, J. Sun, K. Ekena, G. Lazennec, P.G. Martini, E.M. McInerney, R. Delage-Mourroux, K. Weis, and J.A. Katzenellenbogen. 2000. Estrogen receptors: selective ligands, partners, and distinctive pharmacology. *Recent Prog. Horm. Res.* **55**:163–193.
70. Kewon, J.F. and R.W. Everett. 1986. Effect of days carried calf, days dry, and weight of first calf heifers on yield. *J. Dairy Sci.* **69**:1891–1896.
71. Kim, H., J. Xu, Y. Su, H. Xia, L. Li, G. Peterson, J. Murphy-Ullrich, and S. Barnes. 2001. Actions of the soy phytoestrogen genistein in models of human chronic disease: potential involvement of transforming growth factor beta. *Biochem. Soc. Trans.* **29**:216–222.
72. Kleinberg, D.L. 1997. Early mammary development: growth hormone and IGF-1. *J. Mammary Gland Biol. Neoplasia* **2**:49–57.
73. Kleinberg, D.L., M. Feldman, and W. Ruan. 2000. IGF-I: an essential factor in terminal end bud formation and ductal morphogenesis. *J. Mammary Gland Biol. Neoplasia* **5**:7–17.
74. Knabel, M., S. Kolle, and F. Sinowatz. 1998. Expression of growth hormone receptor in the bovine mammary gland during prenatal development. *Anat. Embryol. (Berl)* **198**:163–169.

75. Kordon, E.C. and GH. Smith. 1998. An entire functional mammary gland may comprise the progeny from a single cell. *Development* **125**:1921–1930.
76. Kostelac, D., G. Rechkemmer, and K. Briviba. 2003. Phytoestrogens modulate binding response of estrogen receptors alpha and beta to the estrogen response element. *J. Agric. Food Chem.* **51**:7632–7635.
77. Lamartiniere, C.A. 2002. Timing of exposure and mammary cancer risk. *J. Mammary Gland Biol. Neoplasia* **7**:67–76.
78. Lammers, B.P., A.J. Heinrichs, and R.S. Kensinger. 1999. The effects of accelerated growth rates and estrogen implants in prepubertal Holstein heifers on estimates of mammary development and subsequent reproduction and milk production. *J. Dairy Sci.* **82**:1753–1764.
79. Lammers, B.P., A.J. Heinrichs, and R.S. Kensinger. 1999. The effects of accelerated growth rates and estrogen implants in prepubertal Holstein heifers on growth, feed efficiency, and blood parameters. *J. Dairy Sci.* **82**:1746–1752.
80. Li, R.W., M.J. Meyer, C.P. Van Tassell, T.S. Sonstegard, E.E. Connor, M.E. Van Amburgh, Y.R. Boisclair, and A.V. Capuco. 2006. Identification of estrogen-responsive genes in the parenchyma and fat pad of the bovine mammary gland by microarray analysis. *Physiol. Genomics* **27**:42–53.
81. Lin, C.L. and H.L. Buttle. 1991. Progesterone receptor in the mammary tissue of pregnant and lactating gilts and the effect of tamoxifen treatment during late gestation. *J. Endocrinol.* **130**:251–257.
82. Little, W. and R.M. Kay. 1977. The effects of rapid rearing and early calving on the subsequent performance of dairy heifers. *Anim. Prod.* **29**:131–142.
83. Lucas, A. 1998. Programming by early nutrition: an experimental approach. *J. Nutr.* **128**:401S–406S.
84. Lucas, A., B.A. Baker, M. Desai, and C.N. Hales. 1996. Nutrition in pregnant or lactating rats programs lipid metabolism in the offspring. *Br. J. Nutr.* **76**:605–612.
85. Lyons, W.R. 1958. Hormonal synergism in mammary growth. *Proc. Royal. Soc.* **B149**: 303–325.
86. Mallepell, S., A. Krust, P. Chambon, and C. Brisken. 2006. Paracrine signaling through the epithelial estrogen receptor alpha is required for proliferation and morphogenesis in the mammary gland. *Proc. Natl. Acad. Sci. U S A* **103**:2196–2201.
87. Markusfeld, O. and E. Ezra. 1993. Body measurements, metritis, and postpartum performance of first lactation cows. *J. Dairy Sci.* **76**:3771–3777.
88. McCann, A.H., N. Miller, A. O'Meara, I. Pedersen, K. Keogh, T. Gorey, and P.A. Dervan. 1996. Biallelic expression of the IGF2 gene in human breast disease. *Hum. Mol. Genet.* **5**:1123–1127.
89. Mellor, D.J. 1987. Nutritional effects on the fetus and mammary gland during pregnancy. *Proc. Nutr. Soc.* **46**:249–257.
90. Meyer, M.J., A.V. Capuco, Y.R. Boisclair, and M.E. Van Amburgh. 2006. Estrogen-dependent responses of the mammary fat pad in prepubertal dairy heifers. *J. Endocrinol.* **190**:819–827.
91. Meyer, M.J., A.V. Capuco, D.A. Ross, L.M. Lintault, and M.E. Van Amburgh. 2006. Developmental and nutritional regulation of the prepubertal bovine mammary gland: II. Epithelial cell proliferation, parenchymal accretion rate, and allometric growth. *J. Dairy Sci.* **89**: 4298–4304.
92. Meyer, M.J., A.V. Capuco, D.A. Ross, L.M. Lintault, and M.E. Van Amburgh. 2006. Developmental and nutritional regulation of the prepubertal heifer mammary gland: I. Parenchyma and fat pad mass and composition. *J. Dairy Sci.* **89**:4289–4297.
93. Michels, K.B. and F. Xue. 2006. Role of birthweight in the etiology of breast cancer. *Int. J. Cancer* **119**:2007–2025.
94. Miller, D.M., B.P. Stuart, W.A. Crowel, R.J. Cole, A.J. Goven, and J. Brown. 1978. Aflatoxin in swine: its effect on immunity and relationship to salmonellosis. *Am. Assoc. Vet. Lab. Diagn.* **21**:135–142.

95. Moallem, U., G.E. Dahl, E.K. Duffey, A.V. Capuco, and R.A. Erdman. 2004. Bovine somatotropin and rumen-undegradable protein effects on skeletal growth in prepubertal dairy heifers. *J. Dairy Sci.* **87**:3881–3888.
96. Moallem, U., G.E. Dahl, E.K. Duffey, A.V. Capuco, D.L. Wood, K.R. McLeod, R.L. Baldwin 6th, and R.A. Erdman. 2004. Bovine somatotropin and rumen-undegradable protein effects in prepubertal dairy heifers: effects on body composition and organ and tissue weights. *J. Dairy Sci.* **87**:3869–3880.
97. Morris, R.J., S.M. Fischer, and T.J. Slaga. 1985. Evidence that the centrally and peripherally located cells in the murine epidermal proliferative unit are two distinct cell populations. *J. Invest. Dermatol.* **84**:277–281.
98. Oftedal, O.T. 2002. The mammary gland and its origin during synapsid evolution. *J. Mammary Gland Biol. Neoplasia* **7**:225–252.
99. Oftedal, O.T. 2002. The origin of lactation as a water source for parchment-shelled eggs. *J. Mammary Gland Biol. Neoplasia* **7**: 253–266.
100. Oommen, A.M., J.B. Griffin, G. Sarath, and J. Zempleni. 2005. Roles for nutrients in epigenetic events. *J. Nutr. Biochem.* **16**:74–77.
101. Park, C.S. 2005. Role of compensatory mammary growth in epigenetic control of gene expression. *FASEB J.* **19**:1586–1591.
102. Park, C.S., M.G. Baik, W.L. Keller, I.E. Berg, and G.M. Erickson. 1989. Role of compensatory growth in lactation: a stair-step nutrient regimen modulates differentiation and lactation of bovine mammary gland. *Growth Dev. Aging* **53**:159–166.
103. Park, C.S., R.B. Danielson, B.S. Kreft, S.H. Kim, Y.S. Moon, and W.L. Keller. 1998. Nutritionally directed compensatory growth and effects on lactation potential of developing heifers. *J. Dairy Sci.* **81**:243–249.
104. Peters, R.R., L.T. Chapin, K.B. Leining, and H.A. Tucker. 1978. Supplemental lighting stimulates growth and lactation in cattle. *Science* **199**:911–912.
105. Petitclerc, D., R.D. Kineman, S.A. Zinn, and H.A. Tucker. 1985. Mammary growth response of Holstein heifers to photoperiod. *J. Dairy Sci.* **68**:86–90.
106. Plath, A., R. Einspanier, F. Peters, F. Sinowatz, and D. Schams. 1997. Expression of transforming growth factors alpha and beta-1 messenger RNA in the bovine mammary gland during different stages of development and lactation. *J. Endocrinol.* **155**:501–511.
107. Plaut, K. 1993. Role of epidermal growth factor and transforming growth factors in mammary development and lactation. *J. Dairy Sci.* **76**:1526–1538.
108. Potten, C.S., W.J. Hume, P. Reid, and J. Cairns. 1978. The segregation of DNA in epithelial stem cells. *Cell* **15**:899–906.
109. Potten, C.S., G. Owen, and D. Booth. 2002. Intestinal stem cells protect their genome by selective segregation of template DNA strands. *J. Cell Sci.* **115**:2381–2388.
110. Prelusky, D.B., P.M. Scott, H.L. Trenholm, and G.A. Lawrence. 1990. Minimal transmission of zearalenone to milk of dairy cows. *J. Environ. Sci. Health B* **25**:87–103.
111. Purup, S., K. Sejrsen, J. Foldager, and R.M. Akers. 1993. Effect of exogenous bovine growth hormone and ovariectomy on prepubertal mammary growth, serum hormones and acute in-vitro proliferative response of mammary explants from Holstein heifers. *J. Endocrinol.* **139**:19–26.
112. Purup, S., M. Vestergaard, and K. Sejrsen. 2000. Involvement of growth factors in the regulation of pubertal mammary growth in cattle. *Adv. Exp. Med. Biol.* **480**:27–43.
113. Radcliff, R.P., M.J. VandeHaar, L.T. Chapin, R.E. Pilbeam, D.K. Beede, E.P. Stanisiewski, and H.A. Tucker. 2000. Effects of diet and injection of bovine somatotropin on prepubertal growth and first-lactation milk yields of Holstein cows. *J. Dairy Sci.* **83**:23–29.
114. Radcliff, R.P., M.J. VandeHaar, A.L. Skidmore, L.T. Chapin, B.R. Tadke, J.W. Lloyd, E.P. Stanisiewski, and H.A. Tucker. 1997. Effects of diet and bovine somatotropin on heifer growth and mammary development. *J. Dairy Sci.* **80**:1996–2003.
115. Ryan, F.P. 2006. Genomic creativity and natural selection. *Biol. J. Linnean Soc.* **88**:655–672.
116. Santos, F. and W. Dean. 2004. Epigenetic reprogramming during early development in mammals. *Reproduction* **127**:643–651.

117. Schams, D., S. Kohlenberg, W. Amselgruber, B. Berisha, M.W. Pfaffl, and F. Sinowatz. 2003. Expression and localisation of oestrogen and progesterone receptors in the bovine mammary gland during development, function and involution. *J. Endocrinol.* **177**:305–317.
118. Schwertfeger, K.L., J.M. Rosen, and D.A. Cohen. 2006. Mammary gland macrophages: pleiotropic functions in mammary development. *J. Mammary Gland Biol. Neoplasia* **11**:229–238.
119. Sejrsen, K. 1978. Mammary development and milk yield in relation to growth rate in dairy and dual-purpose heifers. *Acta. Agriculturae Scandinavica* **28**:41–46.
120. Sejrsen, K. 1994. Relationships between nutrition, puberty and mammary development in cattle. *Proc. Nutr. Soc.* **53**:103–111.
121. Sejrsen, K. and J. Foldager. 1992. Mammary growth and milk production capacity of replacemment heifers in relation to diet energy concentration and plasma hormone levels. *Acta Agric. Scand. Sect. A. Anim. Sci.* **42**:99–105.
122. Sejrsen, K., J. Foldager, M.T. Sorensen, R.M. Akers, and D.E. Bauman. 1986. Effect of exogenous bovine somatotropin on pubertal mammary development in heifers. *J. Dairy Sci.* **69**:1528–1535.
123. Sejrsen, K., J.T. Huber, and H.A. Tucker. 1983. Influence of amount fed on hormone concentrations and their relationship to mammary growth in heifers. *J. Dairy Sci.* **66**:845–855.
124. Sejrsen, K., J.T. Huber, H.A. Tucker, and R.M. Akers. 1982. Influence of nutrition on mammary development in pre- and postpubertal heifers. *J. Dairy Sci.* **65**:793–800.
125. Sejrsen, K. and S. Purup. 1997. Influence of prepubertal feeding level on milk yield potential of dairy heifers: a review. *J. Anim. Sci.* **75**:828–835.
126. Sejrsen, K., S. Purup, M. Vestergaard, M.S. Weber, and C.H. Knight. 1999. Growth hormone and mammary development. *Domest. Anim. Endocrinol.* **17**:117–129.
127. Shackleton, M., F. Vaillant, K.J. Simpson, J. Stingl, G.K. Smyth, M.L. Asselin-Labat, L. Wu, G.J. Lindeman, and J.E. Visvader. 2006. Generation of a functional mammary gland from a single stem cell. *Nature* **439**:84–88.
128. Shinin, V., B. Gayraud-Morel, D. Gomes, and S. Tajbakhsh. 2006. Asymmetric division and cosegregation of template DNA strands in adult muscle satellite cells. *Nat. Cell Biol.* **8**: 677–687.
129. Sieber, M., A.E. Freeman, and P.N. Hinz. 1988. Comparison between factor analysis from a phenotypic and genetic correlation matrix using linear type traits of Holstein dairy cows. *J. Dairy Sci.* **71**:477–484.
130. Silva, L.F., J.S. Liesman, M.S. Nielsen Weber, and M.J. VandeHaar. 2003. Intramammary infusion of leptin decreases proliferation of mammary epithelial cells in prepubertal heifers. *J. Dairy Sci.* (Suppl. 1) **82**:166.
131. Silva, L.F., M.J. VandeHaar, M.S. Weber Nielsen, and G.W. Smith. 2002. Evidence for a local effect of leptin in bovine mammary gland. *J. Dairy Sci.* **85**:3277–3286.
132. Sinha, Y.N. and H.A. Tucker. 1969. Mammary development and pituitary prolactin level of heifers from birth through puberty and during the estrous cycle. *J. Dairy Sci.* **52**:507–512.
133. Smith, G.H. 2005. Stem cells and mammary cancer in mice. *Stem Cell Rev.* **1**:215–223.
134. Smith, G.H. 2005. Label-retaining epithelial cells in mouse mammary gland divide asymmetrically and retain their template DNA strands. *Development* **132**:681–687.
135. Smith, G.H. and G. Chepko. 2001. Mammary epithelial stem cells. *Microsc. Res. Tech.* **52**:190–203.
136. Smith, G.H. and D. Medina. 1988. A morphologically distinct candidate for an epithelial stem cell in mouse mammary gland. *J. Cell Sci.* **90**:173–184.
137. Srinivasan, M., S.G. Laychock, D.J. Hill, and M.S. Patel. 2003. Neonatal nutrition: metabolic programming of pancreatic islets and obesity. *Exp. Biol. Med.* **228**:15–23.
138. Sternlicht, M.D., S.W. Sunnarborg, H. Kouros-Mehr, Y. Yu, D.C. Lee, and Z. Werb. 2005. Mammary ductal morphogenesis requires paracrine activation of stromal EGFR via ADAM17-dependent shedding of epithelial amphiregulin. *Development* **132**: 3923–3933.

139. Stingl, J., P. Eirew, I. Ricketson, M. Shackleton, F. Vaillant, D. Choi, H.I. Li, and C.J. Eaves. 2006. Purification and unique properties of mammary epithelial stem cells. *Nature* **439**:993–997.
140. Suetsugi, M., L. Su, K. Karlsberg, Y.C. Yuan, and S. Chen. 2003. Flavone and isoflavone phytoestrogens are agonists of estrogen-related receptors. *Mol. Cancer Res.* **1**:981–991.
141. Swanson, E.W. 1967. Optimum growth patterns for dairy cattle. *J. Dairy Sci.* **50**:244–252.
142. Ting, A.H., K.M. McGarvey, and S.B. Baylin. 2006. The cancer epigenome-components and functional correlates. *Genes Dev.* **20**:3215–3231.
143. Tomlinson, I.P. 2001. Mutations in normal breast tissue and breast tumours. *Breast Cancer Res.* **3**:299–303.
144. Tsai, Y.C., Y. Lu, P.W. Nichols, G. Zlotnikov, P.A. Jones, and H.S. Smith. 1996. Contiguous patches of normal human mammary epithelium derived from a single stem cell: implications for breast carcinogenesis. *Cancer Res.* **56**:402–404.
145. Tsubura, A., N. Uehara, Y. Kiyozuka, and N. Shikata. 2005. Dietary factors modifying breast cancer risk and relation to time of intake. *J. Mammary Gland Biol. Neoplasia* **10**:87–100.
146. Van Amburgh, M.E., D.M. Galton, D.E. Bauman, R.W. Everett, D.G. Fox, L.E. Chase, and H.N. Erb. 1998. Effects of three prepubertal body growth rates on performance of Holstein heifers during first lactation. *J. Dairy Sci.* **81**:527–538.
147. Van Amburgh, M.E., D.M. Galton, D.G. Fox, and D.E. Bauman, 1991. Optimizing heifer growth. p. 85–93. *In* Proceedings of Cornell Nutrition Conference. Department of Animal Science and Division of Nutritional Sciences, Cornell University.
148. van Heugten, E., J.W. Spears, M.T. Coffey, E.B. Kegley, and M.A. Qureshi. 1994. The effect of methionine and aflatoxin on immune function in weanling pigs. *J. Anim. Sci.* **72**:658–664.
149. VandeHaar, M.J. Feeding heifers for lifelong profit. p. 101–109. *In* Proceedings of Southwest Nutrition and Management Conference, Department of Animal Science, University of Arizona.
150. Wagner, K.U. and G.H. Smith. 2005. Pregnancy and stem cell behavior. *J. Mammary Gland Biol. Neoplasia* **10**:25–36.
151. Waldo, D.R., A.V. Capuco, and C.E. Rexroad Jr. 1998. Milk production of Holstein heifers fed either alfalfa or corn silage diets at two rates of daily gain. *J. Dairy Sci.* **81**:756–764.
152. Waldo, D.R., H.F. Tyrrell, A.V. Capuco, and C.E. Rexroad Jr. 1997. Components of growth in Holstein heifers fed either alfalfa or corn silage diets to produce two daily gains. *J. Dairy Sci.* **80**:1674–1684.
153. Wallace, C. 1953. Observations on mammary development in calves and lambs. *J. Agric. Sci.* **43**:413–421.
154. Weaver, G.A., H.J. Kurtz, J.C. Behrens, T.S. Robison, B.E. Seguin, F.Y. Bates, and C.J. Mirocha. 1986. Effect of zearalenone on the fertility of virgin dairy heifers. *Am. J. Vet. Res.* **47**:1395–1397.
155. Weaver, G.A., H.J. Kurtz, J.C. Behrens, T.S. Robison, B.E. Seguin, F.Y. Bates, and C.J. Mirocha. 1986. Effect of zearalenone on dairy cows. *Am. J. Vet. Res.* **47**:1826–1828.
156. Whitacre, M.D. and W.R. Threlfall. 1981. Effects of ergocryptine on plasma prolactin, luteinizing hormone, and progesterone in the periparturient sow. *Am. J. Vet. Res.* **42**:1538–1541.
157. Wiesen, J.F., P. Young, Z. Werb, and G.R. Cunha. 1999. Signaling through the stromal epidermal growth factor receptor is necessary for mammary ductal development. *Development* **126**:335–44.
158. Wogan, G.N. 1966. Chemical nature and biological effects of the aflatoxins. *Bacteriol. Rev.* **30**:460–470.
159. Zinedine, A., J.M. Soriano, J.C. Molto, and J. Manes. 2007. Review on the toxicity, occurrence, metabolism, detoxification, regulations and intake of zearalenone: An oestrogenic mycotoxin. *Food Chem. Toxicol.* **45**:1–18.

Index

P.L. Greenwood et al. (eds.), *Managing the Prenatal Environment to Enhance Livestock Productivity*, DOI 10.1007/978-90-481-3135-8,